THEORY AND APPLICATION OF ANTENNA ARRAYS

THEORY AND APPLICATION OF ANTENNA ARRAYS

M. T. Ma

Senior Member of the Technical Staff
Institute for Telecommunication Sciences
Office of Telecommunications
U. S. Department of Commerce
Boulder, Colorado

and

Professor-Adjoint of Electrical Engineering
University of Colorado

A Wiley-Interscience Publication
John Wiley & Sons
New York · London · Sydney · Toronto

Library of Congress Cataloging in Publication Data:
Ma, M. T.
Theory and application of antenna arrays.

"A Wiley-Interscience publication."
Includes bibliographies.
1. Antenna arrays. I. Title.

TK7871.6.M3 621.38′0283 73-15615
ISBN 0-471-55795-1

Printed in the United States of America

10 9 8 7 6 5 4 3 2 1

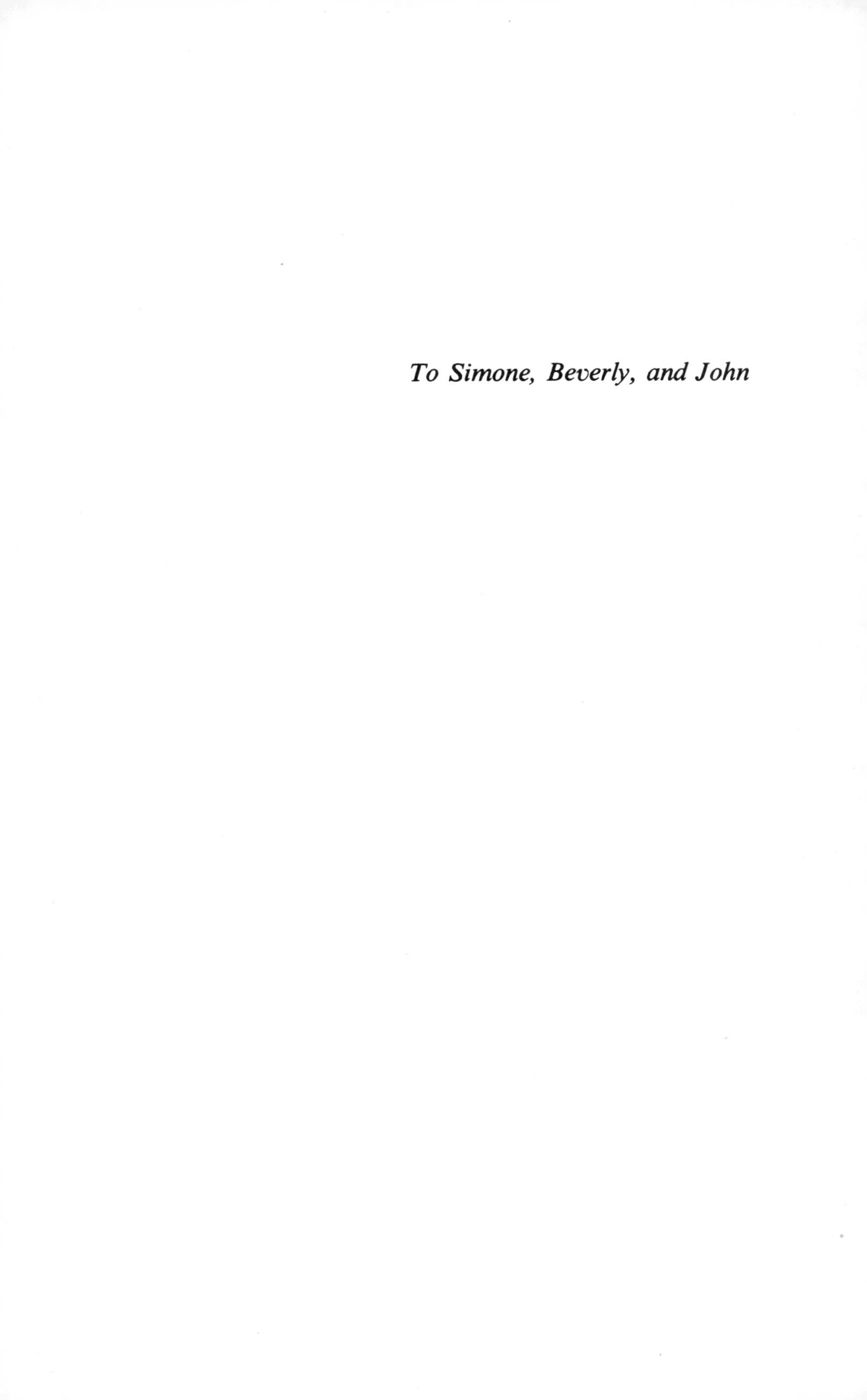

To Simone, Beverly, and John

PREFACE

Since early 1959 when I was first engaged in research on antenna arrays at Syracuse University, my interest in this subject has been divided into two major phases. One of these, which occupied most of my attention from 1959 to 1966, was concerned primarily with basic analysis and synthesis techniques pertinent to antenna arrays, which, for the most part, is an applied mathematics problem. The goals then were to produce a particular class of radiation patterns, to reduce sidelobe levels, to maximize the directivity, and to synthesize an array with a relatively broad frequency band, or to achieve some combinations of these. Isotropic elements were mostly used for the sole purpose of developing general mathematical models, which should not be limited to any particular kind of antennas or frequency bands. Thus, mutual impedances and effects from the ground were usually ignored. The arrays considered were either linear or two dimensional with the number of elements in the array, amplitude and phase excitations, or spacing distributions as controlling parameters. The results so obtained constitute the basis of the first three chapters in this book. They form the mathematical foundation for the theory of antenna arrays and should satisfy the general need to present relevant array topics with modern approaches and ample numerical illustrations in a single volume.

Chapter 1 introduces the reader to the fundamentals of linear arrays of discrete elements. It starts with the analysis for simple uniform arrays and then proceeds to arrays with tapered amplitudes, phases, and spacings. A relatively new approach, using finite Z-transforms for analyzing arrays with nonuniform amplitude excitations, is presented in detail with illustrative examples. Difference patterns produced by a monopulse array are also formulated and studied. Chapter 2 offers many recently developed techniques for synthesizing various linear arrays. It is here that power and field patterns are equally emphasized. Considerable effort is devoted to the understanding of arrays with equal sidelobes. In addition to the synthesis of array patterns, optimization of the array directivity is also thoroughly discussed. As an extension to the material presented in the first two chapters, both analysis and synthesis of two-dimensional arrays are given in Chapter 3. Greater attention is placed on ring and elliptical arrays.

My interest since 1966 has gradually shifted to the "real world" with emphasis on applications. In particular, the ionospheric prediction program was then an activity of major importance at the Central Radio Propagation Laboratory, then a part of the National Bureau of Standards. In this program, there was a strong need to develop reasonably accurate theoretical models for those HF antennas frequently involved in various communication systems. Specifically, expressions for input impedances, mutual impedances, radiated fields, and power gains were formulated with some simplified assumption for current distribution on antennas and many other necessary approximations. These imperfections, of course, will show some effects on the final antenna performance, but were considered adequate in view of the other uncertainties associated with the ionosphere. Major results from this activity were compiled and issued as a laboratory technical report in early 1969. Overwhelming response from a large number of users has since been received in the form of requests for reprints and inquiries about possible computer programs suitable for numerical results. In fact, over one hundred copies of this report have been requested. Because of this strong demand, I was prompted to make further improvements on formulations and to produce more quantitative data. These are essentially reflected in the last three chapters. A significant difference between the book form and the previous report form is that the three-term current distribution proposed originally by Professors R. W. P. King and T. T. Wu of Harvard University has been used in Chapters 4 and 5 to replace the simple sinusoidal current distribution that had been assumed in the earlier report.

In Chapter 4, currents, impedances, fields, and power gains for simple antennas such as dipoles, monopoles, and sleeve antennas above lossy ground are presented in detail. Equal attention is also given to arrays of these antennas in the form of the Yagi-Uda antenna, the curtain array, or the Wullenweber antenna. The basic principle and analytical formulation for a class of broadband antennas, namely, the log-periodic dipole array above lossy ground, are given in Chapter 5. Placing this array in alternative geometric positions (relative to the ground) to yield maximum radiations at low or high angles from the ground plane is also explored for different possible applications. In Chapter 6, a few commonly used traveling-wave antennas above lossy ground are analyzed.

Throughout this book, numerical examples are always presented for each antenna or array subject. These examples are prepared not only to serve the illustrative purpose, but also to give some design insight as to how the various parameters will affect the result. The contents presented in the last three chapters should directly benefit field engineers, design engineers, and other users in their applications, although the first three chapters also provide them with fundamental principles. This entire volume could be used as textbook for a course sponsored by companies for their staffs engaging in array and antenna designs. On the other hand, parts of Chapters 1, 2, 4, and 5 are also suitable for classroom use for a graduate course on arrays or antennas, or as self-study materials for students interested in research on antenna arrays. In fact, a major portion of the first three chapters was once given in a special course on "Antenna Array Theory" at the University of Colorado. For this latter purpose, selected problems are attached at the end of each of the first three chapters as student exercises and as reviews of the material discussed. For each chapter, references are listed to guide the reader to related topics, although there has been no attempt to make the reference list complete.

To the management of my governmental agency I wish to express my thanks for its encouragement and support throughout this undertaking. In particular, D. D. Crombie (Director), W. F. Utlaut (Deputy Director), and F. W. Smith (Executive Officer) of the Institute for Telecommunication Sciences (ITS) at Boulder offered many valuable suggestions, administrative assistance, and initial approval for this manuscript to be published by a nongovernmental publisher. J. M. Richardson (Director), R. C. Kirby (Associate Director), R. Gary (Special Assistant to the Director), and D. M. Malone (Attorney-Adviser) of the Office of Telecommunications at Washington, D. C. also rendered their blessing and frequent services in administrative regards. I wish to express my heartfelt appreciation to Mrs. Lillie C. Walters for her masterful skill in developing computer programs to produce such extensive numerical results for antennas presented in Chapters 4, 5, and 6, without which the value of the book would diminish substantially. I am also indebted to my colleagues at Boulder—E. L. Crow, H. T. Doughtery, E. C. Hayden, R. B. Stoner, Lillie C. Walters (all of ITS), and C. O. Stearns (National Oceanic and Atmospheric Administration), who reviewed parts or all of the chapters and made constructive suggestions for improvement. To Miss Ruth B. Hansen I offer my sincere gratitude for her expert and patient typing of the entire manuscript. Last but not least, my special thanks go to my wife and children for their understanding of my temporary lapse of participation in family affairs, school work, and evening football practice while I was busy preparing this volume.

M.T.Ma

Boulder, Colorado
July 1973

CONTENTS

THEORY AND APPLICATION OF ANTENNA ARRAYS

CHAPTER **1**

ANALYSIS OF DISCRETE LINEAR ARRAYS

An antenna is a device ordinarily used for transmitting and receiving elecrtromagnetic energy. In some circumstances these purposes may well be served by an antenna consisting of a single element, which may be of various types depending on operating frequency range, environment, economy, and many other factors. The single element may be as simple as a dipole or loop antenna or as complex as a parabolic reflector antenna. When a particular application demands higher gain, a more directive pattern, steerability of the main beam, or other performance that a single element antenna cannot provide, an antenna made up of an array of discrete elements may offer a solution to the problem. Although the individual elements of an array may differ among themselves, they ordinarily are identical and are similarly oriented for analytical and operational convenience. A *linear array* of discrete elements is an antenna consisting of several individual and distinguishable elements whose centers are finitely separated and fall on a straight line. The arrays having this particular geometric arrangement will be discussed in this chapter and Chapter 2.

In general, four parameters are accessible for variation in a linear array with a given kind of element, namely, the total number of elements, the spatial distribution of elements, the amplitude excitation function, and the phase excitation function. From the analysis viewpoint, these four parameters would be specified. From them we would determine the appropriate radiation characteristics, such as *pattern, directivity, power gain,* and *impedances.* On the other hand, the synthesis problem is to determine these four parameters in such a manner that the array response will approximate a desired one as closely as necessary under certain criteria.

In this chapter, we will analyze the fundamental properties of various kinds of linear array by several mathematical approaches. The synthesis problem will be treated in Chapter 2. Also, the element to be considered first in this chapter will be the rather idealized *isotropic element.* The analysis will then be extended to some simple, commonly used elements in free space. The analysis of arrays with practical antennas over a realistic flat lossy ground, especially in the high-frequency band, will be dealt with in Chapters 4–6.

1.1 Radiation Characteristics To Be Studied

Consider a typical antenna element located arbitrarily in free space with the coordinates shown in Fig. 1.1. Its distant field can be represented by

$$E_i(\theta,\varphi)=f(\theta,\varphi)I_i\exp[j(kr_i\cos\psi_i+\alpha_i)], \tag{1.1}$$

where $f(\theta,\varphi)$ represents the far-field function associated with the particular element considered, $k=2\pi/\lambda$ (λ being the free-space wavelength), I_i and α_i are, respectively, the amplitude and phase excitations, $j=\sqrt{-1}$,

$$\cos\psi_i=\cos\theta\cos\theta_i+\sin\theta\sin\theta_i\cos(\varphi-\varphi_i), \tag{1.2}$$

and (r_i,θ_i,φ_i) identifies the element location. If n such identical and similarly oriented elements form a linear array coincident with, say, the ζ axis (see Fig. 1.2), we then have

$$\theta_i=0,\quad r_i=\zeta_i,\quad i=1,2,\ldots,n \tag{1.3}$$

and the total field contributed from the array will be

$$E(\theta,\varphi)=\sum_{i=1}^{n}E_i(\theta,\varphi)$$

$$=f(\theta,\varphi)\sum_{i=1}^{n}I_i\exp[j(k\zeta_i\cos\theta+\alpha_i)]. \tag{1.4}$$

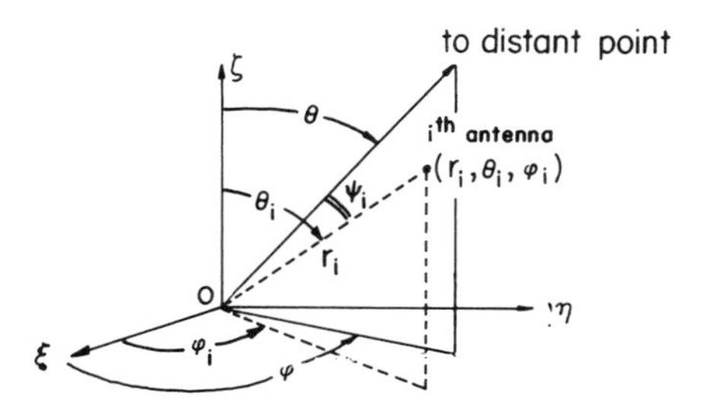

Fig. 1.1 A representative antenna element in freespace. (Here we use $\xi\eta\zeta$ to replace the conventional set of coordinates xyz so that the symbol z can be used later for other purposes.)

The *radiation pattern*, also known as the polar diagram, is obtained by taking the magnitude of $E(\theta,\varphi)$; that is,

$$|E(\theta,\varphi)|=|f(\theta,\varphi)|\cdot|S|, \tag{1.5}$$

where

$$S=\sum_{i=1}^{n} I_i \exp\left[j(k\zeta_i\cos\theta+\alpha_i)\right] \tag{1.6}$$

is usually called the space factor or array factor, because it depends solely on the space distribution of the elements in the array when the element excitations are specified.

Mathematically, (1.5) represents the well-known principle of pattern multiplication, which states that the radiation pattern of the array is given by the product of the *element pattern* $|f(\theta,\varphi)|$ and the magnitude of the array factor.

In general, the radiation pattern of a linear array is a function of θ and φ, consisting of a *main beam* and *several sidelobes* in both the $\theta=$constant and $\varphi=$constant surfaces. Therefore, the principal task of studying the fundamental characteristics of linear arrays is, upon specifications of array geometry and excitation, to determine the location of the *main beam* and *nulls*, if any, the *beamwidth*, the distribution of *sidelobes* and their relative levels, and the directivity.

The term "array gain in a given direction" is defined as the ratio of the radiation intensity in that direction to the average intensity. That is,

$$G(\theta,\varphi)=\frac{|E(\theta,\varphi)|^2}{W_0/4\pi}=\frac{4\pi|E(\theta,\varphi)|^2}{W_0}, \tag{1.7}$$

where W_0 is the total power. When W_0 is taken as the total power delivered to the array, $G(\theta,\varphi)$ in (1.7) is proportional to *power gain* (details

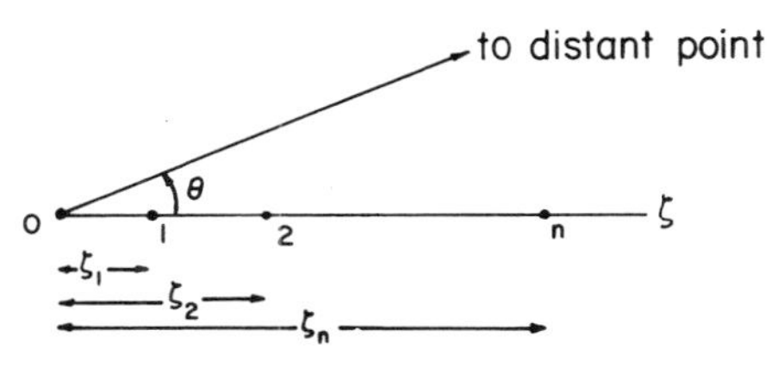

Fig. 1.2 A linear array of n elements along the ζ axis.

are given in Chapter 4). It is called *directive gain* if W_0 represents the total power radiated from the array, in which case

$$W_0 = \int_0^{2\pi} \int_0^{\pi} |E(\theta,\varphi)|^2 \sin\theta \, d\theta \, d\varphi. \tag{1.8}$$

The *directivity* of the array is defined as the maximum value of the directive gain:

$$D = G(\theta,\varphi)_{\max} = \frac{4\pi |E(\theta,\varphi)_{\max}|^2}{\int_0^{2\pi} \int_0^{\pi} |E(\theta,\varphi)|^2 \sin\theta \, d\theta \, d\varphi}. \tag{1.9}$$

We will study only the directivity in the first three chapters and postpone the analysis of power gain until Chapters 4–6, where practical antennas over a lossy ground will be considered.

1.2 Uniform Arrays

We start the study of linear arrays by considering special cases first. One of these is the equally spaced array, where

$$\zeta_i = id, \quad i = 0, 1, \ldots, n-1 \tag{1.10}$$

with d being the distance between two adjacent elements. Here we have chosen the location of the first element to coincide with the coordinate origin. If the elements of the array are excited with uniformly progressive phases,

$$\alpha_i = -ikd\cos\theta_0, \tag{1.11}$$

where θ_0 determines the direction of the maximum of the array factor, the expression (1.6) may be written as a polynomial in z:

$$S = \sum_{i=0}^{n-1} I_i \exp[jikd(\cos\theta - \cos\theta_0)] = \sum_{i=0}^{n-1} I_i z^{-i}, \tag{1.12}$$

where

$$z = e^{-ju}, \quad u = kd(\cos\theta - \cos\theta_0). \tag{1.13}$$

In addition, when $I_i = 1$, $i = 0, 1, 2, \ldots, n-1$, a *uniform array* results. Note that the definition for z in (1.13) is different from the usual notation, which does not include the negative sign in the exponent. This change of notation is necessary in order to conform to the conventional form of the Z

transform of a function to be introduced in Section 1.4. The magnitude of z in (1.13) always equals unity, although its locus, moving in a counterclockwise direction on the circumference of the *unit circle* when θ varies from 0 to π, depends on d and θ_0.

For *broadside arrays,* where the beam maximum occurs in the direction perpendicular to the array axis, $\theta_0=\pi/2$ and $z=\exp(-jkd\cos\theta)$. As the physical angle θ increases from 0 to π, z moves from e^{-jkd} to e^{jkd}. The entire range of variation of z (called the *visible range*) is $2kd$ radians. The locus of z describes a complete circle when $d=\lambda/2$, a portion of a circle when $d<\lambda/2$, and more than a circle when $d>\lambda/2$. A set of examples showing the locus of z is presented in Fig. 1.3. There is an overlap of z on the unit circle for the case $d>\lambda/2$. To explain this point more specifically, let us refer to Fig. 1.3(*c*), where $d=3\lambda/4$. It is clear that two portions of the array factor will have the same value. Since the overlap occurs on the left half-circle in this case, the portion of the array factor between $\theta=0°$ and 70.52° ($z=e^{-j3\pi/2}$ and $e^{-j\pi/2}$) will be identical to that between $\theta=109.48°$ and 180° ($z=e^{j\pi/2}$ and $e^{j3\pi/2}$).

For *endfire arrays,* where the beam maximum occurs in the direction of the array axis, $\theta_0=0$ and $z=\exp[-jkd(\cos\theta-1)]$. The entire range of z on the unit circle is still $2kd$ radians when θ varies from 0 to π. The description of the variation of z is similar to the broadside case except that the starting point of the locus in this case is changed to $z=1$ regardless of d. A corresponding set of examples is shown in Fig. 1.4.

For a finite uniform array of n elements, the expression (1.12) for the array factor S can be written in the convenient form

$$S=\sum_{i=0}^{n-1} z^{-i}=\frac{1-z^{-n}}{1-z^{-1}}. \tag{1.14}$$

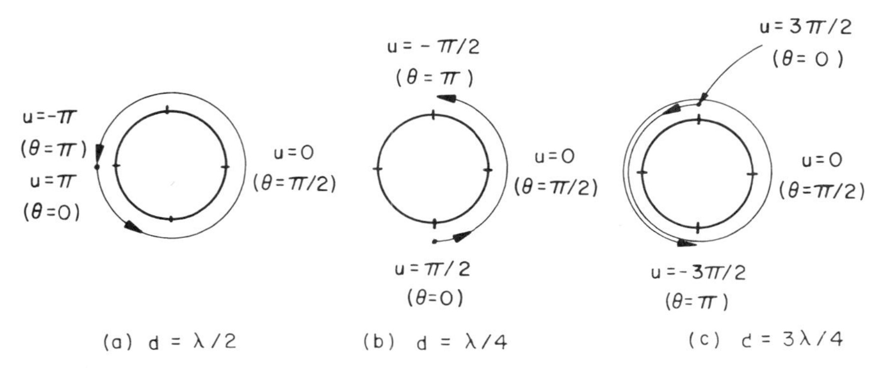

Fig. 1.3 Visible range of z for $\theta_0=\pi/2, z=e^{-ju}, u=kd\cos\theta$.

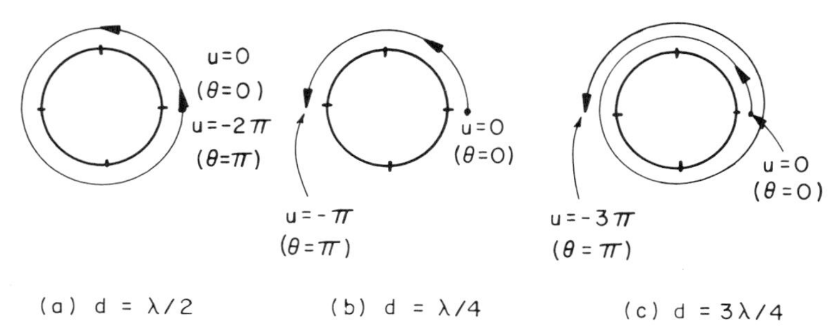

Fig. 1.4 Visible range of z for $\theta_0=0, z=e^{-ju}, u=kd(\cos\theta-1)$.

Since the complex conjugate of z is equal to the inverse of z, the magnitude of S can be obtained in the following manner:

$$|S|^2=\frac{(1-z^{-n})(1-z^{n})}{(1-z^{-1})(1-z)}=\frac{2-(z^{-n}+z^{n})}{2-(z^{-1}+z)}$$

$$=\frac{1-\cos nu}{1-\cos u}=\frac{\sin^2(nu/2)}{\sin^2(u/2)}, \tag{1.15}$$

or

$$|S|=\left|\frac{\sin(nu/2)}{\sin(u/2)}\right|, \tag{1.16}$$

where

$$u=kd(\cos\theta-\cos\theta_0). \tag{1.17}$$

A word about (1.16) is in order. Mathematically, $|S|$ is a periodic function of u for the entire range $-\infty \leqslant u \leqslant \infty$. Physically, however, the visible range of u is fixed by d and θ_0 because $|\cos\theta| \leqslant 1$. Therefore, the radiation pattern in $0 \leqslant \theta \leqslant \pi$ utilizes only a limited range of the expression in (1.16). Examples for $n=5$ and $d=\lambda/2$ are given in Fig. 1.5 to explain this point.

It is instructive to examine the behavior of the array factor S. In particular, it can be manipulated to yield information about

a. Locations of *major maxima* or beams,
b. Locations of *nulls*,
c. Angular widths of major maxima (*beamwidths*), and
d. Locations of minor maxima or *sidelobes*.

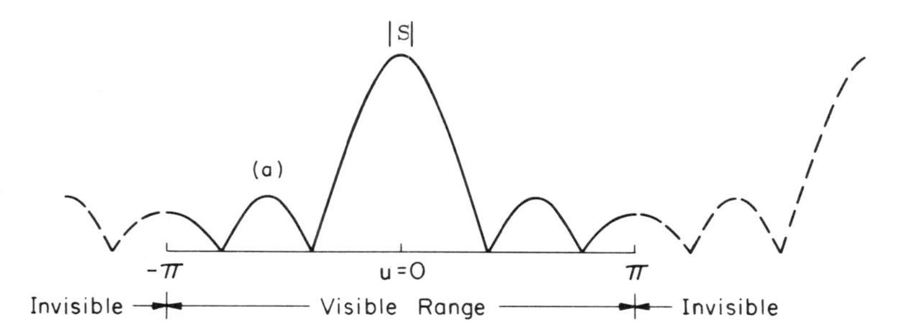

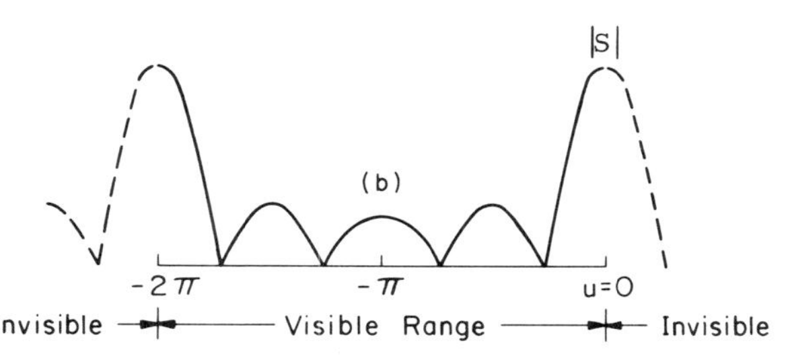

Fig. 1.5 Sketches of $|S|$ in (1.16) for $n=5$ and $d=\lambda/2$: (*a*) $\theta_0=\pi/2$, (*b*) $\theta_0=0$.

a. Major maxima, according to (1.14) or (1.16), occur at $z=1$ or $u=0$, $\pm 2\pi$, $\pm 4\pi$, ... The maximum at $u=0$ is the principal one; those at $u=\pm 2\pi$, $\pm 4\pi,\ldots$,are known as the *grating lobes*. For ordinary directive applications which limit the array pattern to only one maximum, grating lobes are unwanted. They can easily be avoided by choosing $d<\lambda$ in order to leave $u=\pm 2\pi$, $\pm 4\pi,\ldots$,in the invisible range. In terms of the real variable θ, the *principal maximum* occurs at $\theta=\pi/2$ for broadside ($\theta_0=\pi/2$) and $\theta=0$ for endfire ($\theta_0=0$) arrays. The value of $|S|$ at $z=1$ (or $u=0$) is n, the total number of elements in the array.

b. *Null* locations are determined by setting $z^{-n}=1$ or $\sin(nu/2)=0$ (except $z=1$ or $u=0$, which is already identified as the location of the beam maximum):

$$u_m = kd(\cos\theta_m - \cos\theta_0) = \frac{2\pi m}{n}, \tag{1.18}$$

or

$$\cos\theta_m - \cos\theta_0 = \frac{m\lambda}{nd}, \qquad m = \pm 1, 2, \ldots. \tag{1.19}$$

Here again, the total number of nulls in the visible range depends on d and θ_0. For example, when $\theta_0 = \pi/2$, $d = \lambda/2$, the governing equation is $\cos\theta_m = 2m/n$; m can be as large as $\pm n/2$ if n is an even number, or $\pm(n-1)/2$ if n is odd. The total number of nulls will, respectively, be n (even n) and $n-1$ (odd n). When $\theta_0 = 0$, $d = \lambda/2$, the governing equation will be $\cos\theta_m = 1 + (2m/n)$; m can then be $-1, -2, \ldots, -(n-1)$. The total number of nulls will be $n-1$ regardless of whether n is even or odd. For $d > \lambda/2$ and $d < \lambda/2$, the total number of nulls may, respectively, be more and less than that for $d = \lambda/2$. It should also be noted that $u = \pm\pi, \pm 3\pi, \ldots,$ are always possible nulls when n is even.

For broadside arrays, the first null on one side of the main beam (at $\theta = \pi/2$) is given by $m = 1$,

$$\cos\theta_1 = \frac{\lambda}{nd}. \tag{1.20}$$

For endfire arrays, the first null on one side of the main beam (at $\theta = 0$) is given by $m = -1$,

$$\cos\theta_1 = 1 - \frac{\lambda}{nd}. \tag{1.21}$$

The angle θ_1 in (1.20) and (1.21) can be determined exactly once the number of elements n and the element spacing d in terms of wavelength are specified.

c. The *beamwidth* is sometimes defined as the angular space between the first nulls on each side of the main beam. For broadside arrays the expression for the beamwidth is then

$$(\mathrm{BW})_b = 2\left(\frac{\pi}{2} - \theta_1\right) \qquad \text{in radians.} \tag{1.22}$$

When n is very large, θ_1 in (1.20) will be very close to $\pi/2$; we then have

$$\cos\theta_1 = \sin\left(\frac{\pi}{2} - \theta_1\right) \cong \frac{\pi}{2} - \theta_1.$$

Thus,

$$(\mathrm{BW})_b \cong \frac{2\lambda}{nd}. \tag{1.23}$$

In words, the beamwidth of a large broadside array of discrete elements is roughly inversely proportional to the array length.

For endfire arrays the expression for the beamwidth becomes

$$(\mathrm{BW})_e = 2\theta_1. \tag{1.24}$$

When n is very large, θ_1 is close to 0; we then have

$$\cos\theta_1 \cong 1 - \frac{\theta_1^2}{2}.$$

Thus,

$$\theta_1 \cong \sqrt{\frac{2\lambda}{nd}}$$

and

$$(\mathrm{BW})_e \cong 2\sqrt{\frac{2\lambda}{nd}}\,, \tag{1.25}$$

which is approximately inversely proportional to the square root of the array length.

Comparing (1.23) and (1.25), we can conclude that the beamwidth of a broadside array is always narrower than that of an endfire array of the same size.

Another conventional definition of the beamwidth is the angular space between the half-power points on each side of the main beam. In this case, the position of one of the half-power points, θ_h [assuming $f(\theta,\varphi)=1$], is determined by

$$u_h = kd(\cos\theta_h - \cos\theta_0), \tag{1.26}$$

such that u_h satisfies the following relation:

$$|S(u_h)|^2 = \tfrac{1}{2}|S_{\max}|^2 = \frac{n^2}{2}. \tag{1.27}$$

The beamwidths for broadside and endfire arrays will then, respectively, be $2(\pi/2-\theta_h)$ and $2\theta_h$. We will have opportunity to use both of these beamwidths. To distinguish them, we will call the one determined from (1.22) or (1.24) the *first-null beamwidth*, and that from (1.26) and (1.27) the *half-power beamwidth.*

d. Locations of minor maxima or *sidelobes* can be found by setting the derivatives of (1.15) to zero:

$$\frac{d|S|^2}{du}=\frac{\sin(nu/2)[n\cos(nu/2)\sin(u/2)-\cos(u/2)\sin(nu/2)]}{\sin^3(u/2)}=0$$

or

$$n\tan\left(\frac{u}{2}\right)=\tan\left(\frac{nu}{2}\right). \tag{1.28}$$

It is clear that $u=\pm\pi, \pm 3\pi, \ldots,$ are possible locations of sidelobes when n is odd. Other solutions of (1.28) can be obtained numerically when n is given. If, however, n is a large number, an approximate solution for locations of sidelobes may be determined by assuming that sidelobes are roughly half-way between nulls. That is,

$$u'_l=\pm\frac{2\pi}{n}\left(\frac{1+2l}{2}\right), \qquad l=1,2,\ldots, \tag{1.29}$$

or

$$\cos\theta'-\cos\theta_0=\pm\frac{\lambda}{nd}\left(\frac{1+2l}{2}\right).$$

Obviously, the number of sidelobes is also dependent on d and θ_0. The first two sidelobes (closest to the beam maximum), which are usually the most important ones, are at

$$u'_1=\pm\frac{3\pi}{n} \qquad \text{and} \qquad u'_2=\pm\frac{5\pi}{n}. \tag{1.30}$$

The levels of these sidelobes are, respectively,

$$|S_1|=|S(u'_1)|=\left|\frac{\sin(n/2)(3\pi/n)}{\sin(3\pi/2n)}\right|=\frac{1}{|\sin(3\pi/2n)|}$$

$$\cong\frac{2n}{3\pi}, \tag{1.31}$$

and

$$|S_2|=|S(u'_2)|=\frac{1}{|\sin(5\pi/2n)|}\cong\frac{2n}{5\pi}, \tag{1.32}$$

which, when referred to the major maximum $S(0)=n$, becomes

$$\frac{|S_1|}{S(0)} \cong \frac{2}{3\pi} \qquad \text{or} \qquad -13.5 \text{ dB} \tag{1.33a}$$

and

$$\frac{|S_2|}{S(0)} \cong \frac{2}{5\pi} \qquad \text{or} \qquad -17.9 \text{ dB.} \tag{1.33b}$$

The performance of the array will be influenced not only by the behavior of the array factor, but also by the basic radiation pattern, $f(\theta,\varphi)$, of the elements of which the array is constructed. Arrays made of the following three simple elements are now considered: (A) *isotropic elements*, (B) *short parallel dipoles*, and (C) *short collinear dipoles*.

Case A. Uniform arrays of isotropic elements: In this case, $I_i=1$, $f(\theta,\varphi)=1$, and from (1.5) and (1.16)

$$|E(\theta,\varphi)|=|S|=\left|\frac{\sin(nu/2)}{\sin(u/2)}\right|. \tag{1.34}$$

Evidently, all the properties of the array factor, S, just described apply directly to the array of this class. Furthermore, the *directivity* for this case, according to (1.9), becomes

$$D=\frac{4\pi n^2}{2\pi\int_0^{\pi}|E|^2\sin\theta\, d\theta}=\frac{2n^2}{\int_0^{\pi}|E|^2\sin\theta\, d\theta}. \tag{1.35}$$

The expression for $|E|$ given in (1.34) is not very convenient for computing the denominator of (1.35). A workable alternative expression may be found directly from (1.14),

$$|E|^2=|S|^2=\left(\sum_{i=0}^{n-1} z^{-1}\right)\left(\sum_{i=0}^{n-1} z^{i}\right)=n+\sum_{m=1}^{n-1}(n-m)(z^m+z^{-m})$$

$$=n+2\sum_{m=1}^{n-1}(n-m)\cos mu. \tag{1.36}$$

After substituting (1.36) into the denominator of (1.35) and making a transformation of variables from θ to u, we obtain

$$\int_0^{\pi} |E|^2 \sin\theta\, d\theta = (kd)^{-1} \int_a^b |E|^2 du$$

$$= (kd)^{-1}\left[(b-a)n + 2\sum_{m=1}^{n-1} \frac{n-m}{m} (\sin mb - \sin ma) \right]$$

$$= \frac{2kdn + 4\sum_{m=1}^{n-1} \frac{n-m}{m} \sin(mkd)\cos(mkd\cos\theta_0)}{kd}, \tag{1.37}$$

where

$$a = -kd(1+\cos\theta_0),$$
$$b = kd(1-\cos\theta_0). \tag{1.38}$$

The expression for the directivity now becomes

$$D = \frac{(kd)n^2}{nkd + 2\sum_{m=1}^{n-1} \frac{n-m}{m} \sin(mkd)\cos(mkd\cos\theta_0)}. \tag{1.39}$$

Note that there are two factors, $\sin(mkd)$ and $\cos(mkd\cos\theta_0)$, in the second term of the denominator of (1.39). This term vanishes, because of $\sin(mkd)$, whenever $kd = p\pi$, or $d = p\lambda/2$, $p = 1, 2, \ldots$, no matter what values θ_0 may take. Moreover, the same term can also vanish, because of $\cos(mkd\cos\theta_0)$, for other values of d depending on θ_0. As an example, for endfire arrays where $\theta_0 = 0$, the above mentioned term becomes $[(\mathrm{n}-\mathrm{m})/\mathrm{m}]\sin(2mkd)$, which vanishes when $kd = p\pi/2$, or $d = p\lambda/4$. Under all the conditions which make this term vanish, the directivity is numerically equal to n, the total number of elements in the array.

A graphical method for predicting the relative value of D as a function of d is also possible. Using (1.35), (1.37), and (1.38), we can rewrite the expression for directivity as follows,

$$D = \frac{2kdn^2}{\int_a^b |E|^2 du}, \tag{1.40}$$

where the denominator can be interpreted as the area under the curve of $|E(u)|^2$ between $u=a$ and $u=b$.

Let us consider the broadside case $(\theta_0=\pi/2)$ with $n=5$. $|E(u)|^2$ is presented in Fig. 1.6. Now, if $kd=\pi\,(d=\lambda/2)$, the visible range is from $a=-\pi$ to $b=\pi$; the denominator in (1.40) becomes

$$W_1=\int_{-\pi}^{\pi}|E|^2\,du=2\int_0^{\pi}|E|^2\,du, \tag{1.41}$$

or twice the area under $|E(u)|^2$ between points A and A' of Fig. 1.6. When d increases to, say, $4\lambda/5$ or $kd=8\pi/5$, the corresponding visible range is increased to $16\pi/5$ from $a=-8\pi/5$ to $b=8\pi/5$, and the denominator of (1.40) increases to $2\int_0^{8\pi/5}|E(u)|^2\,du$, or twice the area under $|E(u)|^2$ between points A and A''. Since the numerator of (1.40) is also larger, and the rate of increment in the denominator is not as fast as that in the numerator, we conclude that

$$D|_{d=\lambda/2}<D|_{d=4\lambda/5}. \tag{1.42}$$

If, however, d increases further to, say, $\lambda\,(kd=2\pi)$, or exactly twice the beginning element spacing $(d=\lambda/2)$ considered, the denominator of (1.40) will be

$$W_2=2\int_0^{2\pi}|E(u)|^2\,du, \tag{1.43}$$

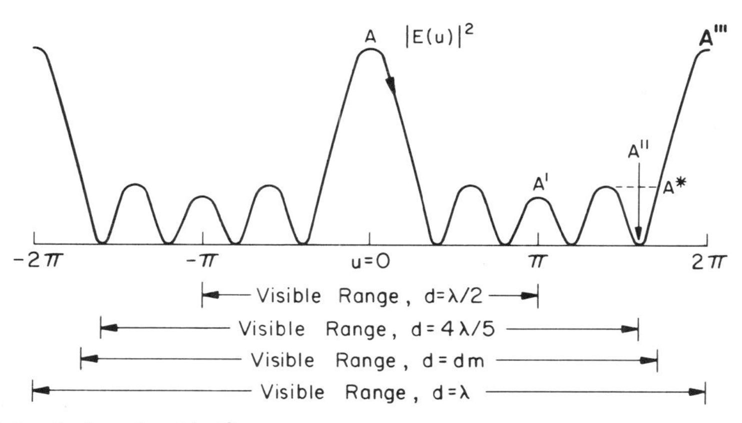

Fig. 1.6 A sketch of $|E|^2$ for $n=5,\theta_0=\pi/2$, and various spacings.

which is twice the area under $|E(u)|^2$ between points A and A'''. The mathematical fact that $W_2 = 2W_1$ is also clearly pictured in Fig. 1.6, since the area under $|E(u)|^2$ between points A and A''' is exactly twice that between points A and A'. In this case, both the numerator and denominator are increased by the same factor, two. This explains why directivities for both spacings ($d=\lambda/2$ and $d=\lambda$) are equal (and numerically equal to n),

$$D|_{d=\lambda/2} = D|_{d=\lambda} = n. \tag{1.44}$$

In reality, of course, d can never be made as large as a full wavelength if the grating lobe located at $u=2\pi$ is to be avoided. Comparing (1.42) and (1.44) reveals that D reaches its maximum somewhere between $d=\lambda/2$ and $d=\lambda$. From Fig. 1.6, we also see that, for single-frequency and fixed-beam operation of a broadside array, d can at most be increased to a value d_m such that the upper limit of integration in the denominator of (1.40), $u=b=kd_m$, will reach as far as the point A^* to avoid a larger sidelobe at $\theta=0$. If the array is designed for capability of scanning by changing the position of its beam maximum, a practical choice of d should be much less than d_m and somewhere near $\lambda/2$. On the other hand, when d decreases to a value below $\lambda/2$, the directivity is always less than n, and decreases monotonically with d. For academic interest, the limiting value of D as d approaches zero can also be evaluated from (1.39), for uniform arrays of isotropic elements, to be unity, because under the condition $d \to 0$ the array reduces to a single element,

$$D|_{d\to 0} \cong \frac{kdn^2}{kdn + 2\sum_{m=1}^{n-1} \frac{n-m}{m}(mkd)} = \frac{n^2}{n + 2\sum_{m=1}^{n-1}(n-m)} = 1. \tag{1.45}$$

For endfire arrays where $\theta_0 = 0$, a similar interpretation is possible. The function $|E(u)|^2$ for this case is given in Fig. 1.7 with $n=5$. Now, $a=-2kd$, $b=0$, and the denominator of (1.40) is

$$W' = \int_{-2kd}^{0} |E(u)|^2\, du. \tag{1.46}$$

When $d=\lambda/4\,(kd=\pi/2)$, (1.46) becomes

$$W'_1 = \int_{-\pi}^{0} |E(u)|^2\, du, \tag{1.47}$$

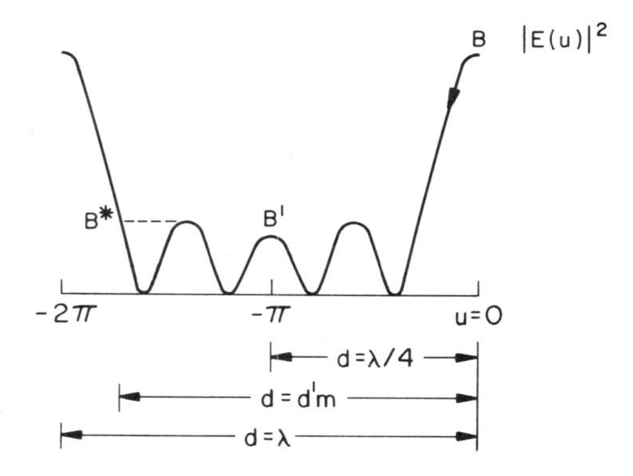

Fig. 1.7 A sketch of $|E|^2$ for $n=5, \theta_0=0$, and various spacings.

representing the area under $|E(u)|^2$ between points B and B'. Since both W_1' and d in this case are exactly half the corresponding values for (1.41), this explains why the directivity for $d=\lambda/4$ and $\theta_0=0$ is also numerically equal to that for $d=\lambda/2$ and $\theta_0=\pi/2$. If we designate the directivity by $D(d, \theta_0)$, signifying that the directivity for an array of n equally spaced isotropic elements is generally a function of element spacing d and main beam position θ_0, we always have

$$D\left(d, \frac{\pi}{2}\right) = D\left(\frac{d}{2}, 0\right). \tag{1.48}$$

To avoid the grating lobe or a sidelobe of higher level, the element spacing must be limited to d'_m such that the lower limit of integration in (1.46), $u=a=-2kd'_m$, does not go beyond the point B^* in Fig. 1.7.

A set of examples for the directivity with $n=$ 2, 3, and 4 is presented in Fig. 1.8 as a function of d for both $\theta_0=0$ and $\pi/2$. Directivities for higher n (up to 20) can be found in an article by Tai.[1] Directivities plotted in contour form for various d and θ_0 with $n=2$, 3, and 6 were discussed by Bach,[2] and those for $n=2$, 3, 4, and 10 were also included in a chapter by Bach and Hansen[3] in a recent book edited by Collin and Zucker.

Case B. Uniform arrays of short parallel dipoles (see Fig. 1.9): In this case, $I_i=1, f(\theta, \varphi)=(1-\sin^2\theta\cos^2\varphi)^{1/2}$, and

$$|E| = \left|\frac{\sin(nu/2)}{\sin(u/2)}\right| \left(1-\sin^2\theta\cos^2\varphi\right)^{1/2}, \tag{1.49}$$

where u is given in (1.17).

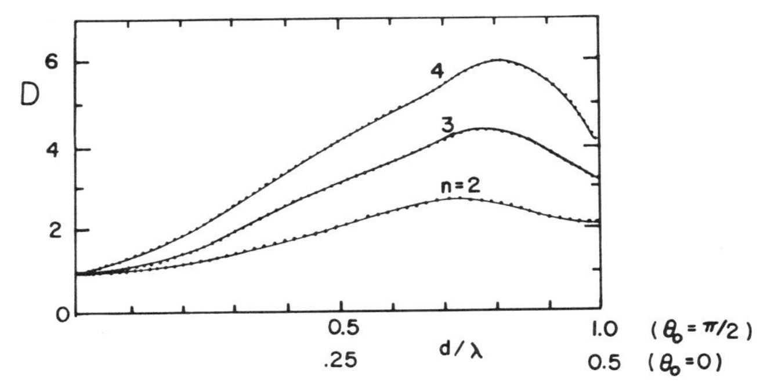

Fig. 1.8 Directivity of uniform arrays of isotropic elements.

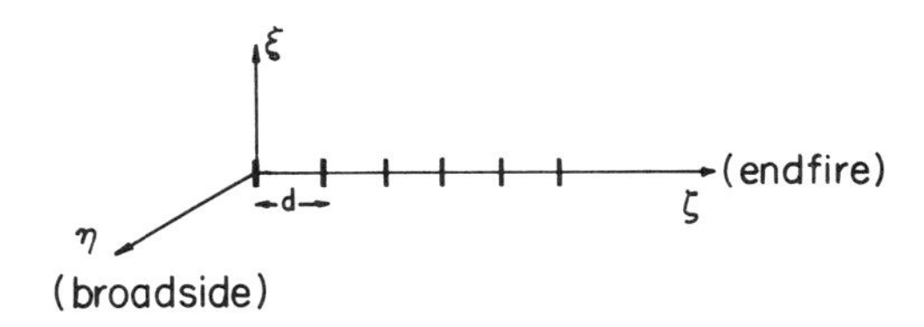

Fig. 1.9 An equally spaced linear array of short parallel dipoles.

Clearly, important characteristics such as beam maximum, nulls, sidelobes, and their levels in the $\eta\zeta$ plane $(\varphi=\pi/2)$ are identical to those for uniform arrays of isotropic elements. Characteristics in the $\xi\zeta$ plane $(\varphi=0)$ will, however, be the product of those for uniform arrays of isotropic elements and the factor $\cos\theta$. Referring to Fig. 1.9, we see that $\theta=\varphi=\pi/2$ for broadside and $\theta=0$ for endfire. Therefore, we always have

$$|E_{\max}|^2=n^2. \tag{1.50}$$

The directivity is also changed because the denominator of (1.9) will now be

$$\int_0^{2\pi}\int_0^{\pi}\left[n+2\sum_{m=1}^{n-1}(n-m)\cos mu\right](1-\sin^2\theta\cos^2\varphi)\sin\theta\,d\theta\,d\varphi,$$

which, after some simple algebraic manipulation, becomes $4\pi W_p$, where

$$W_p = \frac{2n}{3} + \frac{2}{kd}\sum_{m=1}^{n-1}\frac{n-m}{m}\left(1-\frac{1}{m^2k^2d^2}\right)\sin(mkd)\cos(mkd\cos\theta_0)$$

$$+\frac{2}{k^2d^2}\sum_{m=1}^{n-1}\frac{n-m}{m^2}\cos(mkd)\cos(mkd\cos\theta_0). \qquad (1.51)$$

Substituting (1.50) and (1.51) into (1.9), we have the corresponding directivity,

$$D_p = \frac{n^2}{W_p}. \qquad (1.52)$$

While it is rather involved now to predict D_p for finite values of d except through actual calculation according to (1.52), it is still possible to evaluate the limiting values of W_p and D_p as d approaches zero,

$$W_p|_{d\to 0} = \frac{2n}{3} + \frac{4}{3}\sum_{m=1}^{n-1}(n-m)$$

$$= \frac{2n^2}{3}, \qquad (1.53)$$

and

$$D_p|_{d\to 0} = \tfrac{3}{2}, \qquad (1.54)$$

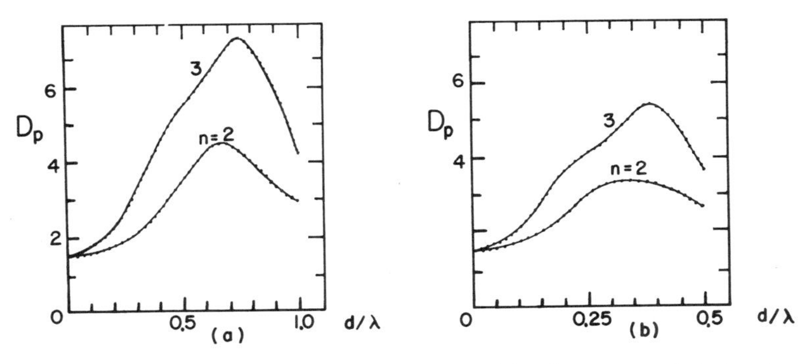

Fig. 1.10 Directivity of uniform arrays with short parallel dipoles: (*a*) broadside ($\theta_0=\pi/2$), (*b*) endfire ($\theta_0=0$).

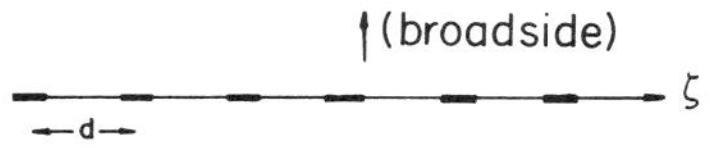

(broadside)

ζ

d

Fig. 1.11 An equally spaced linear array of short collinear dipoles.

which is the same directivity of a single dipole, as it should be. A similar set of examples for $n=2$ and 3 is given in Fig. 1.10 as a function of d for both $\theta_0=0$ and $\pi/2$. Results for a higher number of elements (up to $n=20$) were presented by Tai.[1] Directivities for short parallel dipoles in the form of contour plots can also be found in an article by Bach[2] or the chapter by Bach and Hansen.[3]

Case C. Uniform arrays of short collinear dipoles (see Fig. 1.11): In this case, $I_i=1$, $f(\theta,\varphi)=\sin\theta$, and

$$|E|=\left|\frac{\sin(nu/2)}{\sin(u/2)}\sin\theta\right|. \tag{1.55}$$

Now, the radiation pattern is independent of φ, and is equal to that for uniform arrays of isotropic elements multiplied by $\sin\theta$. The denominator for the directivity becomes

$$\int_0^{2\pi}\int_0^{\pi}\left[n+2\sum_{m=1}^{n-1}(n-m)\cos mu\right]\sin^3\theta\, d\theta\, d\varphi=4\pi W_c, \tag{1.56}$$

where

$$W_c=\frac{2n}{3}+4\sum_{m=1}^{n-1}\frac{n-m}{m^3k^3d^3}\sin(mkd)\cos(mkd\cos\theta_0)$$

$$-4\sum_{m=1}^{n-1}\frac{n-m}{m^2k^2d^2}\cos(mkd)\cos(mkd\cos\theta_0). \tag{1.57}$$

The factor $|E_{\max}|^2$ in the numerator of the directivity is no longer always n^2 in this case. It can be determined according to (1.55). For broadside arrays, $\theta_0=\pi/2, u=kd\cos\theta, |E_{\max}|$ is still n because both factors $\sin(nu/2)/\sin(u/2)$ and $\sin\theta$ assume maximum values at $\theta=\pi/2$ or $u=0$. For $\theta_0=0$, however, $u=kd(\cos\theta-1)$, $|E_{\max}|$ varies since one factor, $\sin(nu/2)/\sin(u/2)$, equals its maximum at $u=0$ or $\theta=0$, and the other factor, $\sin\theta$, vanishes at that point. Actually, under this condition, the pattern is no longer endfire because it has a null at $\theta=0$ rather than a maximum as supposed. Nevertheless, if we still desire to phase the array according to $\alpha_i=-ikd\cos\theta_0$ with $\theta_0=0$, the position at which $|E_{\max}|$ occurs will depend on kd and n. For example, if $kd=\pi$, $n=2$, Eq. (1.55) becomes

$$|E|=\left|\frac{\sin u}{\sin(u/2)}\sin\theta\right|=2\left|\sin\theta\cos\left[\frac{kd}{2}(\cos\theta-1)\right]\right|, \qquad (1.58)$$

which yields

$$|E_{\max}|=1.2981 \qquad \text{approximately at} \qquad \theta=51°. \qquad (1.59)$$

For higher values of n, we can always calculate $|E_{\max}|$ numerically from (1.55). We, therefore, have

$$D_c=\frac{n^2}{W_c}, \qquad \theta_0=\frac{\pi}{2}, \qquad (1.60)$$

or

$$D_c=\left|\frac{\sin(nu/2)}{\sin(u/2)}\sin\theta\right|^2_{\max}\Bigg/W_c, \quad \theta_0=0 \quad (\text{or any other } \theta_0 \text{ except } \theta_0=\pi/2). \qquad (1.61)$$

It is interesting to note that the limiting value of W_c as $d\to0$ is also $2n^2/3$, irrespective of θ_0. Therefore, we again have the expected value

$$D_c|_{d\to0}=\tfrac{3}{2}. \qquad (1.62)$$

A set of curves for D_c with $n=2$ and 3 is presented in Fig. 1.12 as a function of d for both $\theta_0=0$ and $\pi/2$. Similar results for a higher number of elements (with $\theta_0=\pi/2$ only) or plotted in the contour form can also, respectively, be found from articles by Tai,[1] Bach,[2] and Bach and Hansen.[3]

1.3 Improved Uniform Endfire Arrays of Isotropic Elements

In the previous section, we have studied the uniform endfire array of equally spaced isotropic elements where the elements are phased according to

$$\alpha_i=-ikd, \qquad i=0,1,2,\ldots,n-1. \qquad (1.63)$$

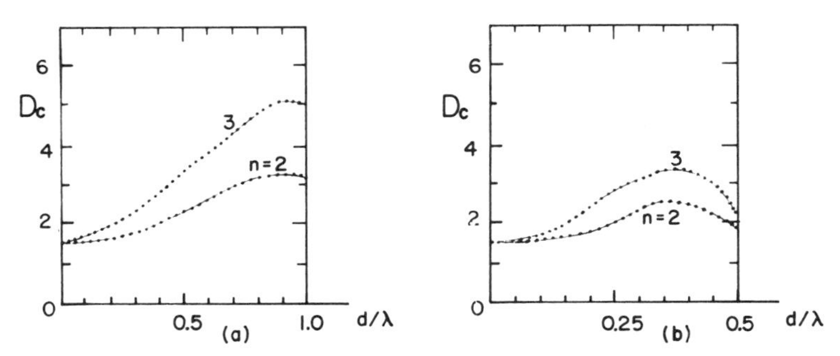

Fig. 1.12 Directivity of uniform arrays with short collinear dipoles: (*a*) $\alpha_i = 0$, (*b*) $\alpha_i = -ikd$.

The pattern for a typical five-element endfire array has been given in Fig. 1.7 and its directivities presented in Fig. 1.8 as a function of the element spacing d for $n = 2$, 3, and 4.

The subject to be considered here is also essentially a uniform array, and therefore can categorically be arranged as a part of the previous section. Since, however, the topic constitutes a significant contribution to array theory, we choose to discuss it in a separate section.

In 1938, Hansen and Woodyard[4] suggested that an increase in directivity can be achieved by increasing the progressive phase lag above that used in (1.63), that is,

$$\alpha'_1 = -i(kd + \delta), \qquad \delta > 0. \tag{1.64}$$

They also concluded that, for large n, the directivity will reach its maximum,

$$D_m = \frac{7.28(n-1)d}{\lambda} \cong \frac{7.28nd}{\lambda}, \tag{1.65}$$

when

$$\delta = \delta_m \cong \frac{2.94}{n-1}. \tag{1.66}$$

Equation (1.66) has since been known as the *Hansen-Woodyard condition* for maximizing the directivity of endfire arrays. However, this condition was derived originally for a continuous excitation distribution. Its validity is therefore subject to the conditions that the number of elements in the discrete array is very large and that the overall array length, $(n-1)d$, is much greater than the wavelength.

Using a somewhat different approach, Eaton, Eyges, and Macfarlane[5] later reached the same condition (1.66) and essentially the same result for directivity, $D_m = 1.82n$ for $d = \lambda/4$.

Since there are approximations and restrictions in both of these derivations, Maher[6] in 1960 formulated an exact approach by determining numerically the optimum value of δ in (1.64), when n and d are given, in order to obtain the maximum directivity.

For the purpose of distinguishing the case being studied from that analyzed in the previous section, the phase distribution in (1.63) is usually called the condition for an *ordinary endfire array* and that in (1.64) is referred to as the condition for an *improved endfire array*.

With (1.64), the array factor for an improved uniform endfire array becomes

$$|S| = \left| \frac{\sin(nu/2)}{\sin(u/2)} \right|, \tag{1.67}$$

where

$$u = kd(\cos\theta - 1) - \delta. \tag{1.68}$$

To understand the basic difference between the ordinary and improved endfire conditions, let us examine again the typical pattern for $n = 5$ and $kd = \pi/2$ in Fig. 1.13, although the principle can be explained with any n and d.

Clearly, the presence of δ shifts the visible range from AB in Fig. 1.13(a) to $A'B'$ in Fig. 1.13(b). As a consequence, the maximum value of $|S|$

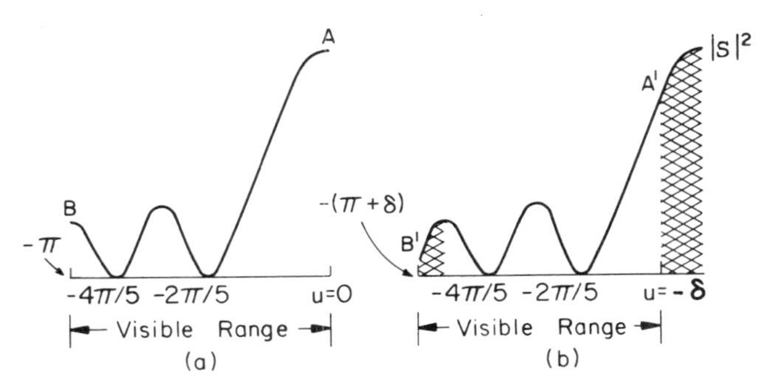

Fig. 1.13 A graphical interpretation for improving directivity of endfire array with $n = 5$ and $d = \lambda/4$: (a) ordinary endfire, $\delta = 0$, (b) improved endfire, $\delta > 0$.

actually decreases from $S_{\max}=5$ for $\delta=0$ to $|S_{\max}|=\sin(5\delta/2)/\sin(\delta/2)$ for $\delta>0$. The area under $|S|^2$ for the entire visible range, represented by

$$\int_0^{2\pi}\int_0^{\pi}|S|^2\sin\theta\,d\theta\,d\varphi=2\pi\int_0^{\pi}|S|^2\sin\theta\,d\theta, \tag{1.69}$$

also decreases because a larger portion of the area between $u=0$ and $u=-\delta$ in Fig. 1.13(*b*) is replaced by a smaller area between $u=-2kd$ and $u=-(2kd+\delta)$. Since the expression for directivity has $|S_{\max}|^2$ in the numerator and (1.69) in the denominator, and $|S_{\max}|^2$ decreases more slowly than the quantity in (1.69) for a small positive δ, we can see that the net effect of having δ is to increase the directivity. From Fig. 1.13(*a*) we also see that half of the beamwidth (between the beam maximum and the first null) is originally

$$\tfrac{1}{2}(\mathrm{BW})_e=\cos^{-1}\left(\tfrac{1}{5}\right)=78.4^\circ,$$

or

$$(\mathrm{BW})_e=156.8^\circ \qquad \text{for} \qquad \delta=0.$$

Now, the half beamwidth for the improved endfire array becomes

$$\tfrac{1}{2}(\mathrm{BW})_{ie}=\cos^{-1}\left(\frac{1}{5}+\frac{2\delta}{\pi}\right)<78.4^\circ \qquad \text{for} \qquad \delta>0.$$

Thus, we have

$$(\mathrm{BW})_{ie}<(\mathrm{BW})_e. \tag{1.70}$$

That is, another favorable effect of having δ is to make the main beamwidth narrower. Disadvantages are to definitely sacrifice relative levels of sidelobes and possibly the level of the backlobe at $\theta=\pi$. While the absolute levels of sidelobes for $\delta>0$ remain the same as those for $\delta=0$, their relative levels are higher because the level of the main beam is lower for the improved endfire than that for the ordinary endfire. The increase or decrease of the backlobe level depends on n, kd, and δ. When n is even and $kd=\pi/2$, there is a null at $\theta=\pi$ for the ordinary endfire array, and generally a sidelobe (backlobe) for the improved endfire array. Therefore, the level of the backlobe is increased by having $\delta>0$. On the other hand, if n is odd and $kd=\pi/2$, there is a sidelobe (backlobe) at $\theta=\pi$ for the ordinary endfire with its level determined by n. The level of the backlobe for the improved endfire may be lower or higher than that of the corresponding ordinary endfire, depending on δ. The details are shown in Fig. 1.14.

Obviously, δ must not be too large. For the example considered in Fig. 1.13(*b*), the entire main beam will be eliminated if δ is allowed to increase

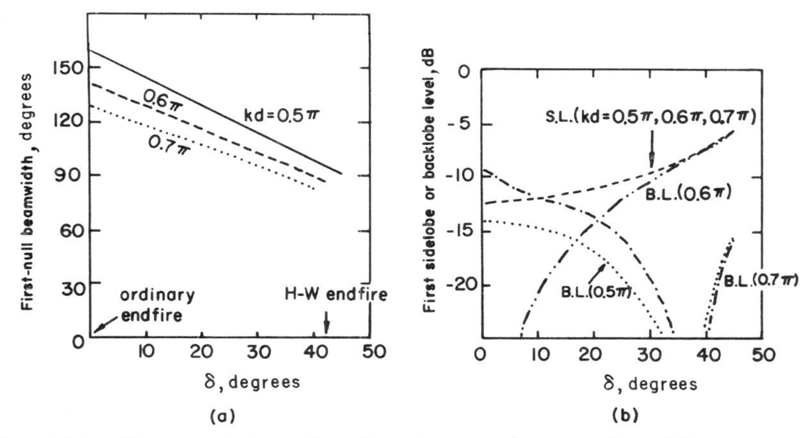

Fig. 1.14 Characteristics of a five-element improved endfire array: (*a*) First-null beamwidth, (*b*) First sidelobe or backlobe level.

to $2\pi/5$ radians. Therefore, the process of maximizing the directivity for endfire arrays works in the following manner: As δ is increased slowly, both the maximum radiation at $\theta=0$ and the total power radiated decrease, but at different rates to yield an increasing directivity. The directivity eventually reaches its maximum at a value δ_m. In what follows, we will try to determine δ_m when n and d are specified.

The directivity for improved endfire arrays can be written as

$$D_i = \frac{2\left|\dfrac{\sin(n\delta/2)}{\sin(\delta/2)}\right|^2}{\displaystyle\int_0^{\pi} |E|^2 \sin\theta \, d\theta}$$

$$= \frac{2kd\left|\dfrac{\sin(n\delta/2)}{\sin(\delta/2)}\right|^2}{\displaystyle\int_{-(\delta+2kd)}^{-\delta} \left[n + 2\sum_{m=1}^{n-1} (n-m)\cos mu \right] du}$$

$$= \frac{kd\left|\dfrac{\sin(n\delta/2)}{\sin(\delta/2)}\right|^2}{kdn + 2\displaystyle\sum_{m=1}^{n-1} \frac{n-m}{m} \sin(mkd)\cos(mkd+m\delta)}. \tag{1.71}$$

As a check, when $\delta=0$, Eq. (1.71) reduces to (1.39) with $\theta_0=0$. Figure 1.15 displays the variation of D_i with δ for $n=5$ and a number of element spacings. Corresponding directivities for the ordinary endfire ($\delta=0$) and that calculated according to (1.71) but using the Hansen-Woodyard condition (1.66) are also indicated therein for the purpose of comparison.

First-null beamwidths, levels of the first sidelobe, and the backlobe are pictured in Fig. 1.14. From these figures, it is clear that the value of δ_m varies with d even when n is kept the same. In general, the directivity increases, the first-null beamwidth decreases, and the level of the first sidelobe increase monotonically with δ in the interval $(0, \delta_m)$. Level of the backlobe can, however, vary widely in the same interval. For similar results with other values of n and d, readers are referred to the report by Maher.[6]

Although we have chosen in this section the uniform endfire array (i.e., $I_i=1$) to discuss the possible improvement in directivity offered by (1.64), the same principle should also apply to arrays with nonuniform amplitude excitations. Material on this latter subject is presented in the next section.

1.4 Finite *Z* Transforms—A Different Approach for Nonuniform Arrays

In Section 1.2, we have analyzed characteristics of uniform arrays where $I_i=1$. The convenience of being able to study the problem with a general term there is mainly due to the possibility of summing the array factor into a finite ratio form such as that shown in (1.14) or (1.16). Although the uniform array offers high directivity, the level of its first sidelobe is, at best, only about 13.5 dB below the main beam [as indicated by (1.33)], which may not be adequate for some applications. For a *nonuniform excitation*,

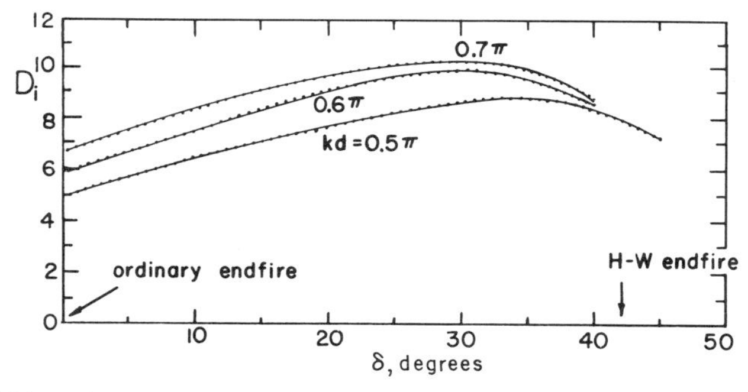

Fig. 1.15 Directivity of a five-element improved endfire array.

the I_i coefficients in (1.12) will not be equal, and a simple form for S such as that of (1.14) is not always obtainable. We would then be faced with the task of having to analyze a polynomial of n terms called the *array polynomial*. Each term in this polynomial consists of both real and imaginary parts. The order of complexity is proportional to the total number of terms involved. Based on a new approach developed by Cheng and Ma,[7,8] which treats linear arrays of discrete elements as a sampled-data system by using *finite Z transforms*, we have found it possible again to sum a large number of array polynomials associated with a variety of nonuniform excitations into a simpler form similar to that of (1.14). This simpler ratio form will be called the *array function*. The distinct facts concerning such an array function are that the denominator has a finite number of terms and is free of n, and that the numerator also has a fixed finite number of terms with n appearing in the exponents or the coefficients. Thus, an increase in n does not change the number of terms involved nor increase the complexity of the expression for S.

Suppose that the envelope of the amplitude excitations in a linear array can be described by a continuous function, $f(\zeta)$, within the interval $0 \leqslant \zeta \leqslant (n-1)d$. Then the excitation coefficients, I_i, in (1.12) can be written as

$$\begin{aligned} I_0 &= f(0), \\ I_1 &= f(d), \\ &\ \ \vdots \\ I_{n-1} &= f[(n-1)d]. \end{aligned} \tag{1.72}$$

The array polynomial (1.12) becomes

$$\begin{aligned} S &= \sum_{i=0}^{n-1} f(id) z^{-i} \\ &= \sum_{i=0}^{\infty} f(id) z^{-i} - \sum_{i=n}^{\infty} f(id) z^{-i}. \end{aligned} \tag{1.73}$$

Here the continuous function, $f(\zeta)$, has been considered the extended envelope function for the entire range $\zeta \geqslant 0$. Now, the first term on the right side of (1.73) is exactly what is called the *Z transform* of the function

$f(\zeta)$, or of the sampled function $f^*(\zeta)$, the sampling period being d[9,10]:

$$\sum_{i=0}^{\infty} f(id)z^{-i} = Z[f(\zeta)] \equiv F(z). \tag{1.74}$$

The second term on the right side of (1.73) can be considered as the Z transform of $f(\zeta)U(\zeta - nd)$:

$$\sum_{i=n}^{\infty} f(id)z^{-i} = Z[f(\zeta)U(\zeta - nd)] \equiv G(z), \tag{1.75}$$

where the shifted unit-step function is defined as

$$U(\zeta - nd)\begin{cases} =1, & \zeta \geqslant nd \\ =0, & \zeta < nd. \end{cases} \tag{1.76}$$

Under conditions that $f(\zeta)$ has an analytic expression and that the Z transform of $f(\zeta)$ exists, both $F(z)$ in (1.74) and $G(z)$ in (1.75), expressed in a ratio form, can be obtained directly from a Z transform table[11] or derived by using the relation between Z transforms and Laplace transforms discussed by Tamburelli.[12] In terms of the unit-gate function used by Christiansen,[13]

$$\gamma_n(\zeta) \equiv U(\zeta) - U(\zeta - nd)\begin{cases} =1, & 0 \leqslant \zeta < nd \\ =0, & \zeta \geqslant nd, \end{cases} \tag{1.77}$$

the array function $S(z)$ may be written more compactly as

$$S(z) = Z[f(\zeta)\gamma_n(\zeta)] = F(z) - G(z) \equiv \frac{Q_1(z)}{Q_2(z)}. \tag{1.78}$$

Since the upper index of summation in the first expression of (1.73) is finite $(n-1)$, $S(z)$ in (1.78) is to be called the *finite Z transform* of $f(\zeta)$.

Before demonstrating the advantage of using finite Z transforms for analyzing linear arrays by showing examples, we first tabulate some simple finite Z transforms and the *shifting theorem*,[9] which are useful for the later analysis. Detailed derivations of these formulas can be found from books covering control theory or sampled-data systems.[10] Extensive tables of finite Z transforms and corresponding numerical results for a large number of $f(\zeta)\gamma_n(\zeta)$ are given in reports by Christiansen,[14] and Ma and Walters.[15]

Table 1.1 A Short Table of Array Functions.

$$S(z) = Z[f(\zeta)\gamma_n(\zeta)] = Q_1(z)/Q_2(z)$$

$f(\zeta)$	$Q_1(z)$	$Q_2(z)$
$U(\zeta)$	$z - z^{-n+1}$	$z-1$
ζ	$d[z - nz^{-n+2} + (n-1)z^{-n+1}]$	$(z-1)^2$
ζ^2	$d^2[z^2 + z - n^2z^{-n+3} + (2n^2 - 2n - 1)z^{-n+2} - (n-1)^2z^{-n+1}]$	$(z-1)^3$
$e^{-a\zeta}$	$1 - e^{-and}z^{-n}$	$1 - e^{-ad}z^{-1}$
$\sin(a\zeta + b)$	$z^2 \sin b + z \sin(ad - b) - z^{-n+2} \sin(nad + b) + z^{-n+1} \sin[(n-1)ad + b]$	$z^2 - (2\cos ad)z + 1$
$\cos(a\zeta + b)$	$z^2 \cos b - z \cos(ad - b) - z^{-n+2} \cos(nad + b) + z^{-n+1} \cos[(n-1)ad + b]$	$z^2 - (2\cos ad)z + 1$
See Fig. 1.16	$z^2 - \dfrac{n+1}{n-1}z + \dfrac{n+1}{n-1}z^{-n+2} - z^{-n+1}$	$(z-1)^2$
See Fig. 1.17	$dz[1 - z^{-(n-1)/2}]^2$	$(z-1)^2$
See Fig. 1.18	$dz(1 - z^{-n/2})[1 - z^{-(n-2)/2}]$	$(z-1)^2$
See Fig. 1.19	$dz(1 - z^{-a})[1 - z^{-(n-1-a)}]$	$(z-1)^2$
Shifting theorem:	$Z[f(\zeta - nd)U(\zeta - nd)] = z^{-n}Z[f(\zeta)]$	

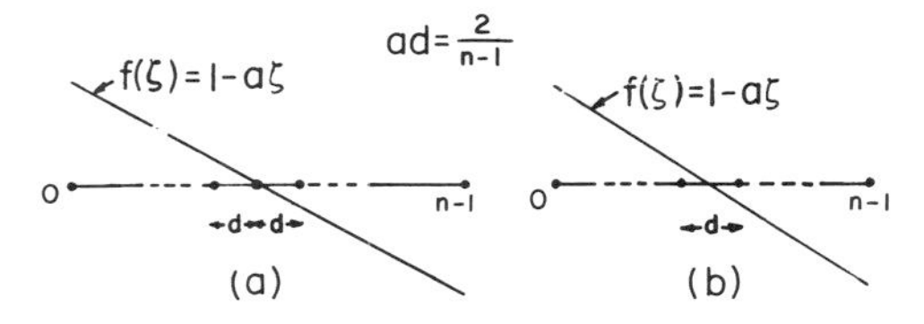

Fig. 1.16 Linear array with linear amplitude distribution of negative slope: (*a*) for odd *n*, (*b*) for even *n*.

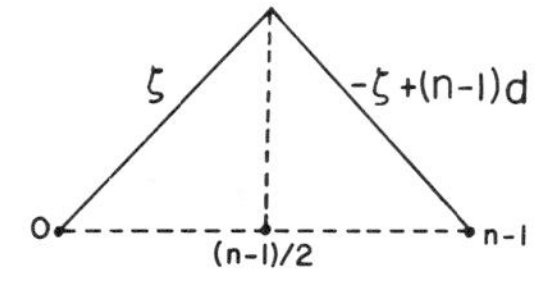

Fig. 1.17 Linear array with triangular amplitude distribution for odd n.

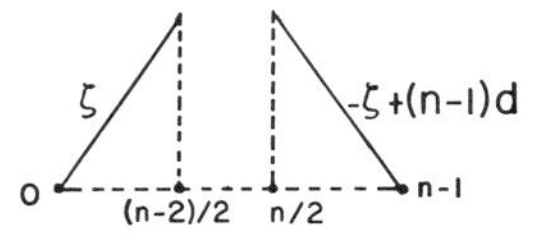

Fig. 1.18 Linear array with triangular amplitude distribution for even n.

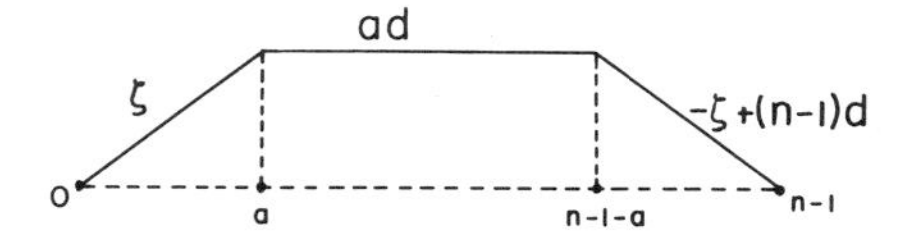

Fig. 1.19 Linear array with trapezoidal amplitude distribution.

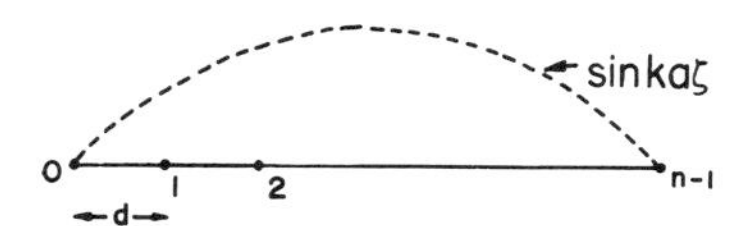

Fig. 1.20 Linear array with sinusoidal amplitude distribution.

Example 1. Analyze the array performance with the excitation function shown in Fig. 1.20. In this case, the envelope function is $f(\zeta)=\sin ka\zeta$, in $0 \leqslant \zeta \leqslant (n-1)d$, where a is a constant. Without using finite Z transforms, the array polynomial would be

$$S(z)=\sin(0)+[\sin(kad)]z^{-1}+\cdots+[\sin(n-1)kad]z^{-(n-1)}. \quad (1.79)$$

We then are confronted with difficulty because it is not obvious that the polynomial in (1.79) can be summed up into a neat form, without which further analysis cannot proceed in general terms.

Now, with the aid of Table 1.1, we have

$$S(z)=Z[\sin ka\zeta\gamma_n(\zeta)]$$

$$=\frac{z\sin(kad)-z^{-n+2}\sin(nkad)+z^{-n+1}\sin[(n-1)kad]}{z^2-2z\cos(kad)+1}, \quad (1.80)$$

which is indeed a convenient expression, and is the ratio form for the polynomial in (1.79). Its denominator is free of n, and the numerator contains n either in the exponent or in a coefficient. Equation (1.80) can be used as the starting point for further analysis of array characteristics.

From Fig. 1.20, we require the following condition for a practical symmetric excitation,

$$(n-1)kad=\pi, \quad (1.81)$$

under which (1.80) becomes

$$S(z)=\frac{(z+z^{-n+2})\sin[\pi/(n-1)]}{z^2-2z\cos[\pi/(n-1)]+1}$$

$$=\frac{(1+z^{-n+1})\sin[\pi/(n-1)]}{z+z^{-1}-2\cos[\pi/(n-1)]}. \quad (1.82)$$

Whenever an array polynomial is transformed into an *array function* such as (1.82), we are always interested in knowing what would happen if the denominator vanishes at a value of z. Physically, we are certain that $S(z)$ cannot be arbitrarily large (although we wish it could be). Mathematically, it is clear that the denominator of (1.82) will vanish when $z=e^{\pm j\pi/(n-1)}$. Since the numerator of $S(z)$, which has been designated as $Q_1(z)$ in (1.78), also vanishes at these particular values by noting

$$Q_1(z)|_{z=e^{\pm j\pi/(n-1)}}=(1+e^{\pm j\pi})\sin\left(\frac{\pi}{n-1}\right)=0,$$

we conclude that $z=e^{\pm j\pi/(n-1)}$ are not *poles*. In fact, the values of $S(z)$ at $z=e^{\pm j\pi/(n-1)}$ are finite because

$$\lim_{z\to e^{\pm j\pi/(n-1)}} S(z)=\lim_{z\to e^{\pm j\pi/(n-1)}}\left[\frac{(-n+1)z^{-n}\sin[\pi/(n-1)]}{1-z^{-2}}\right]$$

$$=\mp j\frac{n-1}{2}. \tag{1.83}$$

The principal maximum of (1.82) occurs at $z=1$, or $u=0$, with

$$S(1)=\frac{\sin[\pi/(n-1)]}{1-\cos[\pi/(n-1)]}=\cot\frac{\pi}{2(n-1)}. \tag{1.84}$$

Nulls are given by $z^{-n+1}=-1$, or

$$z=e^{\pm j(2p-1)\pi/(n-1)}, \qquad p=2,3,4,\ldots, \tag{1.85}$$

where $z=e^{\pm j\pi/(n-1)}$ for $p=1$ have been deleted because they are not zeros of $S(z)$ as evidenced by (1.83). Note that $z=e^{\pm j\pi}$ or $u=\pm\pi$ are always possible nulls when n is even. This same fact was also noted in Section 1.2 in connection with the array factor for uniform arrays.

Locations of sidelobes can be numerically determined by setting the derivative of $|S|$ equal to zero, where $|S|$, according to (1.82) and the relation $z=e^{-ju}$, may be written as

$$|S|=\left|\frac{\cos[(n-1)u/2]\sin[\pi/(n-1)]}{\cos u-\cos[\pi/(n-1)]}\right|. \tag{1.86}$$

For example, when $n=6$, we have

$$\csc 36^\circ\frac{d|S|}{du}=\frac{-2.5\sin 2.5u(\cos u-\cos 36^\circ)+\cos 2.5u\sin u}{(\cos u-\cos 36^\circ)^2},$$

which vanishes at $u'_1=\pm 140.3^\circ$, and

$$S_{u'_1}=0.3675,$$

$$\frac{S_{u'_1}}{S(1)}=0.1194, \qquad \text{or } -18.5\text{ dB}. \tag{1.87}$$

The directivity for $n=6$ can also be evaluated,

$$D=\frac{2kd\cot^2(\pi/10)}{\int_a^b |S|^2\,du}$$

$$=\frac{2kd\cot^2(\pi/10)\csc^2(\pi/5)}{\int_a^b \left\{\cos^2(5u/2)\,du/\left[\cos u-\cos(\pi/5)\right]^2\right\}}, \tag{1.88}$$

where

$$a=-kd(1+\cos\theta_0),\qquad b=kd(1-\cos\theta_0),$$
$$\theta_0=\frac{\pi}{2}\ \text{for broadside},\qquad \theta_0=0\ \text{for ordinary endfire}. \tag{1.89}$$

After some lengthy algebra, the denominator of (1.88) becomes

$$W=\left[\tfrac{2}{3}\sin 3u+4\cos\frac{\pi}{5}\sin 2u+8\cos\frac{\pi}{5}\left(1+\cos\frac{\pi}{5}\right)\sin u+\left(6+4\cos\frac{2\pi}{5}\right)u\right]_a^b$$

$$=2\left[\tfrac{2}{3}\sin 3kd\cos(3kd\cos\theta_0)+4\cos\frac{\pi}{5}\sin 2kd\cos(2kd\cos\theta_0)\right.$$

$$\left.+8\cos\frac{\pi}{5}\left(1+\cos\frac{\pi}{5}\right)\sin kd\cos(kd\cos\theta_0)+\left(6+4\cos\frac{2\pi}{5}\right)kd\right]. \tag{1.90}$$

Substituting (1.90) into (1.88), we can calculate D when d and θ_0 are specified. In particular, when $d=\lambda/2$ or $kd=\pi$, the directivity simplifies to

$$D=\frac{\cot^2(\pi/10)\csc^2(\pi/5)}{6+4\cos(2\pi/5)}=3.7887,$$

regardless of θ_0.

Based on the results calculated above, we can summarize in Table 1.2 the radiation characteristics for the array with $n=6$, $d=\lambda/2$, and the excitation function shown in Fig. 1.20. Note that although $n=6$ was assumed in the above example, the general formulation for $f(\zeta)=\sin ka\zeta$ should work for any larger n with the understanding that the expression in (1.90) will be more complicated. Note also that there are actually only four excited elements in the array discussed in the above example, because excitation coefficients for the first and last elements are nil with $f(\zeta) = \sin ka\zeta$ under the condition (1.81). At this point, it should be instructive to compare the major results for an array with a nonuniform excitation such as the example just considered to those for arrays with a uniform excitation studied previously. For this purpose, a corresponding table (Table 1.3) for $n=4$, $d=\lambda/2$, and $f(\zeta)=U(\zeta)$ is given.

Two important results can be noted by comparing these two tables. First, the directivity for the array with $f(\zeta)=\sin ka\zeta$ is smaller than that for the uniform array. Second, the improvement in the level of the first sidelobe for the array with $f(\zeta)=\sin ka\zeta$ over that for the uniform array is made at the expense of having a wider beamwidth. The first result is true only when $d=\lambda/2$. The fact that the uniform array yields maximum

Table 1.2 Radiation Characteristics for the Linear Array with $n=6$, $d=\lambda/2$, $f(\zeta)=\sin ka\zeta, a=\frac{1}{5}$.

	$\theta_0=\pi/2$ (Broadside)	$\theta_0=0$ (Ordinary Endfire)
Principal maximum	3.0777	3.0777
Location of principal maximum	$u=0$; or $\theta=90°$	$u=0$ and $u=-2\pi$; $\theta=0$ and 180° (bidirectional)
Location of nulls	$u=\pm 3\pi/5$ and $u=\pm\pi$; $\theta=0°, 53.1°$, $126.9°, 180°$	$u=-3\pi/5$, $-\pi$, and $u=-7\pi/5$; $\theta=66.4°, 90°$ $113.6°$
First-null beamwidth	73.8°	132.8°
Location of sidelobes	$u=\pm 140.3°$; $\theta=38.8°$ and $141.2°$	$u=-140.3°$ and $-219.7°$; $\theta=77.3°$ and $102.7°$
Level of first side lobe relative to the principal maximum	−18.5 dB	−18.5 dB
Directivity	3.7887	3.7887

directivity when $d=p\lambda/2$, $p=1, 2, \ldots,$has been proved elsewhere.[5] The second result is, in general, true whenever the array has a *concave downward* type of excitation where the amplitude excitations for the central elements are larger than those for the end elements.

Now, we wish to illustrate, by another example, the opposite effect (narrower beamwidth at the expense of having higher sidelobe level) for arrays having a *concave upward* type of excitation such as that shown in Fig. 1.21. This case not only will serve the purpose of demonstrating once more the easy manner of summing an array polynomial into an array function by using finite Z transforms; it also will reveal the important principle of superposition concerning a composite excitation function. Of course, the reason the principle of superposition applies is that Z transforms and related operations can be used for treating linear systems only.

Table 1.3 Radiation Characteristics for the Linear Array with $n=4$, $d=\lambda/2$, and $f(\zeta)=U(\zeta)$.

	$\theta_0=\pi/2$ (Broadside)	$\theta_0=0$ (Ordinary Endfire)
Principal maximum	4	4
Location of principal maximum	$u=0$; or $\theta=90°$	$u=0$ and -2π; $\theta=0°$ and $180°$
Location of nulls	$u=\pm\pi/2$ and $u=\pm\pi$; $\theta=0°, 60°, 120°, 180°$	$u=-\pi/2, -\pi$, and $-3\pi/2$; $\theta=60°, 90°$ $120°$
First-null beamwidth	60°	120°
Location of sidelobes	$u=\pm 131.8°$; $\theta=42.9°$ and $137.1°$	$u=-131.8°$ and $-228.2°$ $\theta=74.5°$ and $105.5°$
Level of first sidelobe relative to the principal maximum	−11.3 dB	−11.3 dB
Directivity	4	4

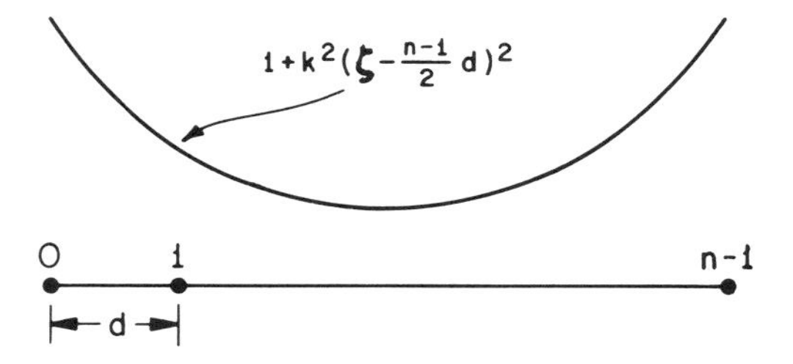

Fig. 1.21 Linear array with $1+k^2[\zeta-(n-1)d/2]^2$ as its amplitude distribution.

If the envelope of the amplitude distribution in a linear array is a linear combination of two or more component functions such that

$$f(\zeta)=\sum_{i=1}^{m} c_i f_i(\zeta), \qquad 0\leqslant\zeta\leqslant(n-1)\,d, \tag{1.91}$$

then the array function in the ratio form after applying finite Z transforms can be expressed as

$$S(z)=Z\,[\,f(\zeta)\gamma_n(\zeta)\,]=\sum_{i=1}^{m} c_i S_i(z), \tag{1.92}$$

where

$$S_i(z)=Z\,[\,f_i(\zeta)\gamma_n(\zeta)\,]$$

is the ith component array function and $\gamma_n(\zeta)$ is defined in (1.77).

Example 2. With the aid of (1.92) and Table 1.1, analyze the array, whose amplitude excitation is shown in Fig. 1.21, in general terms. In this case,

$$f(\zeta)=1+k^2\left(\zeta-\frac{n-1}{2}d\right)^2$$

$$=1+\frac{(n-1)^2}{4}k^2d^2-(n-1)k^2d\zeta+k^2\zeta^2, \tag{1.93}$$

and

$$S(z)=Z[f(\zeta)\gamma_n(\zeta)]$$

$$=\left[1+\frac{(n-1)^2k^2d^2}{4}\right]\frac{z-z^{-n+1}}{z-1}-(n-1)k^2d^2\frac{z-nz^{-n+2}+(n-1)z^{-n+1}}{(z-1)^2}$$

$$+k^2d^2\frac{z^2+z-n^2z^{-n+3}+(2n^2-2n-1)z^{-n+2}-(n-1)^2z^{-n+1}}{(z-1)^3}$$

$$=\frac{Q_1(z)}{Q_2(z)}, \qquad (1.94)$$

where

$$Q_1(z)=4(z-1)^2(z-z^{-n+1})+k^2d^2\big[(n-1)^2(z^3-z^{-n+1})$$
$$-(2n^2-6)(z^2-z^{-n+2})+(n+1)^2(z-z^{-n+3})\big], \qquad (1.95)$$

$$Q_2(z)=4(z-1)^3. \qquad (1.96)$$

The principal maximum occurs at $z=1$ or $u=0$,

$$S(1)=\lim_{z\to 1}\frac{Q_1(z)}{Q_2(z)}=\lim_{z\to 1}\frac{Q_1'''(z)}{Q_2'''(z)}=n+\frac{n(n-1)(n+1)k^2d^2}{12}. \qquad (1.97)$$

Nulls are given by the roots of

$$Q_1(z)=0, \qquad (1.98)$$

excluding $z=1$ which is the location of the principal maximum. From (1.95) it is clear that $z=-1$ or $u=\pm\pi$ are also possible nulls when n is even. In general, the roots of (1.98) depend on n and kd. They are listed as follows for smaller values of n:

$$\cos u_1=-1, \qquad \text{for } n=2: \qquad (1.99)$$

$$\cos u_1=\frac{-1}{2(1+k^2d^2)}, \qquad \text{for } n=3; \qquad (1.100)$$

$$\cos u_1 = \frac{4k^2d^2}{4+9k^2d^2} \text{ and } \cos u_2 = -1, \qquad \text{for } n=4; \tag{1.101}$$

$$\cos u_{1,2} = \frac{-(1+k^2d^2) \pm \sqrt{5+50k^2d^2+129k^4d^4}}{4(1+4k^2d^2)}, \qquad \text{for } n=5. \tag{1.102}$$

The power pattern, $|S|^2$, can be expressed in terms of the real variable u because

$$|S|^2 = \frac{Q_1(z)Q_1(z^{-1})}{Q_2(z)Q_2(z^{-1})}, \tag{1.103}$$

in which the pair, $z^m + z^{-m} = 2\cos mu$, always appears together in both numerator and denominator. Locations of sidelobes can be obtained by solving numerically for

$$\frac{d|S|^2}{du} = 0. \tag{1.104}$$

excluding those roots which have already been identified as locations of the principal maximum and nulls. For $n=4$, the only root in (1.104) for the sidelobe is u'_1, which satisfies

$$\cos u'_1 = \frac{-2(4+7k^2d^2)}{3(4+9k^2d^2)}. \tag{1.105}$$

The directivity for $n=4$ can also be evaluated as

$$D = \frac{2kd(4+5k^2d^2)^2}{W}, \tag{1.106}$$

where

$$\begin{aligned} W &= \int_a^b |S(u)|^2 du \\ &= 2[A_3 \sin 3u + A_2 \sin 2u + A_1 \sin u + A_0 u]_{a=-kd(1+\cos\theta_0)}^{b=kd(1-\cos\theta_0)} \\ &= 4[A_3 \sin 3kd \cos(3kd\cos\theta_0) + A_2 \sin 2kd \cos(2kd\cos\theta_0) \\ &\quad + A_1 \sin kd \cos(kd\cos\theta_0) + A_0 kd], \end{aligned} \tag{1.107}$$

$$
\begin{aligned}
A_3 &= \tfrac{1}{3} + \tfrac{3}{2}k^2d^2 + \tfrac{27}{16}k^4d^4, \\
A_2 &= 1 + \tfrac{5}{2}k^2d^2 + \tfrac{9}{16}k^4d^4, \\
A_1 &= 3 + \tfrac{11}{2}k^2d^2 + \tfrac{19}{16}k^4d^4, \\
A_0 &= 1 + 5k^2d^2 + \tfrac{41}{8}k^4d^4.
\end{aligned}
\tag{1.108}
$$

Substituting (1.107) and (1.108) into (1.106), we can calculate the directivity with known values of d and θ_0. In particular, when $d=\lambda/2$ or $kd=\pi$, D simplifies to

$$
D = \frac{(4+5k^2d^2)^2}{2A_0} = 2.5893 \qquad \left(d=\frac{\lambda}{2},\ n=4\right),
$$

regardless of θ_0. The results for $n=4$, $d=\lambda/2$, and the amplitude excitation function shown in Fig. 1.21 are summarized in Table 1.4. It is clear that the array performance described in Table 1.4 is not at all attractive, because the level of the first sidelobe is only 1.87 dB below that of the main beam. This tells why the "concave upward" type of excitation distribution is very rarely used in practice.

1.5 Nonuniformly Spaced Arrays

In Section 1.4, we have studied, using finite Z transforms, arrays with nonuniform amplitude excitations but constant element spacing and uniformly progressive phase. We also showed therein, by two examples, possibilities for improving some of the array characteristics over those for the uniform array where the amplitude excitation, element spacing, and progressive phase are all kept uniform. Two other alternatives, nonuniform element spacings and nonuniformly progressive phases, are possible approaches to achieving a result better in certain respects than that for the uniform array. In this section, we will analyze the effect on array performance of using variable element spacings, leaving the topic of arrays with nonuniformly progressive phases to be taken up in the next section.

Since the problem of *nonuniformly spaced arrays* was first studied by Unz,[16] many papers on the subject have appeared in the literature.[17–27] Most of the work treats the problem from the synthesis viewpoint, trying to determine a favorable set of spacings in order to improve some of the radiation characteristics. For example, it has been shown that for such an array the level of sidelobes can be lowered, grating lobes eliminated when the average element spacing is large, and the total number of elements

Table 1.4 Radiation Characteristics for the Linear Array with $n=4$, $d=\lambda/2$, and $f(\zeta)=1+k^2[\zeta-(n-1)d/2]^2$.

	$\theta_0=\pi/2$ (Broadside)	$\theta_0=0$ (Ordinary Endfire)
Principal maximum	53.3483	53.3483
Location of principal maximum	$u=0$; or $\theta=90°$	$u=0$, and -2π; $\theta=0°$ and $180°$
Location of nulls	$u=\pm 64.83°$ and $\pm 180°$; $\theta=0°, 68.9°, 111.1°, 180°$	$u=-64.83°, -180°$ and $-295.17°$; $\theta=50.2°, 90°$ $129.8°$
First-null beamwidth	42.2°	100.4°
Location of sidelobes	$u=\pm 121.65$; $\theta=47.5°$, $132.5°$	$u=-121.65°, -238.35°$; $\theta=71.1°, 108.9°$
Level of first sidelobe relative to the principal maximum	−1.87 dB	−1.87 dB
Directivity	2.5893	2.5893

required in the array reduced. These results were achieved by various techniques, some of which will be discussed in detail in Chapter 2 when we present the synthesis work. Here, we are content to analyze two simple arrays only: one of these is to have nonuniform spacings but constant amplitude excitation, and the other is to have both the spacings and amplitude excitations varied. In both cases the number of elements in the array is rather small to reduce the algebraic burden. The purpose of doing this is twofold: to lay the ground work for later dealing with the synthesis problem and to provide the reader with a feeling of how important characteristics of such arrays can be different from those already analyzed.

A. Nonuniformly Spaced but Uniformly Excited Arrays of Isotropic Elements. From the mathematical and practical viewpoints, we require that array elements be symmetrically situated and excited. It is, therefore, more convenient to choose the array center as the coordinate origin. When the total number of elements in the array is odd, $n=2N+1$, such as that shown in Fig. 1.22, the array factor can be written as a real function,

$$S=1+\sum_{i=1}^{N}(e^{-ju_i}+e^{ju_i})=1+2\sum_{i=1}^{N}\cos u_i, \tag{1.109}$$

where

$$u_i=kd_i(\cos\theta-\cos\theta_0), \qquad i=1,2,\ldots,N. \tag{1.110}$$

For $N=2$ or $n=5$, we have

$$E=1+2\cos u_1+2\cos u_2=1+2\cos au_2+2\cos u_2, \tag{1.111}$$

where

$$a=d_1/d_2. \tag{1.112}$$

Naturally, the beam maximum occurs at $u_2=0$ with $S_{\max}=5$. Other characteristics depend on a and d_2. Since

$$\int_0^{2\pi}\int_0^{\pi}|E|^2\sin\theta\,d\theta\,d\varphi=\frac{2\pi}{kd_2}\int_a^b|E|^2\,du_2=\frac{4\pi W}{kd_2},$$

where

$$b=kd_2(1-\cos\theta_0),$$

$$a=-kd_2(1+\cos\theta_0),$$

and

$$\begin{aligned}W={}&5kd_2+4\sin kd_2\cos(kd_2\cos\theta_0)+\sin(2kd_2)\cos(2kd_2\cos\theta_0)\\&+\frac{4}{a}\sin(akd_2)\cos(akd_2\cos\theta_0)+\frac{1}{a}\sin(2akd_2)\cos(2kd_2\cos\theta_0)\\&+\frac{4}{1+a}\sin[(1+a)kd_2]\cos[(1+a)kd_2\cos\theta_0]\\&+\frac{4}{1-a}\sin[(1-a)kd_2]\cos[(1-a)kd_2\cos\theta_0],\end{aligned} \tag{1.113}$$

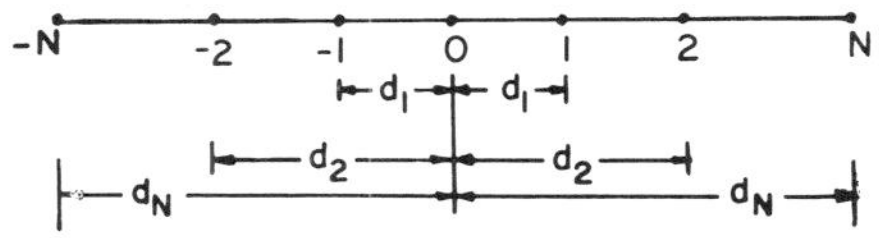

-N -2 -1 0 1 2 N
d_1 d_1
d_2 d_2
d_N d_N

Fig. 1.22 A nonuniformly spaced symmetric array.

the directivity for such an array becomes

$$D=\frac{25kd_2}{W}, \tag{1.114}$$

which is identical to (1.39) when $d_2=2d$ and $a=\frac{1}{2}$.

The directivity as a function of the parameter a representing the ratio of d_1 to d_2 is plotted in Fig. 1.23(a) for $\theta_0=\pi/2$ and $kd_2=3\pi/2$, 2π, $5\pi/2$. The associated first-null beamwidth and level of the first sidelobe are respectively given in Figs. 1.23(b) and (c). It is interesting to see from Fig. 1.23(a) that although the maximum directivity always occurs at $a<0.5$, the value of D_{max} is not much higher than that for the uniform array ($a=0.5$). From Figs. 1.23(b) and (c) we also see that the level of the first sidelobe can be reduced with $a<0.5$ at the expense of having wider beamwidth, and that the beamwidth can be made narrower with $a>0.5$ at the sacrifice of sidelobe level. We should note another important fact that increasing the overall array length by increasing d_2 does not always improve the directivity, as it can clearly be seen from Fig. 1.23(a) for $0.7\leqslant a\leqslant 0.9$. This property is in contrast with that for the uniform array where the directivity is always increased by having a larger element spacing (hence the overall array length) provided, of course, that the element spacing is not too large to cause appearance of the grating lobe. It means that the relative positions of elements in the array should be carefully considered when nonuniform spacings are used.

B. Nonuniformly Spaced and Excited Arrays of Isotropic Elements. When, as before, the array is symmetrically configured and excited, and the number of elements is odd, the array factor may be expressed as

$$S=I_0+2\sum_{i=1}^{N} I_i \cos u_i. \tag{1.115}$$

If we again examine the case with $N=2$ or $n=5$, Eg. (1.115) becomes

$$S=I_0+2(I_1\cos au_2+I_2\cos u_2), \tag{1.116}$$

with $S_{max} = I_0+2(I_1+I_2)$ occurring at $u_2=0$. The expression for the directivity can also be integrated out:

$$D=\frac{kd_2(I_0+2I_1+2I_2)^2}{W'}, \tag{1.117}$$

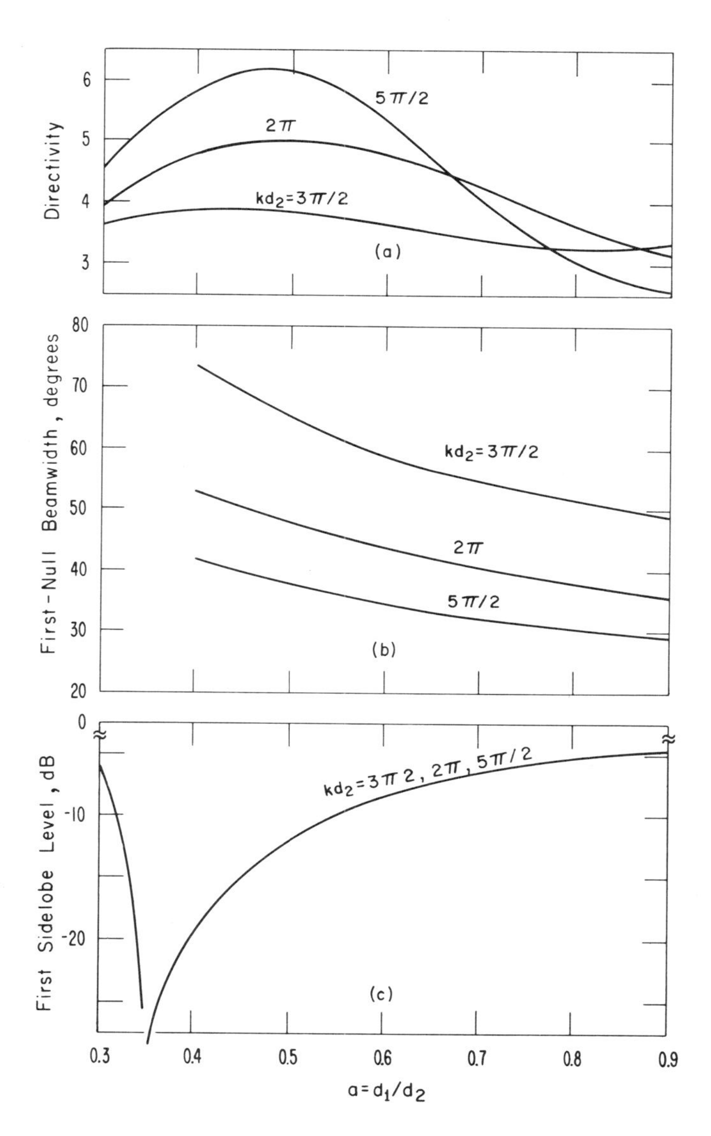

Fig. 1.23 Characteristics of a five-element nonuniformly spaced but uniformly excited symmetric array: (*a*) directivity, (*b*) first-null beamwidth, (*c*) level of the first sidelobe.

where

$$W' = (I_0^2 + 2I_1^2 + 2I_2^2)kd_2 + 4I_2I_0 \sin kd_2 \cos(kd_2 \cos\theta_0)$$
$$+ I_2^2 \sin(2kd_2)\cos(2kd_2\cos\theta_0) + \frac{4I_1I_0}{a}\sin(akd_2)\cos(akd_2\cos\theta_0)$$
$$+ \frac{I_1^2}{a}\sin(2akd_2)\cos(2akd_2\cos\theta_0)$$
$$+ \frac{4I_1I_2}{1+a}\sin[(1+a)kd_2]\cos[(1+a)kd_2\cos\theta_0]$$
$$+ \frac{4I_1I_2}{1-a}\sin[(1-a)kd_2]\cos[(1-a)kd_2\cos\theta_0]. \qquad (1.118)$$

A set of numerical results with $\theta_0 = \pi/2$ is presented in Fig. 1.24 for $I_0 = 0.2704$, $I_1 = 0.2341$, and $I_2 = 0.1302$. This kind of amplitude excitation corresponds roughly to the sinusoidal distribution shown in Fig. 1.20 for the equally spaced counterpart. Basic features remain the same as the previous case displayed in Fig. 1.23. The major differences between these two cases are that the location of maximum directivity for a fixed d_2 is now shifted to $a > 0.5$, and that the level of the first sidelobe is still reasonable when the parameter a is moderately larger than 0.5.

Although we have chosen an odd n to discuss the subject and presented the results for a broadside array only, the technique will apply also to an endfire array and arrays with an even number of elements. When the number of elements in the array is larger, the number of parameters, $a_i = d_i/d_N$, under one's control will also be larger. A greater improvement in resultant characteristics may be possible if a favorable set of a_i can be determined.[18,26] This will be studied in detail in Chapter 2 when the synthesis problem is considered.

1.6 Arrays of Isotropic Elements with Nonuniformly Progressive Phases

In Sections 1.2–1.4 we have studied various arrays in which the element spacing and progressive phase were kept uniform. Mathematically speaking, these restrictions are

$$\zeta_i = id, \qquad \alpha_i = -ikd\cos\theta_0, \qquad i = 0, 1, \ldots, (n-1), \qquad (1.119)$$

with θ_0 determining the position of the beam maximum.

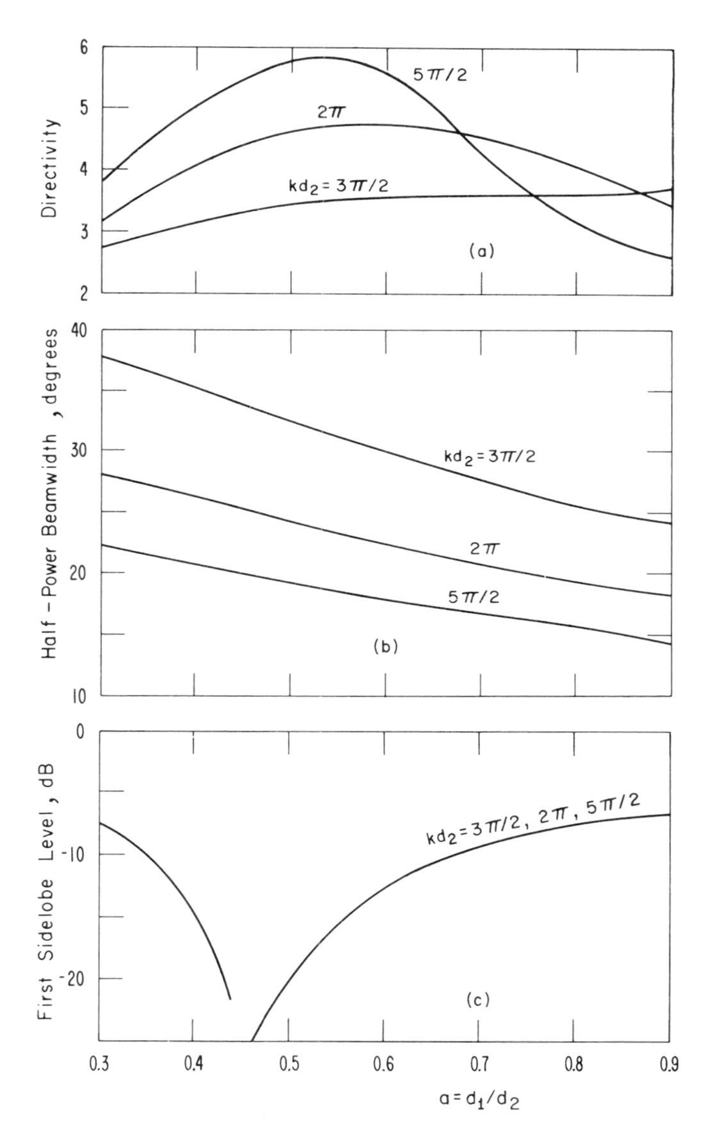

Fig. 1.24 Characteristics of a symmetric five-element array, nonuniformly spaced and excited, with $I_0 = 0.2704$, $I_1 = 0.2341$, $I_2 = 0.1302$; (*a*) directivity, (*b*) H-P beamwidth, (*c*) first S. L.

Now, let us consider the array with a *modified progressive phase* (but still with constant element spacing),

$$\alpha'_i = i\alpha + \delta_i, \tag{1.120}$$

where α and δ_i are arbitrary. The array factor in this case can be written as

$$S_{\text{NU}}(z) = \sum_{i=0}^{n-1} I_i \exp[j(ikd\cos\theta + \alpha'_i)] = \sum_{i=0}^{n-1} a_i z^{-i}, \tag{1.121}$$

where

$$a_i = I_i e^{j\delta_i}, \tag{1.122}$$

$$z = e^{-ju}, \tag{1.123}$$

$$u = kd\cos\theta + \alpha, \tag{1.124}$$

and the subscript NU signifies nonuniformity in progressive phasing.

On the surface, the mathematical form of (1.121) is very similar to that in (1.12), with a_i replacing I_i. However, because of the simple difference that a_i is complex while I_i was real, we find the analysis of arrays with *nonuniformly progressive phases* more challenging. For example, we are not even sure now whether the beam maximum still occurs at $z=1\,(u=0)$ even with positive values for I_i. The facts that the radiation patterm is no longer symmetric with respect to the beam maximum and that the Z transform technique cannot be applied any more give added difficulty. The task of finding the roots of $|S_{\text{NU}}(z)|=0$ and $(d/dz)|S_{\text{NU}}(z)|=0$ to determine the locations of nulls and sidelobes is even more formidable. Fortunately, we can simplify the problem by a special transformation of variables to be introduced later in this section. This special transformation is obtained through a study of the power pattern. Often, it is to our advantage to deal with the power pattern since it involves real quantities only. In order to discover the difference between the array with nonuniformly progressive phases (NUPP) being studied and that with uniformly progressive phases (UPP) considered previously, let us first formulate the power pattern for the UPP array. It is defined as

$$|S_{\text{U}}|^2 = S_{\text{U}}(z)\,\overline{S_{\text{U}}(z)} = \left(\sum_{i=0}^{n-1} I_i z^{-i}\right)\left(\sum_{i=0}^{n-1} I_i z^{i}\right)$$

$$= \sum_{m=0}^{n-1} I'_m(z^m + z^{-m}), \tag{1.125}$$

where

$$I'_m = \sum_{i=0}^{n-1-m} I_i I_{m+i}, \quad 1 \leqslant m \leqslant (n-1), \tag{1.126}$$

$$I'_0 = \frac{1}{2} \sum_{i=0}^{n-1} I_i^2, \tag{1.127}$$

and the subscript U signifies uniformity in progressive phasing. Since the pair z^m and z^{-m} always appears together in (1.125), it is clear that $|S_U|^2$ is a real function of u,

$$|S_U|^2 = 2 \sum_{m=0}^{n-1} I'_m \cos mu. \tag{1.128}$$

While the form in (1.128) is very helpful for finding the denominator of the directivity, it is still rather inconvenient for determining the number of nulls and sidelobes. Alternatively, (1.128) can be expressed as a real polynomial,

$$P_U(y) = \sum_{m=0}^{n-1} A_m y^m, \tag{1.129}$$

if we make the following substitution,

$$z + z^{-1} = 2\cos u = y,$$

$$z^2 + z^{-2} = (z + z^{-1})^2 - 2 = y^2 - 2,$$

$$z^3 + z^{-3} = (z + z^{-1})^3 - 3(z + z^{-1}) = y^3 - 3y, \tag{1.130}$$

$$\ldots .$$

In (1.129), A_m is, of course, the final combined coefficient made of I'_m or I_i through (1.126) and (1.127). Indeed, the form in (1.129) is convenient because it is real and the nulls and sidelobes can be obtained by setting, respectively,

$$P_U(y) = 0 \tag{1.131}$$

and

$$\frac{d}{dy} P_U(y) = 0. \tag{1.132}$$

Now, let us go back to NUPP arrays. The power pattern in this case, based on (1.121), becomes

$$|S_{\mathrm{NU}}|^2=\left(\sum_{i=0}^{n-1} a_i z^{-i}\right)\left(\sum_{i=0}^{n-1} \bar{a}_i z^{i}\right)=\sum_{m=0}^{n-1}\left(b_m z^m+\bar{b}_m z^{-m}\right), \quad (1.133)$$

where

$$b_m=\sum_{i=0}^{n-1-m} a_i \bar{a}_{m+i}, \quad 1\leqslant m\leqslant(n-1), \quad (1.134)$$

$$b_0=\tfrac{1}{2}\sum_{i=0}^{n-1}|a_i|^2, \quad (1.135)$$

and $\bar{a}_i$ and $\bar{b}_m$ are, respectively, complex conjugates of a_i and b_m. In this case b_m is complex and (1.133) is different from (1.125). The expression corresponding to (1.129) after the substitution in (1.130) is used takes the form

$$P_{\mathrm{NU}}(y)=\sum_{m=0}^{n-1} A'_m y^m+\left(\sum_{m=0}^{n-2} A''_m y^m\right)(4-y^2)^{1/2}, \quad (1.136)$$

where the combined coefficients A'_m and A''_m are all real. Comparing (1.136) with (1.129), we see that the main difference between NUPP and UPP arrays is the presence of the extra factor $(4-y^2)^{1/2}$ in NUPP arrays. With (1.136) itself, we still have difficulty in determining the nulls and sidelobes through $P_{\mathrm{NU}}(y)=0$ and $(d/dy)P_{\mathrm{NU}}(y)=0$. Since, however, we can readily verify that[5] if z_i is a zero of (1.133), so also is its conjugate reciprocal $1/\bar{z}_i$, therefore a special transformation of variables such as

$$y=\frac{4x}{1+x^2}, \quad (4-y^2)^{1/2}=\frac{2(1-x^2)}{1+x^2} \quad (1.137)$$

is suggested for (1.136) so that the original irrational polynomial in y will become a rational polynomial in x,

$$p_{\mathrm{NU}}(x)=\frac{Q(x)}{(1+x^2)^{n-1}}, \quad (1.138)$$

where $Q(x)$ is a polynomial of degree $2(n-1)$ and its coefficients depend on a_i in (1.133). In terms of the variable z, the transformation in (1.137) is equivalent to a special *bilinear transformation,*[28]

$$x=\frac{(1+j)z+(1-j)}{(1-j)z+(1+j)} \tag{1.139}$$

or

$$z=\frac{-(1+j)x+(1-j)}{(1-j)x-(1+j)}. \tag{1.140}$$

It is, in general, easier to handle a polynomial such as $Q(x)$ than one like $|S_{\mathrm{NU}}(z)|^2$. The success of this technique lies in the fact that the transformation in (1.139) or (1.140) converts the zero pairs which are inverse complex conjugates of each other into ordinary complex conjugate pairs. Figure 1.25 exhibits detailed one-to-one mappings between the z and x planes.

With this complicated transformation presented, we are now ready to analyze the characteristics of NUPP arrays. First of all, it is not trivial to determine the location of the principal maximum even when the condition $\alpha=0$ is inserted in (1.124) because the main lobe no longer occurs at $\theta=\pi/2$ with the appearance of δ_i in (1.121). However, when

$$\alpha=-(kd+\alpha_1), \qquad \frac{\pi}{2}>\alpha_1\geqslant 0, \qquad kd<\pi, \tag{1.141}$$

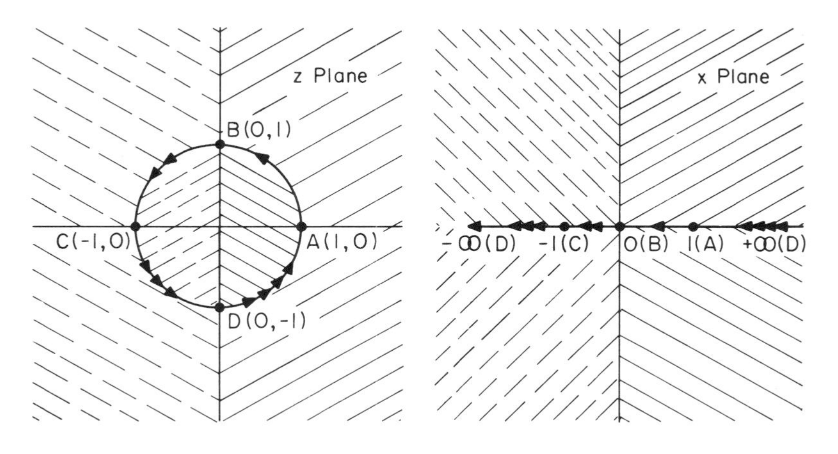

Fig. 1.25 One-to-one mappings between z plane and x plane.

the visible principal maximum still occurs at $\theta=0$, $u=-\alpha_1$, $z=e^{j\alpha_1}$, or $x=x_1$ where $0<x_1\leqslant 1$ (see Fig. 1.25), giving

$$|S_{\mathrm{NU}}|_{\max}=\left|\sum_{i=0}^{n-1} a_i e^{-ji\alpha_1}\right| \quad \text{or} \quad \sqrt{p_{\mathrm{NU}}|_{\max}}=\sqrt{p_{\mathrm{NU}}(x_1)}\ .$$

Note that, in terms of UPP arrays, the cases $\alpha_1=0$ and $\alpha_1>0$ correspond, respectively, to the ordinary and improved endfire conditions.

Since all the physical nulls, minima, and sidelobes should lie on the circumference of the unit circle in the z plane, it is clear from Fig. 1.25 that they all should be on the real axis in the x plane. In fact, they should be in the range $x_l\leqslant x\leqslant 1$, where the lower bound x_l depends on kd. For example, when $kd=\pi/2$ and $\alpha_1=0$, the entire visible range is $-\pi\leqslant u\leqslant 0$, $-1\leqslant z\leqslant 1$, or $-1\leqslant x\leqslant 1$. Thus $x_l=-1$. Therefore, as far as the task of determining nulls, minima, and sidelobes is concerned, all we have to do is to find the real roots of

$$Q(x)=0 \tag{1.142}$$

and

$$\frac{d}{dx}p_{\mathrm{NU}}(x)=0 \tag{1.143}$$

in the range $(1,x_l)$ and then reexpress them in terms of z or θ.The directivity for NUPP arrays, based on the formulation outlined above, can be obtained by modifying (1.133) into a more suitable form,

$$|S_{\mathrm{NU}}|^2=\sum_{m=0}^{n-1}\left(b_m z^m+\bar{b}_m z^{-m}\right)$$

$$=2\sum_{m=0}^{n-1}\left(b_{rm}\cos mu+b_{im}\sin mu\right), \tag{1.144}$$

where b_{rm} and b_{im} are, respectively, the real and imaginary parts of b_m in (1.134). The denominator in the expression for the directivity thus becomes

$$W=2\pi\int_0^{\pi}|S_{\mathrm{NU}}|^2\sin\theta\,d\theta=\frac{2\pi}{kd}\int_a^b|S_{\mathrm{NU}}|^2\,du$$

$$=4\pi W_1, \tag{1.145}$$

where

$$b = kd + \alpha, \tag{1.146}$$

$$a = -kd + \alpha, \tag{1.147}$$

and

$$W_1 = \sum_{i=0}^{n-1} |a_i|^2 + \frac{2}{kd} \sum_{m=1}^{n-1} \frac{\sin(mkd)}{m} (b_{rm} \cos m\alpha + b_{im} \sin m\alpha). \tag{1.148}$$

The directivity for NUPP is, therefore,

$$D_{\text{NU}} = \frac{|S_{\text{NU}}|^2{}_{\max}}{W_1}. \tag{1.149}$$

Two special cases can be noted from (1.148). They are

$$D_{\text{NU}} = \frac{|S_{\text{NU}}|^2_{\max}}{\sum_{i=0}^{n-1} |a_i|^2}, \qquad \text{for} \qquad d = p\frac{\lambda}{2}, p = 1, 2, \ldots, \tag{1.150}$$

and

$$D_{\text{NU}} = \frac{\left| \sum_{i=0}^{n-1} a_i e^{ji\alpha} \right|^2}{W'_1}, \qquad d \to 0 \tag{1.151}$$

where

$$W'_1 = \sum_{i=0}^{n-1} |a_i|^2 + 2 \sum_{m=1}^{n-1} (b_{rm} \cos m\alpha + b_{im} \sin m\alpha). \tag{1.152}$$

Example. Analyze the characteristics of a five-element NUPP array with[29]

$$a_0 = a_1 = 1, \qquad a_2 = e^{j\delta_2}, \qquad a_3 = e^{j\delta_3}, \qquad \text{and} \qquad a_4 = e^{j\delta_4}.$$

In this case,

$$|S_{\text{NU}}|^2 = 5 + \sum_{m=1}^{4} \left(b_m z^m + \bar{b}_m z^{-m} \right),$$

$$b_1 = 1 + e^{-j\delta_2} + e^{j(\delta_2 - \delta_3)} + e^{j(\delta_3 - \delta_4)},$$

$$b_2 = e^{-j\delta_2} + e^{-j\delta_3} + e^{j(\delta_2 - \delta_4)},$$

$$b_3 = e^{-j\delta_3} + e^{-j\delta_4},$$

$$b_4 = e^{-j\delta_4};$$

$$P_{\mathrm{NU}}(y) = (A'_4 y^4 + A'_3 y^3 + A'_2 y^2 + A'_1 y + A'_0) + (A''_3 y^3 + A''_2 y^2 + A''_1 y + A''_0)(4 - y^2)^{1/2},$$

where

$$A'_4 = \cos\delta_4, \qquad A'_3 = \cos\delta_3 + \cos\delta_4,$$

$$A'_2 = \cos\delta_2 + \cos\delta_3 + \cos(\delta_2 - \delta_4) - 4\cos\delta_4,$$

$$A'_1 = 1 + \cos\delta_2 + \cos(\delta_2 - \delta_3) + \cos(\delta_3 - \delta_4) - 3\cos\delta_3 - 3\cos\delta_4,$$

$$A'_0 = 5 + 2\cos\delta_4 - 2\cos\delta_2 - 2\cos\delta_3 - 2\cos(\delta_2 - \delta_4),$$

$$A''_3 = \sin\delta_4, \qquad A''_2 = \sin\delta_3 + \sin\delta_4,$$

$$A''_1 = \sin\delta_2 + \sin\delta_3 - \sin(\delta_2 - \delta_4) - 2\sin\delta_4,$$

$$A''_0 = \sin\delta_2 - \sin(\delta_2 - \delta_3) - \sin(\delta_3 - \delta_4) - \sin\delta_3 - \sin\delta_4,$$

$$p_{\mathrm{NU}}(x) = \frac{Q(x)}{(x^2+1)^4},$$

where

$$\begin{aligned} Q(x) = {} & (A'_0 - 2A''_0)x^8 + (4A'_1 - 8A''_1)x^7 + (4A'_0 + 16A'_2 - 4A''_0 - 32A''_2)x^6 \\ & + (12A'_1 + 64A'_3 - 8A''_1 - 128A''_3)x^5 + (6A'_0 + 32A'_2 + 256A'_4)x^4 \\ & + (12A'_1 + 64A'_3 + 8A''_1 + 128A''_3)x^3 + (4A'_0 + 16A'_2 + 4A''_0 + 32A''_2)x^2 \\ & + (4A'_1 + 8A''_1)x + (A'_0 + 2A''_0), \end{aligned} \tag{1.153}$$

$$|S_{NU}|_{max} = |1 + e^{-j\alpha_1} + e^{j(\delta_2 - 2\alpha_1)} + e^{j(\delta_3 - 3\alpha_1)} + e^{j(\delta_4 - 4\alpha_1)}| \leqslant 5,$$

$$\begin{aligned} W_1 = \frac{1}{2kd}\{ & 10kd + 4\sin kd \cos\alpha + 4\sin kd \cos(\delta_2 + \alpha) \\ & + 4\sin kd \cos(\delta_3 - \delta_2 + \alpha) + 4\sin kd \cos(\delta_4 - \delta_3 + \alpha) \\ & + 2\sin 2kd [\cos(\delta_2 + 2\alpha) + \cos(\delta_3 + 2\alpha) + \cos(\delta_4 - \delta_2 + 2\alpha)] \\ & + \tfrac{4}{3}\sin 3kd [\cos(\delta_3 + 3\alpha) + \cos(\delta_4 + 3\alpha)] \\ & + \sin 4kd \cos(\delta_4 + 4\alpha)\}. \end{aligned} \tag{1.154}$$

It is clear that when $d = \lambda/2$ or $kd = \pi$, $W_1 = 5$. Then, according to (1.149), we have

$$D_{NU}\,(\text{when } kd = \pi) = \frac{|S_{NU}|^2_{max}}{5} \leqslant 5 \;(D_U, \text{ directivity for UPP array}).$$

This result implies that, for $I_i = 1$, $kd = p\pi$, $p = 1, 2, 3, \ldots$, it is impossible to make the directivity for NUPP arrays higher than that for UPP arrays. However, when $kd < \pi$, we may have many sets of values for α and δ's such that $D_{NU} > D_U$. Specific results for $d = \lambda/4$ ($kd = \pi/2$), which is the element spacing commonly used for an endfire array, are shown in Figs. 1.26–1.28. In Fig. 1.26, where $\alpha = -90°$ ($\alpha_1 = 0°$, corresponding to the ordinary endfire condition) and $\delta_2 = 0°$, D_{NU} is plotted as a function of δ_4 with δ_3 as the parameter. It is clear that D_{NU} can be as large as 6.80 when $\delta_3 = -40°$ and $\delta_4 = -80°$, which is much greater than $D_U = 5$ for the corresponding UPP array. Actually, as far as the case $\alpha_1 = 0$ is concerned, D_{NU} reaches its maximum of 7.43 approximately when $\delta_2 = -30°$, $\delta_3 = -55°$, and $\delta_4 = -95°$, as indicated by curve I in Fig. 1.27. Curves II and III in Fig. 1.27 give the same information for $\alpha_1 = 10°$ ($\alpha = -100°$) and $\alpha_1 = 20°$ ($\alpha = -110°$), respectively. Curve I in Fig. 1.28 shows that the absolute maximum of D_{NU} for the case considered occurs approximately when $\alpha_1 = 55°$ ($\alpha = -145°$), $\delta_2 = +30°$, $\delta_3 = +60°$, and $\delta_4 = +60.5°$ with $D_{NU}|_{max} = 9.81$, a 96.2% increase over that for a UPP ordinary endfire array with the same element spacing, or an 11.9% increase over that for a UPP improved endfire array with the same element spacing and with $\delta = 35°$ as presented in Fig. 1.15.

To gain insight into this rather dramatic improvement in directivity, following are the details for calculating radiation patterns, beamwidths, nulls, and sidelobes for two of the five cases just considered.

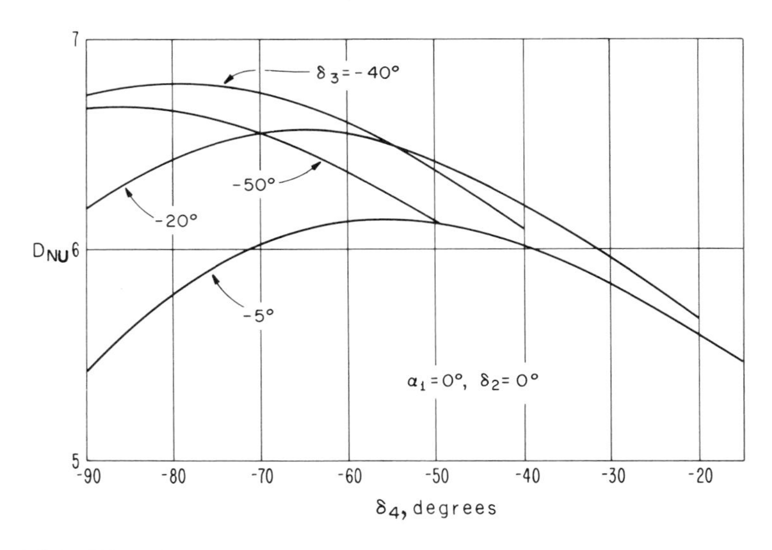

Fig. 1.26 Directivity of a five-element NUPP array with $d = \lambda/4$.

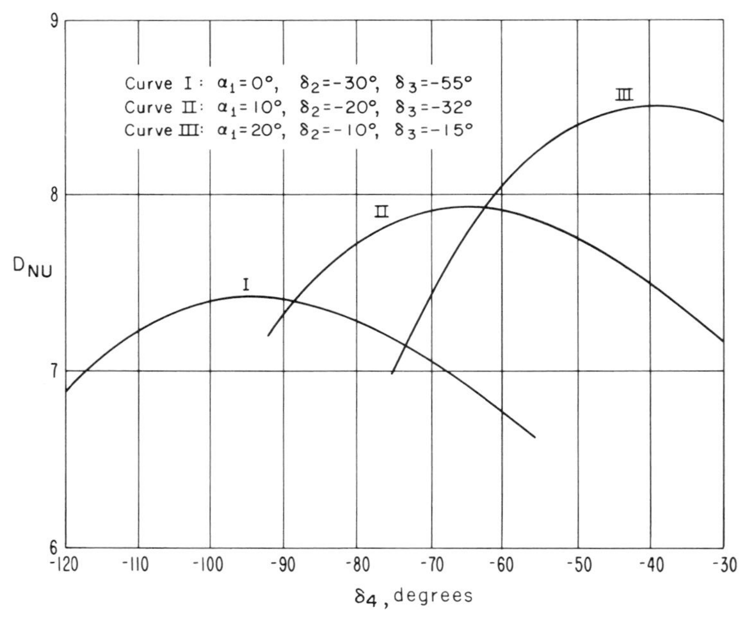

Fig. 1.27 Further results for directivity of a five-element NUPP array.

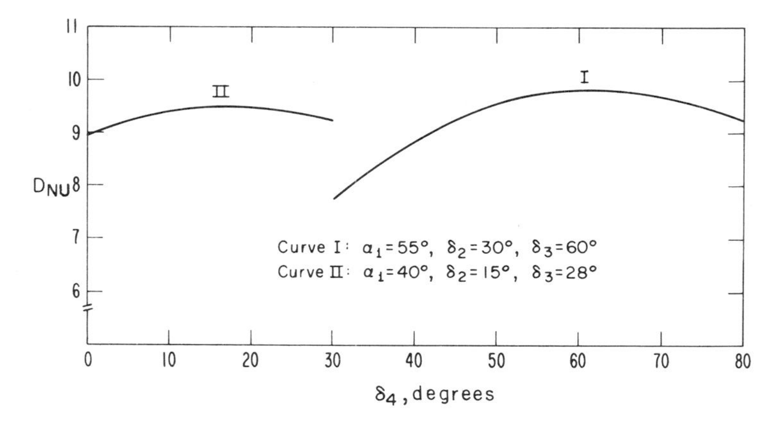

Fig. 1.28 More results for directivity of a five-element NUPP array.

Case A. $\alpha_1 = 0$, $\delta_2 = -30°$, $\delta_3 = -55°$, $\delta_4 = -95°$, and $kd = 90°$. This case corresponds to curve I in Fig. 1.27. From (1.153), we have

$$Q(x) = 0.6012x^8 + 10.1812x^7 + 96.8736x^6 + 185.4572x^5 + 55.0360x^4$$
$$- 73.2996x^3 - 17.3120x^2 + 6.4516x + 1.6012,$$

$$p_{\mathrm{NU}}(x) = \frac{Q(x)}{(1+x^2)^4}.$$

Visible range: $-1 \leqslant x \leqslant 1$.

Principal maximum: at $x = 1$ ($\theta = 0$) with $p_{\mathrm{NU}}(1) = 16.5996$.

The roots of $(d/dx)p_{\mathrm{NU}}(x) = 0$ are approximately

$$x_1 = 0.3985 \ (\theta_1 = 61.2°),$$

$$x_2 = 0.0943 \ (\theta_2 = 83.1°),$$

$$x_3 = -0.2667 \ (\theta_3 = 109.4°),$$

$$x_4 = -0.6646 \ (\theta_4 = 138.7°),$$

$$x_5 = 1.5548,$$

$$x_6 = -1.3217,$$

$$x_7 = -3.0183,$$

$$x_8 = -15.3340,$$

of which only the first four are in the visible range. Their corresponding values of θ are also indicated.

There are no nulls for this case.

Minima are given by x_1 and x_3, with $p_{NU}(x_1)=0.2444$ or 18.32 dB below the principal maximum, and $p_{NU}(x_3)=0.0773$ (-23.32 dB).

Sidelobes are given by x_2 and x_4, with $p_{NU}(x_2)=1.9304$ (-9.34 dB) and $p_{NU}(x_4)=1.3107$ (-11.02 dB).

The level at the backlobe direction ($x=-1$ or $\theta=180°$) is given by $p_{NU}(-1)=0.5006$ or -15.20 dB.

The half-power point can be determined by solving for $p_{NU}(x_h) = \frac{1}{2}p_{NU}(1)$, which yields $x_h=0.7660$ or $\theta_h=33.5°$.

With the important data listed above, the radiation pattern for this case can easily be plotted, but is omitted here.

Case B. $\alpha_1=55°$, $\delta_2=30°$, $\delta_3=60°$, $\delta_4=60.5°$, and $kd=90°$. This case corresponds to curve I in Fig. 1.28.

$$Q(x)=2.9850x^8+1.9576x^7-42.4076x^6-39.9016x^5+143.4880x^4$$
$$+185.0440x^3+62.9004x^2+4.0808x+0.0742,$$

$$p_{NU}(x)=\frac{Q(x)}{(1+x^2)^4}.$$

Visible range: $-3.1721 \leqslant x \leqslant 0.3152$.

Principal maximum: at $x=0.3152$ ($\theta=0°$) with $p_{NU}(0.3152)=10.0340$.

The roots of $(d/dx)p_{NU}(x)=0$ are approximately:

$$x_1=-0.0388 \ (\theta_1=55.8°),$$

$$x_2=-0.2691 \ (\theta_2=73.9°),$$

$$x_3=-0.5799 \ (\theta_3=93.3°),$$

$$x_4=-1.1974 \ (\theta_4=120.1°),$$

$$x_5 = -2.5051\ (\theta_5 = 154.9°),$$

$$x_6 = 0.7276,$$

$$x_7 = 2.7938,$$

$$x_8 = 56.5936,$$

of which the first five are in the visible range.

Nulls are given by x_1, x_3, and x_5 as evidenced by

$$Q(x_1) = Q(x_3) = Q(x_5) = 0.$$

Sidelobes are given by x_2 and x_4, with $p_{NU}(x_2) = 0.5426$ (-12.67 dB) and $p_{NU}(x_4) = 1.1854$ (-9.27 dB).

The backlobe is at $x = -3.1721$ ($\theta = 180°$), with $p_{NU}(-3.1721) = 0.2087$ (-16.82 dB).

The half-power point is at $x_h = 0.1986$ or $\theta_h = 30.5°$.

These characteristics can now be readily compared with those for the UPP ordinary and improved endfire arrays discussed previously.

Details for the other three cases in Figs. 1.27 and 1.28 ($\alpha_1 = 10°$, $20°$, and $40°$) are left to the reader as an exercise.

Although the condition $I_i = 1$ is used in the examples shown, the general formulation for NUPP arrays described in this section also applies to arrays whose amplitude excitations are not uniform.

1.7 Monopulse Arrays

When an antenna is used in a tracking radar, its beam is usually positioned by a servo-mechanism actuated by an error signal. There are many techniques for producing this error signal. One is the well-known *simultaneous lobing*,[30] where all necessary information on tracking can be obtained on the basis of a single pulse from an amplitude-comparison monopulse system. In such a system, the same antenna with two offset feeds or two identical antennas illuminated separately are used for both transmitting and receiving. Work on monopulse theory has, in the past, been done primarily on aperture antennas[31–33] with only very limited attention paid to arrays with discrete elements,[34] because of the cost and complexity involved. Recent new development in high-power solid-state phasing devices has raised performance capacities of arrays to the point that it may be feasible to reconsider the use of arrays for radar tasks.

In this section, we will give a mathematical formulation for discrete *monopulse arrays*. Basically, two arrays or one array with dual excitations are involved for producing *sum* and *difference* radiation patterns. The former is responsible for range detection of moving targets, and the latter is important for determining the angular tracking accuracy. As will be seen later, the sum pattern from such an array is essentially the same as that discussed before, and therefore is not repeated here in detail. Only the difference pattern will be studied in addition. Both equally and unequally spaced arrays are considered.

Suppose we have an unequally spaced but symmetric UPP broadside array of $2N+1$ isotropic elements, as shown in Fig. 1.22, which constitutes a component array of the monopulse system. The far-field can be represented by (1.115), or

$$E(u)=I_0+2\sum_{i=1}^{N} I_i \cos u_i = I_0+2\sum_{i=1}^{N} I_i \cos(a_i u), \tag{1.155}$$

where

$$u_i = kd_i \cos\theta, \tag{1.156}$$

$$u = kd_N \cos\theta, \tag{1.157}$$

$$a_i = \frac{d_i}{d_N}. \tag{1.158}$$

Note that in (1.158) we have chosen to normalize d_i with respect to d_N, the same manner as in (1.111).

If an array with dual excitations is used to replace the two antennas in a conventional monopulse tracking radar, the excitations have to be so phased that one gives $E(u+u_s)$ and the other $E(u-u_s)$, as shown in Fig. 1.29, where the sidelobe region is suppressed for clarity. The *sum* and *difference* pattern functions are given, respectively, by

$$\begin{aligned}\Sigma(u) &= E(u-u_s)+E(u+u_s)\\ &= 2I_0+4\sum_{i=1}^{N} I_i \cos(a_i u_s)\cos(a_i u)\end{aligned} \tag{1.159}$$

and

$$\begin{aligned}\Delta(u) &= E(u-u_s)-E(u+u_s)\\ &= 4\sum_{i=1}^{N} I_i \sin(a_i u_s)\sin(a_i u),\end{aligned} \tag{1.160}$$

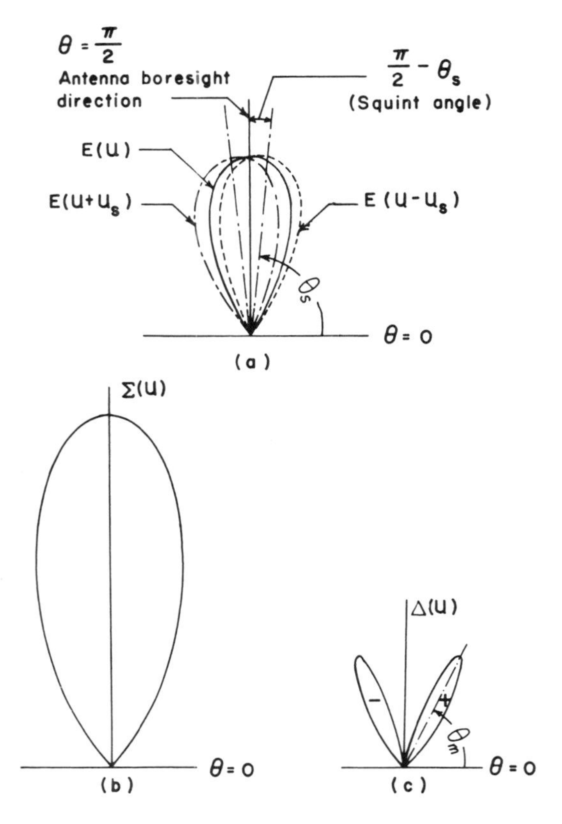

Fig. 1.29 Sketches of monopulse array patterns: (*a*) overlapping, (*b*) sum pattern, (*c*) difference pattern.

where

$$u_s = kd_N \cos\theta_s, \tag{1.161}$$

and the angle $(\pi/2) - \theta_s$ is usually called the *squint angle*.

It is clear that upon a specification of θ_s and a_i, Eq. (1.159) reduces essentially to (1.155) if $I_i \cos(a_i u_s)$ is considered as the new set of excitation coefficients. Thus, the basic properties learned before in studying the single array still apply to the sum pattern here.

The sum and difference directivities are defined, respectively, as

$$D_s = \frac{kd_N \Sigma^2(0)}{\int_0^{kd_N} \Sigma^2(u)\, du} \tag{1.162}$$

and

$$D_d = \frac{kd_N \Delta^2(u_m)}{\int_0^{kd_N} \Delta^2(u)\, du}, \tag{1.163}$$

where

$$u_m = kd_N \cos\theta_m \tag{1.164}$$

and θ_m is the position of the beam maximum for $\Delta(u)$.

In addition, the quantity known as the *difference slope* at the *boresight* ($\theta = \pi/2$, or $u = 0$) is also important for determining the tracking accuracy. It is defined below:

$$\Delta'(0) = \left.\frac{d\Delta(u)}{du}\right|_{u=0} = 4 \sum_{i=1}^{N} I_i a_i \sin(a_i u_s). \tag{1.165}$$

From the design point of view, we wish to make those quantities defined in (1.162), (1.163), and (1.165) as large as possible. The problem of determining a set of array parameters such as N, θ_s, or I_i to maximize some of the characteristics will be discussed in Chapter 2. Here, to analyze the effects of array parameters on the general performance of monopulse arrays, we will specifically assume certain different excitations and spacings. For each excitation and spacing, the directivities and difference slope can be calculated as functions of N and θ_s. Normally, when θ_s is below a certain value, depending on d_i and I_i, the sum pattern will have more than one peak (pattern splitting) resulting in an ambiguity for target detection. This fact may be anticipated by examining Fig. 1.29. Therefore, the value of θ_s is usually limited to $80° \leqslant \theta_s \leqslant 89°$ to avoid the pattern splitting.

A. Uniformly Spaced and Excited Monopulse Arrays of Isotropic Elements. In this case, $I_i = 1$ and $a_i = i/N$. Typical curves showing the variation of D_s and D_d for $n = 9$ ($N = 4$) as functions of θ_s and λ are given in Fig. 1.30. The corresponding result for $\Delta'(0)$ with $d = \lambda/2$ is shown in Fig. 1.31. It is apparent that both D_s and D_d for a given element spacing are relatively insensitive to changes in θ_s for $82° \leqslant \theta_s \leqslant 90°$, while $\Delta'(0)$ decreases almost linearly when θ_s is increased from 82° to 90°. For $d = \lambda/2$, $\Delta'(0)$ reaches its maximum value approximately at $\theta_s = 81°$.

Figure 1.32 shows D_s as a function of N with $d = \lambda/2$ and θ_s as a parameter, from which we can conclude that, for a given $\theta_s < 89°$, the sum directivity cannot always be imporved by merely increasing the total

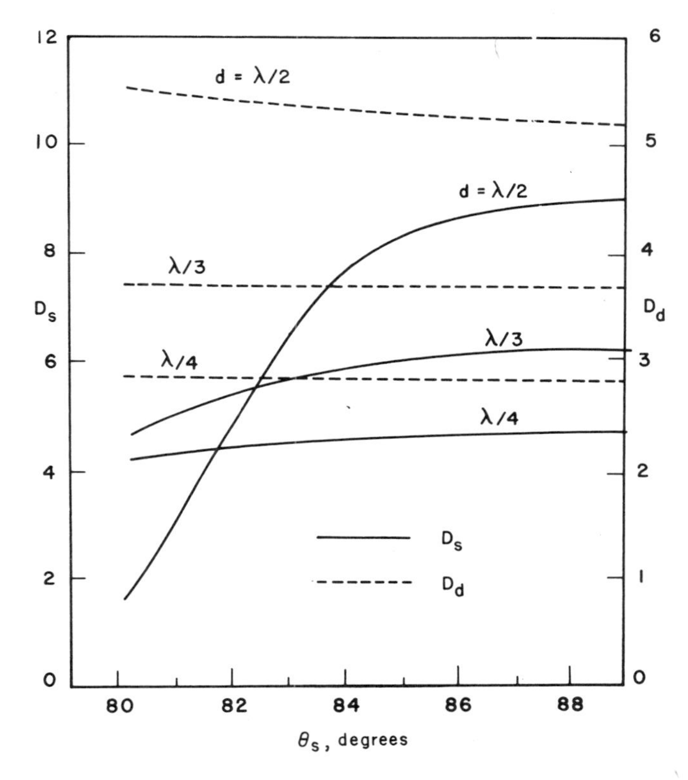

Fig. 1.30 Variations of D_s and D_d as functions of θ_s with $N=4, I_i=1$, and $a_i=i/4$.

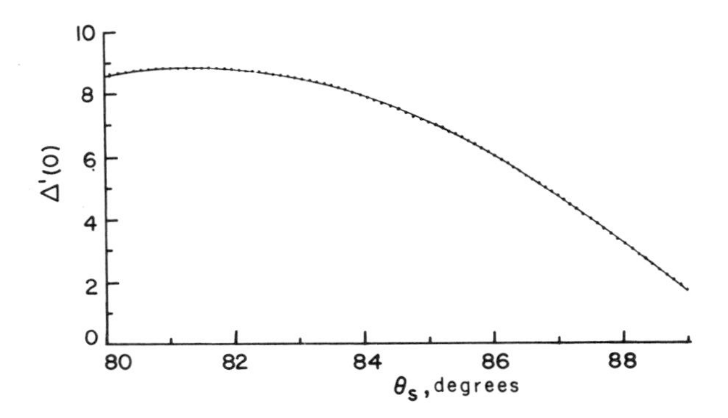

Fig. 1.31 Difference slope of a nine-element symmetric monopulse array as a function of θ_s with $I_i=1, a_i=i/4, d=\lambda/2$.

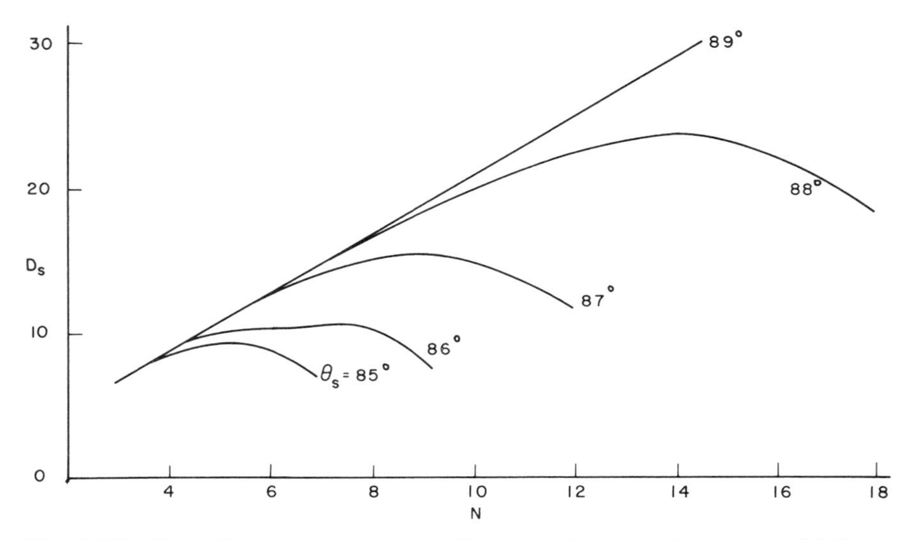

Fig. 1.32 Sum directivity of an equally spaced monopulse array of $2N+1$ elements with $d=\lambda/2$ and $I_i=1$.

number of elements in the array. This fact is quite contrary to the conventional single broadside array where the directivity (sum) is always proportional to the number of elements for a given d. On the other hand, the difference directivity when $d=\lambda/2$ is almost linearly proportional to N, as can be seen from Fig. 1.33.

B. Uniformly Spaced but Nonuniformly Excited Monopulse Arrays of Isotropic Elements. In this case, $a_i=i/N$, $I_i \neq 1$. Here we choose to consider only two special excitations, one of which is the *concave-downward* type with $I_i=\cos(i\pi/2N)$, $i=0,1,2,\ldots,N$, and the other the *concave-upward* type of excitation with $I_i=(i/N)^2$. Numerical results for D_s, D_d, and $\Delta'(0)$ as a function of θ_s with both of these excitations when $N=3$ and $d=\lambda/2$ are presented in Fig. 1.34. From 1.34(a) and (b), we see another opposite effect for these two special excitations: when θ_s is increased from 80° to 89°, both the sum and difference directivities increase for $I_i=\cos(i\pi/2N)$, while they generally decrease for $I_i=(i/N)^2$ after D_s reaches its maximum approximately at $\theta_s=82°$. When θ_s is fixed at 88°, the directivities, however, increase with respect to N (up to $N=9$ at least) for both excitations, as evidenced by Fig. 1.35(a) and (b). They begin to decrease when N is larger than 15, as can be seen from Fig. 1.32. Corresponding results for u_m [see (1.164)] where the difference pattern

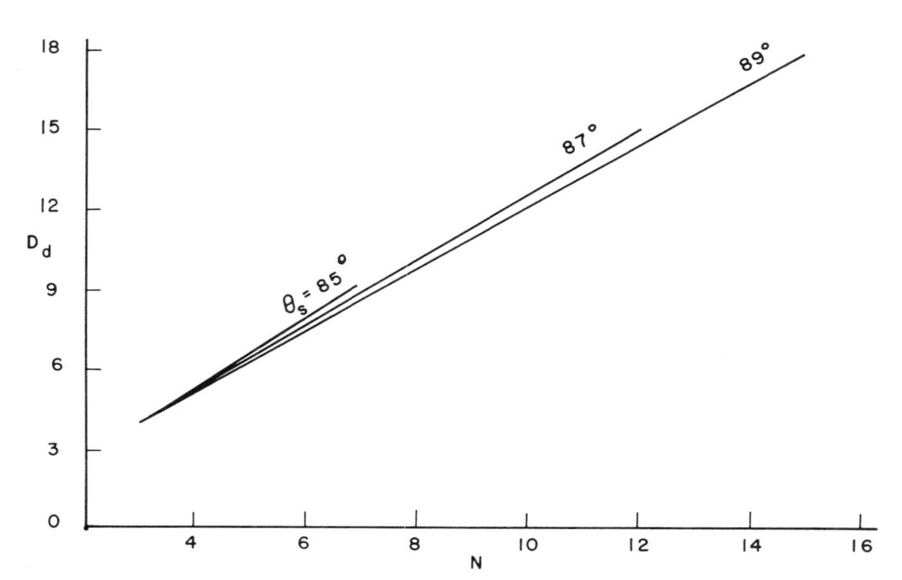

Fig. 1.33 Difference directivity of an equally spaced monopulse array of $2N+1$ elements with $d=\lambda/2$ and $I_i=1$.

reaches its maximum are respectively given in Figs. 1.34(*c*) and 1.35(*c*) as a function of θ_s and N.

As far as the difference slope is concerned, it decreases monotonically for both excitations in the range $80° \leqslant \theta_s \leqslant 89°$. In practice, in addition to using $\Delta'(0)$ defined in (1.165), a somewhat different factor called the slope of the error signal at the boresight[35] may be used to locate the true boresight direction when lost in the presence of noise. In his book on monopulse, Rhodes[30] shows that this slope of the error signal, depending on the type of phase comparators employed, is directly proportional to either the product or ratio of the difference slope $\Delta'(0)$ and the sum pattern evaluated at the boresight $\Sigma(0)$. Mathematically, these are, respectively,

$$\Delta'(0)\Sigma(0), \qquad \text{slope-sum product,}$$
$$\frac{\Delta'(0)}{\Sigma(0)}, \qquad \text{slope-sum ratio.}$$

Detailed analysis and discussion on both of these quantities can be found in a report by Pang and Ma,[36] but are omitted here.

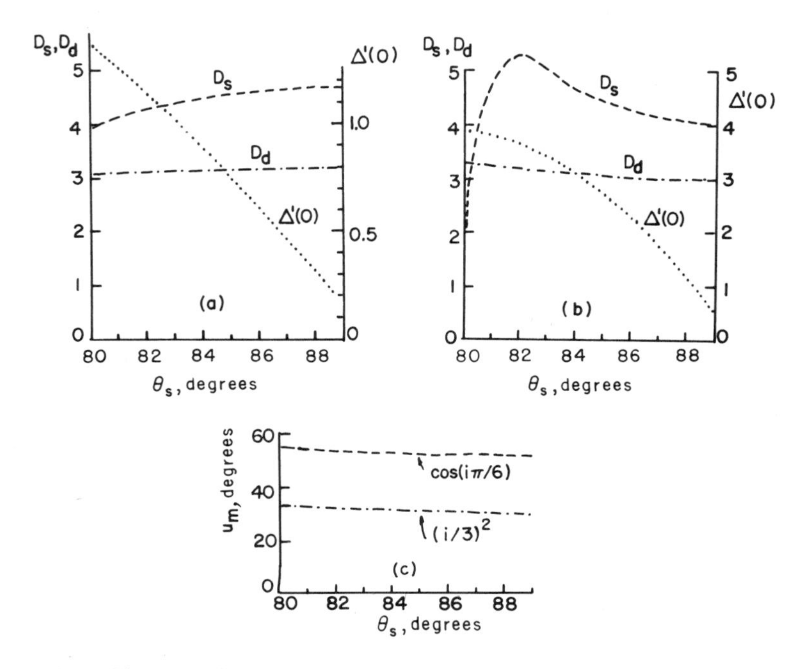

Fig. 1.34 Characteristics of an equally spaced monopulse array as a function of θ_s with $N=3$: (*a*) D_s, D_d, $\Delta'(0)$ for $I_i=\cos(i\pi/6)$, (*b*) $D_s, D_d, \Delta'(0)$ for $I_i=(i/3)^2$, and (*c*) u_m.

C. Nonuniformly Spaced but Uniformly Excited Monopulse Arrays of Isotropic Elements. In this case, $I_i=1$ and $a_i \neq i/N$. For the purpose of simple analysis, we choose once again, as in Section 1.5, $N=2$ (n=5) with a_1 as the varying parameter. In fact, a_i in this section is determined according to two special kinds of spacing arrangement suggested by King et al.[17] They are the logarithmic spacing (LS) and reverse logarithmic spacing (RLS). The LS broadens the spacing between two adjacent elements as the elements of interest are farther away from the array center. The RLS makes the spacing between two adjacent elements progressively smaller. Therefore, we set

$$\text{for LS:} \qquad d_i = A[1-\log(10-i)], \qquad i=1,2,\ldots,N \qquad (N<10)$$

$$(1.166)$$

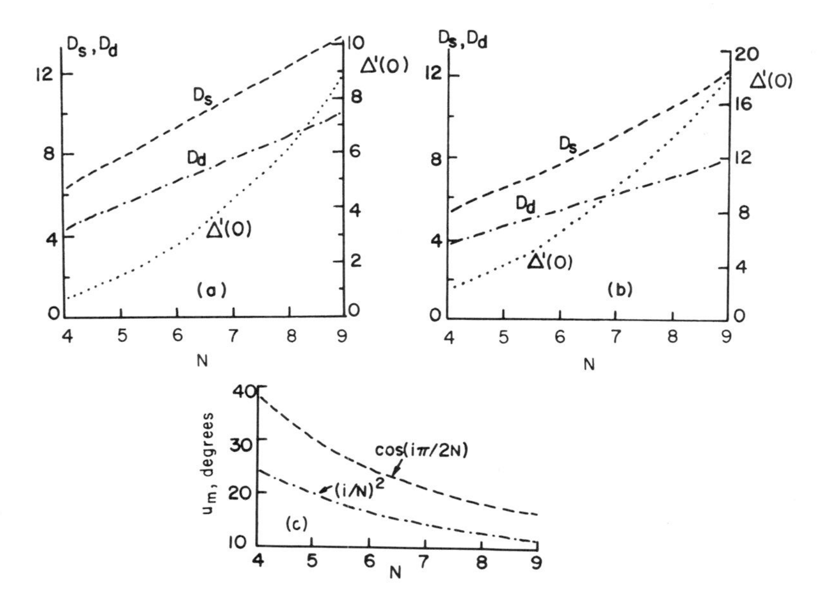

Fig. 1.35 Characteristics of an equally spaced monopulse array as a function of N with $\theta_s = 88°$: (*a*) $D_s, D_d, \Delta'(0)$ for $I_i = \cos(i\pi/2N)$, (*b*) $D_s, D_d, \Delta'(0)$ for $I_i = (i/N)^2$, and (*c*) u_m.

and

$$\text{for RLS:} \qquad d_i = A \log(1+i), \tag{1.167}$$

where log is the common logarithm and A, a common factor for all d_i, is determined by the relation $2d_N = L$, L being the overall array length.

From (1.166) and (1.167), we can compute spacings and associated array characteristics. Generally speaking, for LS geometry, the major improvement over the corresponding equally spaced case is in the level of the first sidelobe of the sum pattern at some sacrifice in difference slope. Results in the difference directivity are also better although the improvement is only marginal. Representative results for LS ($N=2, a_1=0.4731, L=2\lambda$) are listed in Table 1.5. Results for the corresponding equally spaced monopulse array ($N=2, a_1=0.5, L=2\lambda$) are given in Table 1.6 for comparison. It is seen that the sum directivity for LS distribution in this case is higher only

Table 1.5 Partial Characteristics for a Monopulse Array with $N=2, I_i=1, d_1=0.4731\lambda, d_2=\lambda$.

θ_s (deg)	D_s	D_d	$\Delta'(0)$	$\Sigma(0)$	First Sum sidelobe Level (dB)
83	4.8747	3.0717	3.4435	8.6232	−17.47
85	4.9700	3.0533	2.5691	9.2808	−15.34
87	5.0051	3.0412	1.5842	9.7380	−14.11
89	5.0145	3.0366	0.5376	9.9702	−13.56

Table 1.6 Partial Characteristics for an Equally Spaced Monopulse Array with $N=2, I_i=1, d_1=0.5\lambda, d_2=\lambda$.

θ_s (deg)	D_s	D_d	$\Delta'(0)$	$\Sigma(0)$	First Sum Sidelobe Level (dB)
83	4.9103	3.0279	3.5208	8.5928	−15.28
85	4.9783	3.0113	2.6252	9.2650	−13.50
87	4.9974	3.0003	1.6180	9.7328	−12.54
89	4.9999	2.9945	0.5486	9.9699	−12.08

Table 1.7 Partial Characteristics for a Monopulse Array with $N=2, I_i=1, d_1=0.6309\lambda, d_2=\lambda$.

θ_s (deg)	D_s	D_d	$\Delta'(0)$	$\Sigma(0)$	First Sum sidelobe Level (dB)
83	4.7022	2.6199	3.9453	8.4252	−9.28
85	4.6383	2.6134	2.9387	9.1780	−8.38
87	4.5856	2.6095	1.8107	9.7004	−7.86
89	4.5574	2.6077	0.6139	9.9662	−7.62

when $\theta_s \geqslant 87°$, and in fact is greater than 5. As far as the RLS ($N=2, a_1=0.6309, L=2\lambda$) is concerned, the effect is just the opposite; only the difference slope is improved. Similar results are presented in Table 1.7. Before concluding this section, we should note that neither the LS nor the RLS distribution considered here is the best choice. Other nonuniform spacings may offer even greater improvement over the equally spaced monopulse array. The problem of determining analytically and uniquely the best spacing distribution for any N and θ_s to maximize some of the characteristics still remains unsolved.

1.8 Concluding Remarks and Discussion

In this chapter, we have analyzed basic characteristics associated with linear arrays, in which all the elements are assumed to be identical and similarly oriented, the mutual interactions among elements are totally ignored, and the element used in the array is rather idealized or simple. Uniform arrays where the amplitude excitations, element spacing, and the progressive phase distribution are all kept constant are studied in terms of isotropic elements, short parallel, and collinear dipoles. Arrays with nonuniform amplitude excitations are presented by the finite Z-transform method. Improved endfire arrays, arrays with nonuniformly progressive phases, and monopulse arrays are analyzed for isotropic elements only. They can be easily extended to include other simple elements by applying the principle of pattern multiplication.

Arrays of practical elements with mutual impedances also included will be considered in Chapter 4. Work on arrays with nonidentical elements, or with identical elements not similarly oriented, has not been thoroughly investigated in the past.[37] Very limited attention has been given to arrays whose elements are but small distances away from a straight line.[38]

For all the cases covered in this chapter, results are presented in a deterministic manner. That is, the array performance can be evaluated precisely once the required array parameters are specified. When the total number of elements in the array becomes extremely large, say of the order of 10^3, it may be desirable to analyze the characteristics from a probabilistic approach. Excellent work in this respect has been presented by Professor Lo in his unique papers.[39,40] Maher and Cheng have considered a different facet of this subject by randomly removing elements from an equi-spaced linear array.[41]

The analytical methods presented in this chapter should also apply to arrays with acoustic sources.

PROBLEMS

1.1 Analyze the broadside uniform array of five isotopic elements with $d=2\lambda/3$ by determining (a) the directivity, (b) beam maximum, (c) null positions, (d) approximate sidelobe positions and relative levels, and (e) the directivity when d approaches zero.

1.2 Analyze the ordinary endfire uniform array of four short parallel dipoles with $d=\lambda/4$.

1.3 Determine the directivity for the broadside uniform array of four collinear dipoles with $0 \leqslant d \leqslant \lambda$.

1.4 Plot the directivity of an improved endfire uniform array with four isotropic elements as a function of δ. What is δ_m?

1.5 Verify some of the formulas given in Table 1.1.

1.6 From the text we know that (1.79) and (1.80) are equivalent. Therefore, if the division is carried out for (1.80), we should get (1.79). Find some trigonometric identities for $n=7$ in this process.

1.7 Verify (1.90) by using the original polynomial form for $S(z)$ and carrying out the integration in the denominator of (1.88).

1.8 In Section 1.4, we analyzed two arrays whose amplitude excitations were described by $f(\zeta)=\sin ka\zeta$ and $f(\zeta)=U(\zeta)$. Principal results were respectively summarized in Tables 1.2 and 1.3. By comparison, we noted that the sidelobe level for the array with a tapered amplitude excitation $[f(\zeta)=\sin ka\zeta]$ is improved over that with constant amplitude excitation $[f(\zeta)=U(\zeta)]$ at the expense of having a wider beamwidth. Naturally, the degree of improvement in sidelobe level and penalty in beamwidth depends on how strongly the amplitude excitation is tapered. The extreme case is the well-known binomial amplitude excitation which would eliminate all the sidelobes. Analyze this binomial array with $n=4$, $I_0=1$, $I_1=3$, $I_2=3$, $I_3=1$, $d=\lambda/2$ and summarize the result in a table similar to Table 1.2.

1.9 In problem 1.8 we have analyzed the four-element binomial array with $I_0:I_1:I_2:I_3=1:3:3:1$ and $d=\lambda/2$ where the sidelobes are totally eliminated and the beamwidth is very broad. The level of the sidelobe relative to the beam maximum for this case may be considered the lowest ($-\infty$ dB). What amplitude distribution (also for $n=4$, $d=\lambda/2$) will make the level of the sidelobe the highest (same level as the beam maximum) and the beamwidth very narrow?

1.10 Formulate the problem of nonuniformly spaced symmetric arrays when the total number of elements in the array is even, $n=2N$. Calculate the directivity for $N=2$, $I_i=1$, and $kd_2=3\pi/2$ as a function of d_1/d_2.

1.11. Analyze the characteristics of a four-element NUPP array with $a_0=a_1=1$, $a_2=e^{j\delta_2}$, $a_3=e^{j\delta_3}$, $kd=\pi/2$. What α_1, δ_2, and δ_3 will approximately yield the maximum directivity?

1.12 Complete the example given in Section 1.6 for $\alpha_1=10°$, 20°, and 40°.

1.13 How can you explain why the sum directivity of an equally spaced monopulse array with $n=2N+1$, $I_i=1$, $a_i=i/N$ does not always increase with N for a fixed θ_s (other than 90° of course).

1.14. Calculate the sum and difference directivities for a seven-element monopulse array with $I_i=1$, $a_i=i/N$, $L=3\lambda$, and $\theta_s=86°$. What is u_m in this case?

REFERENCES

1. **Tai, Chen T.** The nominal directivity of uniformly spaced arrays of dipoles, *Microwave J.* Vol. 7, No. 9, pp. 51–55, September, 1964.

2. **Bach, Henning.** Directivity of basic linear arrays, *IEEE Trans. Antennas and Propagation,* Vol. AP-18, No. 1, pp. 107–110, January, 1970.

3. **Bach, H. and J. E. Hansen.** Uniformly spaced arrays, Chap. 5 in *Antenna Theory,* Part I, edited by R. E. Collin and Francis J. Zucker, McGraw-Hill Book Company, New York, 1969.

4. **Hansen, W. W. and J. R. Woodyard.** A new principle in directional antenna design, *Proc. IRE,* Vol. 26, No. 3, pp. 333–345, March, 1938.

5. **Eaton, J. E., L. J. Eyges, and G. G. Macfarlane.** Linear-array antennas and feeds, Chap. 9 in *Microwave Antenna Theory and Design,* edited by S. Silver, McGraw-Hill Book Company, New York, 1949.

6. **Maher, Thomas M.** Optimum progressive phase shifts for discrete endfire arrays, *Syracuse University Research Institute Report No. EE 492-6002T8,* Syracuse, N. Y., February, 1960.

7. **Cheng, D. K. and M. T. Ma.** A new mathematical approach for linear array analysis, *IRE Trans. Antennas and Propagation,* Vol. AP-8, No. 3, pp. 255–259, May, 1960.

8. **Ma, M. T.** A new mathematical approach for linear array analysis and synthesis, Ph. D dissertation, Syracuse University, Syracuse, N. Y., 1961.

9. **Cheng, D. K.** *Analysis of Linear Systems,* Addison-Wesley Publishing Company, Inc., Reading, Mass., 1959.

10. **Jury, E. I.** *Sampled-Data Control Systems,* John Wiley and Sons, Inc., New York, 1958.

11. **Truxal, J. G.** *Automatic Feedback Control System Synthesis,* McGraw-Hill Book Company, New York, 1955.

12. **Tamburelli, G.** Some new properties of the Z-transform, *Alta Frequenta,* Vol. 32, pp. 157–161, February, 1963.

13. **Christiansen, P. L.** On the closed form of the array factor for linear arrays, *IEEE Trans. Antennas and Propagation,* Vol. AP-11, No. 2, March, 1963.

14. **Christiansen, P. L.** List of Z-transforms, *Report No. 32, Laboratory of Electromagnetic Theory,* Technical University of Denmark, Lyngby, Denmark, July, 1963.

15. **Ma, M. T. and L. C. Walters.** A table and computation of array functions using finite Z-transforms, *Technical Report No. IER 36-ITSA 36,* ESSA, Boulder, Colo., May, 1967.

16. **Unz, H.** Linear arrays with arbitrarily distributed elements, *IRE Trans. Antennas and Propagation,* Vol. AP-8, No. 2, pp. 222–223, March, 1960.

17. **King, D. D., R. F. Packard, and R. K. Thomas.** Unequally spaced broadband antenna arrays, *IRE Trans. Antennas and Propagation,* Vol. AP-8, No. 4, pp. 380–385, July, 1960.

18. **Harrington, R. F.** Sidelobe reduction by nonuniform element spacing, *IRE Trans. Antennas and Propagation,* Vol. AP-9, No. 2, pp. 187–192, March, 1961.

19. **Andreasen, M. G.** Linear arrays with variable inter-element spacings, *IRE Trans. Antennas and Propagation,* Vol. AP-10, No. 2, pp. 137–143, March, 1962.

20. **Maffett, A. L.** Array factors with nonuniform spacing parameters, *IRE Trans. Antennas and Propagation,* Vol. AP-10, No. 2, pp. 131–136, March, 1962.

21. **Lo, Y. T.** A spacing weighted antenna array, *IRE International Convention Record,* part 1, pp. 191–195, March, 1962.

22. **Ishimaru, A.** Theory of unequally spaced arrays, *IRE Trans. Antennas and Propagation,* Vol. AP-10, No. 6, pp. 691–701, November, 1962.

23. **Ishimaru, A. and Y. S. Chen.** Thinning and broadbanding antenna

arrays by unequal spacings, *IEEE Trans. Antennas and Propagation,* Vol. AP-13, No. 1, pp. 34–42, January, 1965.

24. **Yen, J. L. and Y. L. Chow.** On large nonuniformly spaced arrays, *Can. J. Phys.,* Vol. 41, pp. 1–11, January, 1963.

25. **Brown, F. W.** Note on nonuniformly spaced arrays, *IRE Trans. Antennas and Propagation,* Vol. AP-10, No. 5, pp. 639–640, September, 1962.

26. **Ma, M. T.** Note on nonuniformly spaced arrays, *IEEE Trans. Antennas and Propagation,* Vol. AP-11, No. 4, p. 508, July, 1963.

27. **Tang, C. H.** On a design method for nonuniformly spaced arrays, *IEEE Trans. Antennas and Propagation,* Vol. AP-13, No. 4, pp. 642–643, July, 1965.

28. **Churchill, R. V.** *Introduction to Complex Variables and Applications,* McGraw-Hill Book Company, New York, 1948, p. 59.

29. **Ma, M. T.** Linear arrays with nonuniform progressive phase shift, *IEEE International Convention Record,* part 1, pp. 70–76, March, 1963.

30. **Rhodes, D. R.** Introduction to Monopulse, McGraw-Hill Book Company, New York, 1959.

31. **Hannan, P. W.** Maximum gain in monopulse difference mode, *IRE Trans. Antennas and Propagation,* Vol. AP-9, No. 3, pp. 314–315, May, 1961.

32. **Kirkpatrick, G. M.** Angular accuracy improvement, *Final Report,* General Electric Co., Syracuse, N. Y., 1952.

33. **Powers, E. J.** Utilization of lambda function in the analysis and synthesis of monopulse antenna difference patterns, *IEEE Trans. Antennas and Propagation,* Vol. AP-15, No. 6, pp. 771–777, November, 1967.

34. **Brennan, L. E.** Angular accuracy of a phased array radar, *IRE Trans. Antennas and Propagation,* Vol. AP-9, No. 3, pp. 268–275, May, 1961.

35. **Cohen, W. and C. M. Steinmetz.** Amplitude and phase sensing monopulse system parameters, parts I and II, *Microwave J.* Vol. 2, No. 10, pp. 27–33 and No. 11, pp. 33–38, 1959.

36. **Pang, C. C. and M. T. Ma.** Analysis and synthesis of monopulse arrays with discrete elements, *Technical Report No. ERL 91-ITS 70,* ESSA, Boulder, Colo., November, 1968.

37. **Larson, R. W., J. E. Ferris, and V. M. Powers.** Nonuniformly spaced linear arrays of unequal-width elements, *IEEE Trans. Antennas and Propagation,* Vol. AP-13, No. 6, pp. 974–976, November, 1965.

38. **Galindo, Victor.** Nonlinear antenna arrays, *IEEE Trans. Antennas and Propagation,* Vol. AP-12, No. 6, pp. 782–783, November, 1964.

39. **Lo, Y. T.** A probabilistic approach to the design of large antenna arrays, *IEEE Trans. Antennas and Propagation,* Vol. AP-11, No. 1, pp. 95–97, January, 1963.

40. **Lo, Y. T.** A mathematical theory of antenna arrays with randomly spaced elements, *IEEE Trans. Antennas and Propagation,* Vol. AP-12, No. 3, pp. 257–268, May, 1964.

41. **Maher, T. M. and D. K. Cheng.** Random removal of radiators from large linear arrays, *IEEE Trans. Antennas and Propagation,* Vol. AP-11, No. 2, pp. 106–112, March, 1963.

ADDITIONAL REFERENCES

Blasi, E. A. and R. S. Elliott. Scanning antenna arrays of discrete elements, *IRE Trans. Antennas and Propagation,* Vol. AP-7, No. 4, pp. 435–436, October, 1959.

Cheng, David K. *Z*-transform theory for linear array analysis, *IEEE Trans. Antennas and Propagation,* Vol. AP-11, No. 5, p. 593, September, 1963.

Chow, Y. Leonard. On grating plateaux of nonuniformly spaced arrays, *IEEE Trans. Antennas and Propagation,* Vol. AP-13, No. 2, pp. 208–215, March, 1965.

Chow, Y. Leonard. On the error involved in Poisson's sum formula of nonuniformly spaced antenna arrays, *IEEE Trans. Antennas and Propagation,* Vol. AP-14, No. 1, pp. 101–102, January, 1966.

Christiansen, Peter L. *Z*-transform theory in general array analysis, *IEEE Trans. Antennas and Propagation,* Vol. AP-12, No. 5, p. 647, September, 1964.

Elliott, R. S. A limit on beam broadening for linear arrays, *IEEE Trans. Antennas and Propagation,* Vol. AP-11, No. 5, pp. 590–591, September, 1963.

Elliott, R. S. The theory of antenna arrays, Chap 1, in *Microwave Scanning Antennas* edited by R. C. Hansen, Academic Press, New York, 1966, Vol. II.

Hansen, Robert C. Gain limitations of large antennas, *IRE Trans. Antennas and Propagation,* Vol. AP-8, No. 5, pp. 490–495, September, 1960.

Hansen R. C. Gain limitations of large antennas, *IEEE Trans. Antennas and Propagation,* Vol. AP-13, No. 6, pp. 997–998, November, 1965.

Harrington, Roger F. On the gain and beamwidth of directional antennas, *IRE Trans. Antennas and Propagation,* Vol. AP-6, No. 3, pp. 219–225, July, 1958.

Jordan, Edward C. *Electromagnetic Waves and Radiating Systems,* Chap. 12, Prentice-Hall, Inc., Englewood Cliffs, N.J., 1950.

King, H. E. Directivity of a broadside array of isotropic radiators, *IRE Trans. Antennas and Propagation,* Vol. AP-7, No. 2, pp. 197–198, April, 1959.

King, M. J. and R. K. Thomas. Gain of large scanned arrays, *IRE Trans. Antennas and Propagation,* Vol. AP-8, No. 6, pp. 635–636, November, 1960.

Kinsey, Richard R. Monopulse difference slope and gain standards, *IRE Trans. Antennas and Propagation,* Vol. AP-10, No. 3, pp. 343–344, May, 1962.

Kirkpatrick, George M. A relationship between slope functions for array and aperture monopulse antennas, *IRE Trans. Antennas and Propagation,* Vol. AP-10, No. 3, p. 350, May, 1962.

Kraus, John D. *Antennas,* Chap. 4, McGraw-Hill Book Co., New York, 1950.

Lo, Y. T. Sidelobe level in nonuniformly spaced antenna arrays, *IEEE Trans. Antennas and Propagation,* Vol. AP-11, No. 4, pp. 511–512, July, 1963.

Lo, Y. T. and S. W. Lee. Sidelobe level of nonuniformly spaced antenna arrays, *IEEE Trans. Antennas and Propagation,* Vol. AP-13, No. 5, pp. 817–818, September, 1965.

Lo, Y. T. and S. W. Lee. A study of space-tapered arrays, *IEEE Trans. Antennas and Propagation,* Vol. AP-14, No. 1, pp. 22–30, January, 1966.

Maffett, A. L. and C. T. Tai. Some properties of the gain of uniform and nonuniform arrays, *IEEE Trans. Antennas and Propagation, Vol. AP-18, No. 4, pp. 556–558, July, 1970.*

Nester, William H. A study of tracking accuracy in monopulse phased arrays, *IRE Trans. Antennas and Propagation,* Vol. AP-10, No. 3, pp. 237–246, May, 1962.

Neustadter, Siegfried. The second moment sum method and space tapered arrays, *IEEE Trans. Antennas and Propagation,* Vol. AP-11, No. 6, pp. 706–707, November, 1963.

Pelchat, Guy M. Relationships between squinted, sum and difference radiation patterns of amplitude monopulse antennas with mutual coupling between feeds, *IEEE Trans. Antennas and Propagation,* Vol. AP-15, No. 4, pp. 519–526, July, 1967.

Sandler, Sheldon S. Some equivalences between equally and unequally spaced arrays, *IRE Trans. Antennas and Propagation,* Vol. AP-8, No. 5, pp. 496–500, September, 1960.

Sherman, J. W. and M. I. Skolnik. An upper bound for the sidelobes of an unequally spaced array, *IEEE Trans. Antennas and Propagation,* Vol. AP-12, No. 3, pp. 373–374, May, 1964.

Stegen, Robert J. The null depth of a monopulse tracking antenna, *IEEE Trans. Antennas and Propagation,* Vol. AP-12, No. 5, pp. 645–646, September, 1964.

Strait, Bradley J. Antenna arrays with partially tapered amplitudes, *IEEE Trans. Antennas and Propagation,* Vol. AP-15, No. 5, pp. 611–617, September, 1967.

CHAPTER 2

SYNTHESIS OF DISCRETE LINEAR ARRAYS

In Chapter 1, we analyzed linear arrays and monopulse arrays with different approaches where the total number of elements in the array, space distribution, amplitude, and phase excitations were employed as four sets of varying parameters. In this chapter, we will study the reverse problem—synthesis. This is done by establishing various techniques to determine a favorable combination of the above-mentioned four sets of parameters to achieve certain specifications. One simple kind of specification is to express a general requirement on sidelobes and beamwidth without concerning details of the pattern or the directivity. This subject will be studied in Sections 2.2 and 2.3.

A second kind of synthesis is to achieve a prespecified pattern shape. It is here that the problem of approximation is usually involved because the specifications are, almost without exception, of an ideal nature or are arbitrary, and therefore do not necessarily correspond to a physically realizable finite array. The approximation involved is to obtain a realizable array factor approximating a prescribed pattern within some tolerable limit of error. The synthesis procedure is then to determine the number of elements required, space distribution, and excitations in order to yield exactly this realizable array factor. Whenever possible, attention is always focused on the point of devising a "best" array factor in the sense that the mean-square error or the maximum error committed by replacing the specified pattern by the approximating array factor is a minimum. Alternatively, an upper bound of error, in the sense of either mean-square or maximum deviation, is given so that the worst case we may expect in the process of approximation is known beforehand. These topics are to be included in Sections 2.4–2.7.

A third kind of synthesis is to achieve desired results from a known case. This method, which is usually referred to as the perturbation approach, is discussed in Section 2.8. A fourth kind of synthesis we consider in this chapter is to maximize the directivity of an array with a given number of elements without duly paying attention to details of other characteristics. This subject will be studied in Section 2.9 and part of Section 2.10.

Finally, because of the similarity in mathematics used, the analogy between theories of antenna arrays and passive networks will also be noted in Section 2.11.

Since we intend to present the synthesis techniques in general terms without giving due attention to applications of any particular antenna element, the idealized isotropic element in free space is assumed throughout this chapter. Studies of various practical HF antennas above a lossy ground plane will be given in the last three chapters.

2.1 Power Patterns and Relations to Excitation Coefficients and Others

In Section 1.6 when equally spaced arrays with nonuniformly progressive phases were discussed, we had the occasion to note there that sometimes it is more advantageous to deal with the power pattern because it involves only real quantities. In essence, the power pattern for an equally spaced UPP array of isotropic elements is related to the array polynomial, $E(z)$, by

$$|E(z)|^2 = E(z)\,\overline{E(z)} = \left(\sum_{i=0}^{n-1} I_i z^{-i}\right)\left(\sum_{i=0}^{n-1} I_i z^{i}\right)$$

$$= \sum_{m=0}^{n-1} I'_m (z^m + z^{-m}), \tag{2.1}$$

where

$$I'_m = \sum_{i=0}^{n-1-m} I_i I_{m+i}, \qquad 1 \leqslant m \leqslant (n-1), \tag{2.2}$$

$$I'_0 = \tfrac{1}{2} \sum_{i=0}^{n-1} I_i^2. \tag{2.3}$$

In (2.1) we have assumed that $I_i, i = 0, 1, \ldots, (n-1)$ are real coefficients. Since the pair z^m and z^{-m} always appears together in (2.1), the power pattern can be converted into a polynomial of y of degree $(n-1)$ with real coefficients,

$$P(y) = \sum_{i=0}^{n-1} A_i y^i, \tag{2.4}$$

if the following substitution of variables is made:

$$y = z + z^{-1} = 2\cos u, \tag{2.5}$$

where

$$u = kd(\cos\theta - \cos\theta_0), \tag{2.6}$$

with d the element spacing, k the phase constant, θ the observation angle measured from the array axis, and θ_0 the position of the principal maximum. It is clear from (2.5) that the maximum visible range for y is from -2 to 2.

From the analysis viewpoint, $I_i, i=0, 1, \ldots, (n-1)$ are given first. It is then straightforward to calculate A_i in (2.4) via (2.2), (2.3), and (2.5). On the other hand, when the synthesis problem is presented, we often try to obtain a physically realizable $P(y)$ from the specification through certain approximation techniques. Here, by the adjective "realizable" we mean that $P(y)$ should satisfy

$$P(y) \geqslant 0 \quad \text{in} \quad -2 \leqslant y \leqslant 2. \tag{2.7}$$

The question is, then, how can the excitation coefficients I_i be determined from that realizable $P(y)$? The answer to this question can be found through a factorization process. Since $P(y)$ is a polynomial satisfying (2.7), it must only contain a combination of the following possible elementary factors:

$$\begin{array}{lll} \text{(i)} & (y+c_i) \quad \text{or} \quad (c_i - y) & \text{with } c_i \text{ real and } \geqslant 2, \\ \text{(ii)} & (y+c_i)^2 & \text{with } c_i \text{ real and } |c_i|<2, \\ \text{(iii)} & (y^2+2c_{i1}y+c_{i1}^2+c_{i2}^2) & \text{with } c_{i1} \text{ and } c_{i2} \text{ real.} \end{array} \tag{2.8}$$

It can be readily verified that the elementary array polynomials corresponding to these factors are, respectively,

$$\text{(i)} \quad E_i(z) = \frac{1}{\sqrt{c'_i}}(1+c'_i z^{-1}), \quad \text{or} \quad \frac{1}{\sqrt{c'_i}}(1-c'_i z^{-1})$$

with $\quad c'_i = \frac{1}{2}\left(c_i \pm \sqrt{c_i^2 - 4}\right); \quad c'_i = 1 \text{ when } c_i = 2,$

$$\text{(ii)} \quad E_i(z) = 1 + c_i z^{-1} + z^{-2}, \tag{2.9}$$

$$\text{(iii)} \quad E_i(z) = \frac{1}{\sqrt{c'_{i1}c'_{i2}}}\left[1+(c'_{i1}+c'_{i2})z^{-1}+c'_{i1}c'_{i2}z^{-2}\right]$$

with $\quad c'_{i1} = \frac{1}{2}\left[c_{i1} + jc_{i2} \pm \sqrt{(c_{i1}+jc_{i2})^2 - 4}\right],$

$$c'_{i2} = \overline{c'_{i1}} \quad \text{(complex conjugate of } c'_{i1}).$$

Therefore the procedures of determining I_i from $P(y)$ are (a) to find the elementary factors as those in (2.8), (b) to calculate the corresponding array polynomials according to (2.9), and (c) multiply all these array polynomials to obtain the final coefficients.

Note that the factor (i) in (2.8) gives a possible physical null at $y=-2$ or $y=2$ only when $c_i=2$. It is also when $c_i=2$ that the solution for $E_i(z)$ is unique and symmetric, as can be seen from the factor (i) in (2.9). The factor (ii) in (2.8) always identifies a null at $y=-c_i$, and there always exists a unique and symmetric solution for $E_i(z)$ corresponding to this particular factor. The factor (iii) in (2.8) gives, however, no nulls but only a minimum. Hence, in order that arrays have the maximum possible number $(n-1)$ of physical nulls in the visible range $-2 \leqslant y \leqslant 2$ (and thus that there is a unique and symmetric solution for the array polynomial), the only factors that can appear in $P(y)$ are $(y+2)$, $(2-y)$, and $(y+c_i)^2$ with c_i real and $|c_i|<2$.

Furthermore, although the point $y=2$ $(z=1, u=0)$ can be a null in principle, it normally is the position of the beam maximum, especially when all the excitation coefficients are non-negative, as evidenced many times in the previous chapter. Thus, the power pattern of an array having symmetric and non-negative excitations and the maximum possible number of nulls must take either of the following two forms, depending on whether n is odd or even:

$$P_0(y) = \prod_{i=1}^{(n-1)/2} (y+c_i)^2 \qquad \text{for odd } n, \tag{2.10}$$

$$P_e(y) = (y+2) \prod_{i=1}^{(n-2)/2} (y+c_i)^2 \qquad \text{for even } n, \tag{2.11}$$

where all c_i's are real, distinct, and $|c_i|<2$.

The corresponding array polynomials are, respectively, according to (2.9),

$$E_0(z) = \prod_{i=1}^{(n-1)/2} (1+c_i z^{-1}+z^{-2}) \qquad \text{for odd } n \tag{2.12}$$

and

$$E_e(z) = (1+z^{-1}) \prod_{i=1}^{(n-2)/2} (1+c_i z^{-1}+z^{-2}) \qquad \text{for even } n. \tag{2.13}$$

Excitation coefficients in array elements can then be determined by expanding (2.12) or (2.13).

The directivity in (1.9), which has been discussed extensively in Chapter 1, can also be expressed in terms of the power pattern

$$D = \frac{4\pi P(y)_{\max}}{\int_0^{2\pi}\int_0^{\pi} P(y)\sin\theta\, d\theta\, d\varphi} = \frac{2kdP(y)_{\max}}{W}, \tag{2.14}$$

where

$$W = \int_{y_a}^{y_b} \frac{P(y)}{\sqrt{4-y^2}}\,dy, \tag{2.15}$$

$$\begin{aligned} y_b &= 2\cos[kd(1+\cos\theta_0)], && \text{corresponding to } \theta=\pi,\\ y_a &= 2\cos[kd(1-\cos\theta_0)], && \text{corresponding to } \theta=0. \end{aligned} \tag{2.16}$$

The maximum value of the power pattern, $P(y)_{\max}$, becomes $P(2)$ when $y=2$ is the position of the beam maximum.

For broadside arrays, $\theta_0=\pi/2, y_a=y_b$, Eq. (2.15) should be replaced by

$$W = 2\int_{y_c}^{2} \frac{P(y)}{\sqrt{4-y^2}}\,dy, \tag{2.17}$$

where

$$y_c = 2\cos kd, \qquad \text{corresponding to } \theta=0 \text{ and } \theta_0=\frac{\pi}{2}. \tag{2.18}$$

In (2.14), (2.15), and (2.17), we have treated $P(y)$ as a general power pattern which can take any form as long as it satisfies the realization condition (2.7). In particular, it can be in the form of either (2.10) or (2.11) if those condition on excitations and number of nulls are satisfied. Note that the directivity has been expressed directly in terms of $P(y)$; it is therefore not necessary to determine the amplitude distribution in the array polynomial first. Since $P(y)$ is a finite polynomial in y, it can always be rearranged in terms of the factor $(4-y^2)$ so that the exact analytic expression for W and hence that for D can be obtained by integrating (2.15) or (2.17) term by term. The integration itself is straightforward; no manipulations with complicated transcendental functions, as are usually involved, are required.

Equations (2.10) to (2.13), though simple, are very useful and general enough to cover most of the practical cases. It should be instructive to give an example to demonstrate its usefulness, and at the same time show some interesting interrelationships among the power pattern, amplitude excitations, positions of nulls and sidelobes, and the directivity.

For an array of five elements, the power pattern and the resulting array polynomial can, according to (2.10) and (2.12), be expressed as

$$P_0(y)=(y+c_1)^2(y+c_2)^2, \tag{2.19}$$

$$\begin{aligned} E_0(z)&=(1+c_1z^{-1}+z^{-2})(1+c_2z^{-1}+z^{-2}) \\ &=1+(c_1+c_2)z^{-1}+(2+c_1c_2)z^{-2}+(c_1+c_2)z^{-3}+z^{-4}. \end{aligned} \tag{2.20}$$

It is clear that $E_0(z)$ does give a symmetric amplitude excitation. In order that all the excitation coefficients be non-negative, we require

$$c_1+c_2 \geqslant 0 \qquad \text{and} \qquad c_1c_2 \geqslant -2, \tag{2.21}$$

which ensures that the beam maximum is at $y=2$. In addition, when $kd=\pi$ and $\theta_0=\pi/2$, we have, according to (2.17) and (2.14),

$$W=2\pi[6+2(c_1^2+c_2^2)+8c_1c_2+c_1^2c_2^2] \tag{2.22}$$

and

$$D=\frac{(2+c_1)^2(2+c_2)^2}{6+2(c_1^2+c_2^2)+8c_1c_2+c_1^2c_2^2}. \tag{2.23}$$

Positions of nulls and sidelobes are as follows:

$$\text{nulls:} \qquad y=-c_1 \qquad \text{and} \qquad -c_2, \tag{2.24}$$

$$\text{sidelobes:} \qquad y=-(c_1+c_2)/2 \qquad \text{and} \qquad -2. \tag{2.25}$$

In (2.25), the position of the first sidelobe is obtained from the roots of

$$\frac{d}{dy}P_0(y)=0, \tag{2.26}$$

excluding $y=-c_1$ and $y=-c_2$, which have already been identified a nulls. The point $y=-2$ is taken as the position of the second sidelobe although it does not satisfy (2.26), because $P_0(-2)$ is not the principa maximum and has a non zero value. Levels of the sidelobes relative to the principal maximum are

$$\text{first sidelobe} = \frac{(c_1 - c_2)^4}{16(2+c_1)^2(2+c_2)^2} \tag{2.27}$$

and

$$\text{second sidelobe} = \frac{(2-c_1)^2(2-c_2)^2}{(2+c_1)^2(2+c_2)^2}. \tag{2.28}$$

It is important to note that (2.20) through (2.28) clearly exhibit the interrelationships among characteristics of this simple array.

Now, we are ready to discuss some special cases from the synthesis viewpoint.

Case A. If the amplitude excitations are required to be equal (uniform array), we then have from (2.20),

$$c_1 + c_2 = 1 \qquad \text{and} \qquad c_1 c_2 = -1$$

yielding $c_1 = (1-\sqrt{5})/2$ and $c_2 = (1+\sqrt{5})/2$. Substituting these values into appropriate expressions, we obtain the following for $\theta_0 = \pi/2$ and $kd = \pi$:

Principal maximum: 25 at $y=2$, $u=0$, $\theta=\pi/2$;

Nulls: at $y=0.618$ and -1.618, $u=\pm 72°$ and $\pm 144°$,$\theta=36.9°$, 66.4°, 113.6°, and 143.1°;

First-null beamwidth: 47.2°;

Sidelobes: at $y=-0.5$ and -2, $u=\pm 104.4°$ and $\pm 180°$,$\theta=0°$, 54.5°, 125.5°, and 180°;

First sidelobe $= \frac{1}{16}$, or -12.04 dB;

Second sidelobe $= \frac{1}{25}$, or -13.98 dB;

Directivity = 5.

All of these results, of course, agree with what we learned from Section 1.2. In addition, the position of the first sidelobe at $u = \pm 104.4°$ is exact and obtained easily. No approximation of estimating it or solving a transcendental equation for it is necessary. The approximate position of the first sidelobe, according to (1.29), would have been at $u = \pm 108°$. The results of c_1 and c_2 determined for this particular case also satisfy both $\partial D/\partial c_1 = 0$ and $\partial D/\partial c_2 = 0$ obtained from (2.23), giving further assurance that an equally spaced array does yield the maximum directivity when $I_i = 1$ and $d = \lambda/2$.

Case B. If we are not satisfied with the first-null beamwidth obtained above, we can reassign c_1 in such a way that the first null at $y = -c_1$ be moved closer to the beam maximum at $y = 2$. From previous experience we know that this can only be achieved by sacrificing levels of the sidelobes. In general, levels of both of the sidelobes will be changed when c_1 and c_2 are allowed to vary. Detailed variations of sidelobes levels are displayed in Fig. 2.1, where we have restricted $c_1 + c_2 \geqslant 0$ so that none of the excitation coefficients will become negative. It can be seen there that when $c_1 = -0.8$ and $c_2 = 1.142$, we still can maintain the first sidelobe (now at $y = -0.171$) at the same level as case A by allowing the second sidelobe (still at $y = -2$) to bear the full burden for improving the beamwidth. Of course, in the practical situation when a physical element used in the array has a null at $y = -2$ or nearby (e.g., the short collinear dipole), the unusually high level of the second sidelobe will be reduced to an acceptable level in view of the principle of pattern multiplication. It is also clear that the first-null beamwidth cannot be made arbitrarily narrow by choosing a more negative value for c_1, because this process will eventually raise the first sidelobe to a very high level [see both Fig. 2.1(*a*) and Eq. (2.27)] or even change the

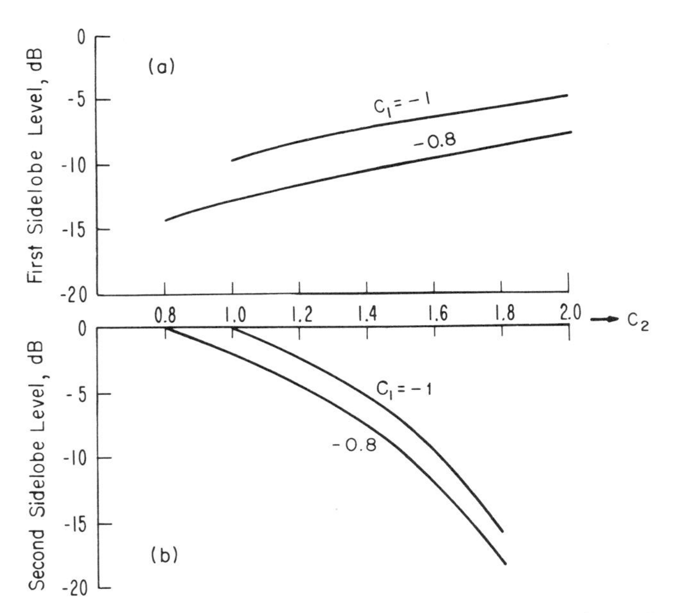

Fig. 2.1 Sidelobe levels of a five-element array as a function of amplitude excitations (or null positions).

location of the principal maximum. When the extreme situation $c_1+c_2=0$ occurs, the excitation coefficients for the second and fourth elements, according to (2.20), will vanish. The resultant array will have only three elements with a doubled spacing $d'=2d=\lambda$, and therefore the second sidelobe will become a grating lobe, as clearly indicated in Fig. 2.1(b).

Case C. If we wish to make both of the sidelobes at a same level, say -20 dB, we then can set, from (2.27) and (2.28),

$$(2+c_1)^2(2+c_2)^2=100(2-c_1)^2(2-c_2)^2 \tag{2.29}$$

and

$$(c_1-c_2)^4=16(2-c_1)^2(2-c_2)^2, \tag{2.30}$$

which yield $c_1=-0.0413$ and $c_2=1.6498$. The required excitation coefficients will be 1 : 1.6085 : 1.9318 : 1.6085 : 1. We also have the following, assuming the same values for d and θ_0 as those in case A:

Principal maximum: 51.1062 at $y=2$, $u=0$, or $\theta=\pi/2$;

Nulls: at $y=0.0413$ and -1.6498, $u=\pm 88.8°$ and $\pm 145.6°$;$\theta=36°$, 60.4°, 119.6°, and 144°;

First-null beamwidth: 59.2°;

Sidelobes: at $y=-0.8042$ and -2, $u=\pm 113.7°$ and $\pm 180°$,$\theta=0°$, 50. 8°, 129.2°, and 180°, all at -20 dB;

Directivity: 4.6857.

Comparing these results with those for case A, we find naturally that the first-null beamwidth is wider and the directivity is lower for the present case—a price we have to pay in order to bring the sidelobe level down to -20 dB. Should we specify a higher or lower sidelobe level, the corresponding first-null beamwidth will, respectively, be narrower or wider. As a special case, we can choose $c_1=c_2=2$. Then, positions of the nulls and sidelobes all coincide with the point $y=-2$, and the sidelobes are totally eliminated. It is equivalent to saying that the sidelobes are still equal but at $-\infty$ dB. The required excitation coefficients will then be 1 : 4 : 6 : 4 : 1, a binomial distribution[1] (also see Probelm 1.8).

Note that the results obtained from the present case are identical to those for the commonly known Dolph-Chebyshev array,[2,3] although a totally different approach is used here. A general synthesis procedure of obtaining arrays of any number of elements with equal sidelobes is presented in the next section.

2.2 Arrays with Equal Sidelobes

In the previous section, we demonstrated the usefulness and advantage of dealing with the power pattern and illustrated a few cases for a simple five-element array from different considerations. One of these considerations is to equalize all the sidelobes to the same level, as shown in case C. Now, we wish to formulate the general problem of synthesizing an array with equal sidelobes, also starting from an expression for the power pattern.

Referring to (2.10) or (2.11) we can set the following procedures: (a) to solve for the positions, y_l, of all the sidelobes from $(d/dy)\ P(y_l)=0$; (b) to equate all $P(y_l)$'s; and (c) to set the ratio $P(2)/P(y_1)$ equal to some constant, say K^2, which is fixed by a desired sidelobe level relative to the principal maximum. Alternatively, step (c) may be replaced by a prespecified first null, $y=-c_1$, which is tantamount to specifying the beamwidth. The above-mentioned procedures should result in a right number of independent equations for the complete determination of $P(y)$, from which the required array polynomial $E(z)$ can then readily be derived according to (2.12) or (2.13).

For an odd n, we have $(n-1)/2$ distinct values of c_i to be determined for $P_0(y)$ in (2.10). We then do the following:

(a) Set $(d/dy)P_0(y)=0$ giving $(n-3)/2$ distinct values of y_l, namely, y_1, $y_2,\ldots,y_{(n-3)/2}$. Note that $y_l \neq -c_i$, $i=1, 2,\ldots,(n-1)/2$, which have been identified as the nulls. Note further that $y=-2$ is always a sidelobe in this case, as evidenced by the example shown in Section 2.1.

(b) Set

$$\left.\begin{aligned} P_0(y_1)&=P_0(y_2),\\ P_0(y_2)&=P_0(y_3),\\ &\cdots\\ P_0(y_{(n-5)/2})&=P_0(y_{(n-3)/2}),\\ P_0(y_{(n-3)/2})&=P_0(-2). \end{aligned}\right\} \text{a total of } (n-3)/2 \text{ independent equations} \tag{2.31}$$

(c) Set

$$\frac{P_0(2)}{P_0(y_1)}=\frac{P_0(2)}{P_0(-2)}=K^2, \tag{2.32}$$

or alternatively, set $c_1=$ a specified value. (2.33)

Steps (b) and (c) give the required $(n-1)/2$ equations for solving for c_i.

For an even n, we have $(n-2)/2$ distinct values of c_i to be determined from $P_e(y)$ in (2.11). We then do the following:

(a) Set $(d/dy)P_e(y)=0$ giving $(n-2)/2$ values of y_l, namely $y_1, y_2, \ldots, y_{(n-2)/2}$. Note that $y_l \neq -c_i$. Note further that $y=-2$ is a null in this case.

(b) Set

$$\left.\begin{aligned} P_e(y_1) &= P_e(y_2), \\ P_e(y_2) &= P_e(y_3), \\ &\cdots \\ P_e(y_{(n-4)/2}) &= P_e(y_{(n-2)/2}). \end{aligned}\right\} \text{a total of } (n-4)/2 \text{ independent equations} \tag{2.34}$$

(c) Set

$$\frac{P_e(2)}{P_e(y_1)} = K^2, \tag{2.35}$$

$$\text{or alternatively, set } c_1 = \text{a specified value.} \tag{2.36}$$

The system of (2.34) and (2.35) or (2.36) again gives the required $(n-2)/2$ equations.

When n is very large, it appears that we would have to solve for a large set of simultaneous equations in (2.31)–(2.33) or (2.34)–(2.36). This, however, is fortunately not true. The actual probelm-solving procedures are really simple. Since we have imposed the condition that all the sidelobes are equal in level through the system of equations (2.31) or (2.34), it is clear that all c_i's and y_l's are related to each other. The simple example given in the previous section shows that, for $n=5$, c_1, c_2, and y_1 are related by (2.25) and (2.30). Therefore, if a general and simple relationship among all c_i and y_l can also be found for an arbitrary n, we can solve the problem directly in the case where (2.33) or (2.36) is specified; alternatively, we express $c_i (i \geqslant 2)$ and y_l (all l) in terms of c_1 only and then solve (2.32) or (2.35) for c_1 when the constant K^2 there is given. Thus, after finding the general relationship for c_i and y_l, we have only one equation, either (2.32) or (2.35), to solve numerically for c_1. Furthermore, since $y=-c_1$ determines the first null, and since its position should be close to the principal maximum at $y=2$, its approximate value is already known. The larger the n, the closer the value of c_1 is to -2. Therefore it should not be a difficult task to determine c_1 numerically. This is the reason we have chosen here to express all c_i and y_l in terms of c_1 first and then solve for it. Actually, we can, with equal ease, express all the unknowns in terms of the last $c_i[c_{(n-1)/2}$ for odd n, $c_{(n-2)/2}$ for even $n]$ and then solve for it because the value of this last c_i should be close to 2.

Now, let us find the general and simple relationship among c_i's and y_l's. As we noted at the end of Section 2.1, the results obtained for arrays with equal sidelobes by our approach of dealing directly with $P(y)$ should be identical to those for the well-known Dolph–Chebyshev array.[2] This is true because the entire array pattern is representable by a polynomial and there exists only one kind of polynomial (Chebyshev polynomial[4]) that displays equal ripples (sidelobes). Therefore, the power pattern for an array of n isotropic elements, $P(y)$, discussed so far is but the square of the normalized Chebyshev polynomial of order $n-1$, $T_{n-1}(x)$. The normalization constant is necessary to make the absolute value of sidelobes (ripples) equal to unity as required by the ordinary way of defining $T_{n-1}(x)$. This extra normalization constant does not, however, produce any effect on null positions. It is with this note that a simple relationship among c_i's and y_l's can be derived by the known properties associated with a Chebyshev polynomial. In his paper, Dolph[2] used the transformation of variable,

$$x = x_0 \cos\frac{u}{2}, \tag{2.37}$$

where x_0 is the positon of the principal maximum and u is the same as (2.6), although only the broadside array was discussed by Dolph. The symbol y used in our work here is related to u by (2.5). It is, therefore, straghtforward to establish the following:

$$x^2 = x_0^2 \cos^2\left(\frac{u}{2}\right) = \frac{x_0^2}{2}(1+\cos u) = \frac{x_0^2}{4}(y+2), \tag{2.38}$$

which checks with the obvious fact that $x = x_0$ when $y = 2$.

Positions of nulls and sidelobes, in terms of x, are given respectively by

$$T_{n-1}(x) = \cos[(n-1)\cos^{-1}x] = 0 \tag{2.39}$$

and

$$\sin[(n-1)\cos^{-1}x] = 0, \tag{2.40}$$

which yield, respectively,

$$x_i = \pm \cos\frac{(2i-1)\pi}{2(n-1)} \tag{2.41}$$

and

$$x_l = \pm \cos\frac{l\pi}{n-1}, \tag{2.42}$$

where

$$i = 1,2,\ldots,\frac{n-1}{2}; \qquad l = 1,2,\ldots,\frac{n-3}{2} \qquad \text{for odd}\, n,$$

and

$$i,l = 1,2,\ldots,\frac{n-2}{2} \qquad \text{for even}\, n.$$

Equating (2.41) to $y = -c_i$ and (2.42) to $y = y_l$ through the use of (2.38), we have

$$\cos^2\left[\frac{(2i-1)\pi}{2(n-1)}\right] = \frac{x_0^2}{4}(2-c_i) \tag{2.43}$$

and

$$\cos^2\left[\frac{l\pi}{n-1}\right] = \frac{x_0^2}{4}(2+y_l). \tag{2.44}$$

Thus, the following simple and important relations can be established:

$$(2-c_1):(2-c_2):\cdots:(2-c_{(n-1)/2}):(2+y_1):(2+y_2):\cdots:(2+y_{(n-3)/2})$$

$$= \cos^2\left[\frac{\pi}{2(n-1)}\right]:\cos^2\left[\frac{3\pi}{2(n-1)}\right]:\cdots:\cos^2\left[\frac{(n-2)\pi}{2(n-1)}\right]$$

$$:\cos^2\left(\frac{\pi}{n-1}\right):\cos^2\left(\frac{2\pi}{n-1}\right):\cdots:\cos^2\left[\frac{(n-3)\pi}{2(n-1)}\right] \qquad \text{for odd}\, n, \tag{2.45}$$

$$(2-c_1):(2-c_2):\cdots:(2-c_{(n-2)/2}):(2+y_1):(2+y_2):\cdots:(2+y_{(n-2)/2})$$

$$=\cos^2\left[\frac{\pi}{2(n-1)}\right]:\cos^2\left[\frac{3\pi}{2(n-1)}\right]:\cdots:\cos^2\left[\frac{(n-3)\pi}{2(n-1)}\right]$$

$$:\cos^2\left(\frac{\pi}{n-1}\right):\cos^2\left(\frac{2\pi}{n-1}\right):\cdots:\cos^2\left[\frac{(n-2)\pi}{2(n-1)}\right] \quad \text{for even } n, \qquad (2.46)$$

where

$$2>-c_1>y_1>-c_2>y_2>\cdots>y_{(n-3)/2}>-c_{(n-1)/2}>-2 \qquad \text{for odd } n \qquad (2.47)$$

and

$$2>-c_1>y_1>-c_2>y_2>\cdots>-c_{(n-2)/2}>y_{(n-2)/2}>-2 \qquad \text{for even } n. \qquad (2.48)$$

It is seen clearly that, when c_1 (or, equivalently, the first null) is specified, all the c_i's ($i \geqslant 2$) can be easily determined from the first half of (2.45) or (2.46) no matter how large n may be. We can then calculate (a) y_l's by the second half of (2.45) or (2.46), (b) the sidelobe level relative to the principal maximum by evaluating $P(y_1)/P(2)$,(c) the directivity by (2.14) and (2.15) with a known element spacing and θ_0, and (d) the required amplitude excitations in the array from (2.12) or (2.13).

On the other hand, if K^2 is specified, we can (a) find the relations expressing c_i ($i \geqslant 2$) and y_l in terms of c_1, (b) substitute them into $P(2)/P(y_1)=K^2$, (c) solve for c_1 numerically, (d) calculate the first-null beamwidth from c_1, and (e) evaluate other characteristics according to the same procedure stated in the above paragraph.

The array so synthesized is known to have optimum characteristics in the sense that the sidelobe level is the lowest for a specified first-null beamwidth or that the first-null beamwidth is the narrowest for a specified sidelobe level. This means that it is impossible to find another set of excitation coefficients yielding a better performance in both the first-null beamwidth and sidelobe level, for a given n, d, and θ_0. This is actually true only when $d \geqslant \lambda/2$ for $\theta_0=\pi/2$ and $d \geqslant \lambda/4$ for $\theta_0=0$, as pointed out by Riblet.[5] The reason for this is that if $d<\lambda/2$ for $\theta_0=\pi/2$ or $d<\lambda/4$ for

$\theta_0=0$, a portion of the pattern in the sidelobe region will be shifted to the invisible range of u to make the final beamwidth wider although the sidelobe level is not changed. Under that condition, we can synthesize another array with the same set of n, d, and θ_0 to give a narrower beamwidth for the same desired sidelobe level. Details on this particular point, which is important from the optimization viewpoint, will be presented in the next section. For the remaining material in this section, we will consider $d \geqslant \lambda/2$ for $\theta_0=\pi/2$ and $d \geqslant \lambda/4$ for $\theta_0=0$ only.

Although the general relations between various c_i's and y_l's have been derived in (2.45) and (2.46), some more special relations, which are interesting and useful, are presented below:

A. For an array with equal sidelobes and an odd number of elements ($n=2N+1$), the entire region of sidelobes is symmetric with respect to a central point.

A-1. If N is odd, this central point is a null at $y=-c_{(N+1)/2}$, and the following relations hold true:

$$c_{(N+1)/2}=\tfrac{1}{2}(c_i+c_{N+1-i}), \qquad i=1,2,\ldots,\frac{N-1}{2} \tag{2.49}$$

$$-c_{(N+1)/2}=\tfrac{1}{2}(y_i+y_{N-i}). \tag{2.50}$$

Proof. For $n=2N+1$, the power pattern becomes

$$P_0(y)=\prod_{i=1}^{N}(y+c_i)^2.$$

With an odd N, we have, by using (2.45),

$$\frac{2-c_i}{2-c_{(N+1)/2}}=\frac{\cos^2[(2i-1)\pi/2(n-1)]}{\cos^2[N\pi/2(n-1)]}=\frac{\cos^2[(2i-1)\pi/4N]}{\cos^2(\pi/4)}, \tag{2.51}$$

$$\frac{2-c_{N+1-i}}{2-c_{(N+1)/2}}=\frac{\cos^2[(2N-2i+1)\pi/4N]}{\cos^2(\pi/4)}=\frac{\sin^2[(2i-1)\pi/4N]}{\cos^2(\pi/4)}, \tag{2.52}$$

$$\frac{2+y_i}{2-c_{(N+1)/2}}=\frac{\cos^2[i\pi/(n-1)]}{\cos^2(\pi/4)}=\frac{\cos^2(i\pi/2N)}{\cos^2(\pi/4)}, \tag{2.53}$$

and

$$\frac{2+y_{N-i}}{2-c_{(N+1)/2}}=\frac{\cos^2[(N-i)\pi/2N]}{\cos^2(\pi/4)}=\frac{\sin^2(i\pi/2N)}{\cos^2(\pi/4)}. \tag{2.54}$$

The addition of (2.51) and (2.52) gives

$$\frac{4-(c_i+c_{N+1-i})}{2-c_{(N+1)/2}}=\frac{1}{\cos^2(\pi/4)}=2$$

or

$$c_{(N+1)/2}=\tfrac{1}{2}(c_i+c_{N+1-i}). \tag{2.55}$$

The additon of (2.53) and (2.54) yields

$$\frac{4+(y_i+y_{N-i})}{2-c_{(N+1)/2}}=2$$

or

$$-c_{(N+1)/2}=\tfrac{1}{2}(y_i+y_{N-i}). \tag{2.56}$$

Thus the proof for (2.49) and (2.50) is completed.

An example for $n=11$ ($N=5$) is shown in Fig. 2.2, where we can see that the null, $y=-c_3$, is the symmetric point, and that positions of nulls and sidelobes have the following interesting relations:

$$c_3=\tfrac{1}{2}(c_1+c_5)=\tfrac{1}{2}(c_2+c_4) \tag{2.57}$$

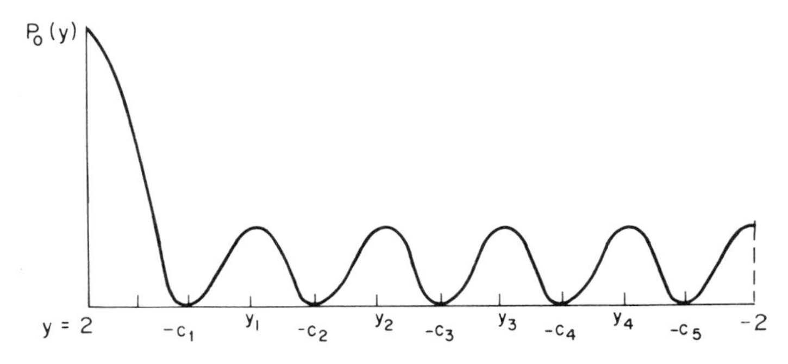

Fig. 2.2 Relative positions of nulls and sidelobes for an eleven-element array with equal sidelobes.

and

$$-c_3=\tfrac{1}{2}(y_1+y_4)=\tfrac{1}{2}(y_2+y_3). \qquad (2.58)$$

A-2. If N is an even number, the central point is a sidelobe at $y=y_{N/2}$ and the following relations hold true:

$$y_{N/2}=\tfrac{1}{2}(y_{N/2-i}+y_{N/2+i}), \qquad i=1,2,\ldots,\left(\frac{N}{2}-1\right), \qquad (2.59)$$

$$-y_{N/2}=\tfrac{1}{2}(c_i+c_{N+1-i}), \qquad i=1,2,\ldots,\frac{N}{2}. \qquad (2.60)$$

The formal proof for (2.59) and (2.60), which can be done by procedures similar to those for case A-1, is left to the reader, who should also draw a graphical presentation as that in Fig. 2.2 in order to appreciate the detailed symmetry.

B. For an array with equal sidelobes and an even number of elements ($n=2N$) there is no particular point in the entire region of y with respect to which the sidelobes are symmetric. Instead, the following relations hold true:

$$y_1-y_i=c_{N-1}-c_{N-i}, \qquad i=1,2,\ldots,(N-1). \qquad (2.61)$$

The proof for the above can be found elsewhere,[6] and is omitted here.

The expression for directivity given in (2.14) is still good for arrays discussed in this section once $P_0(y)$ or $P_e(y)$ is determined.

Before giving a numerical example to illustrate the synthesis of arrays with equal sidelobes, we are interested to note, once again, that if the parameter c_1 should become 2 (equivalently, if the first null of the array is moved to the point $y=-2$), then, by using the general relation (2.45) or (2.46) we conclude that

$$c_i=2, \text{ and } y_l=-2 \qquad \text{for all } i \text{ and } l \text{ (both odd and even } n).$$

Thus, the power pattern and the corresponding array polynomial will be, respectively,

$$P(y)=(y+2)^{n-1} \qquad \text{for both odd and even } n \qquad (2.62)$$

and

$$E(z)=(1+z^{-1})^{n-1}=\sum_{i=0}^{n-1}\frac{(n-1)!}{i!(n-1-i)!}z^{-i}, \qquad (2.63)$$

which is the binomial distribution.

Since the power pattern has only one null at $y=-2$ and no sidelobes, a sidelobeless array results. As we already mentioned in Section 2.1 when case C of the example was discussed, this array still can be considered as a special array with equal sidelobes whose level is at $-\infty$ dB relative to the principal maximum [equivalently, $P(2)/P(y_1)=K^2=\infty$].

Example. Let it be desired to synthesize a seven-element, equally spaced, broadside array with $d=\lambda/2$, which will have -20 dB sidelobes.

The given conditions are then $n=7$ $(N=3)$, $K^2=100$, $\theta_0=\pi/2$, and $kd=\pi$. Since n is an odd number, we must start with (2.10),

$$P_0(y)=(y+c_1)^2(y+c_2)^2(y+c_3)^2, \tag{2.64}$$

where c_1, c_2, c_3, and the positions of sidelobes are related according to (2.45). That is,

$$\begin{aligned} c_2&=0.9282+0.5359c_1,\\ c_3&=1.8564+0.0718c_1, \end{aligned} \tag{2.65}$$

$$\begin{aligned} y_1&=-0.3922-0.8039c_1,\\ y_2&=-1.4640-0.2680c_1. \end{aligned} \tag{2.66}$$

The other condition, $P(2)/P(y_1)=P(2)/P(-2)=100$, yields

$$(2+c_1)(2+c_2)(2+c_3)=10(2-c_1)(2-c_2)(2-c_3). \tag{2.67}$$

Solving (2.65) and (2.67), we obtain approximately

$$\begin{aligned} &c_1=-0.9382, \qquad c_2=0.4254, \qquad c_3=1.7890,\\ &y_1=0.3620, \qquad y_2=-1.2128, \end{aligned} \tag{2.68}$$

which, of course, check with (2.49) and (2.50).

The principal maximum for this array is at $y=2$, $u=0$, or $\theta=90°$, with $P_0(2)=95.2141$.

Nulls occur at $y=0.9382,\ -0.4254,\ -1.7890$;
$u=\pm 62°,\ \pm 102.3°,\ \pm 153.5°$;
$\theta=31.5°,\ 55.4°,\ 69.9°,\ 110.1°,\ 124.6°,\ 148.5°$.

Sidelobes are at $y_1=0.3620$, $y_2=-1.2128$, and $y_3=-2$;
$u=\pm 79.6°,\ \pm 127.3°,\ \pm 180°$;
$\theta=0°,\ 45°,\ 63.8°,\ 116.2°,\ 135°,\ 180°$,

with $P_0(0.3620)=P_0(-1.2128)=P(-2)=0.9521$.

First-null beamwidth is $(\mathrm{BW})_1=40.2°$.

The half-power point, if desired, can also be determined by solving $P_0(y_h) = \frac{1}{2}P_0(2)$, or

$$y_h^3 + 1.2762y_h^2 - 1.3165y_h - 7.6138 = 0. \tag{2.69}$$

There is only one change of sign in (2.69), yielding one positive real root, $y_h = 1.8015$, $u_h = \pm 25.7°$, or $\theta_h = 81.8°$ and $98.2°$. The half-power beamwidth is, therefore, $(\mathrm{BW})_h = 16.4°$.

The required array polynomial is

$$\begin{aligned} E_0(z) &= (1 - 0.9382z^{-1} + z^{-2})(1 + 0.4254z^{-1} + z^{-2})(1 + 1.7890z^{-1} + z^{-2}) \\ &= 1 + 1.2762z^{-1} + 1.6835z^{-2} + 1.8384z^{-3} + 1.6835^{-4} \\ &\quad + 1.2762z^{-5} + z^{-6}. \end{aligned} \tag{2.70}$$

It is seen that for the case being considered ($d = \lambda/2$, $\theta_0 = \pi/2$) all the excitation coefficients are positive and symmetric with respect to the central element.

For the purpose of calculating the directivity we rearrange the power pattern given in (2.64) in terms of $(y^2 - 4)$ so that the integration required by (2.17) can be carried out term by term. That is,

$$\begin{aligned} P_0(y) &= (y^2-4)^3 + 2.5524y(y^2-4)^2 + 10.9957(y^2-4)^2 \\ &\quad + 15.6309y(y^2-4) + 39.8763(y^2-4) + 23.5654y + 48.0833, \end{aligned}$$

$$W = 2\int_{-2}^{2} \frac{P_0(y)\,dy}{\sqrt{4-y^2}} = 2\pi(14.3050). \tag{2.71}$$

Substituting (2.71) into (2.14), we obtain

$$D = 6.6560. \tag{2.72}$$

A caution should be borne in mind that the characteristics obtained above can be expected exactly only if the accuracy in excitation coefficients is maintained to the fourth digit after the decimal point, as shown in (2.70). From the practical consideration, it may not be possible to get such accurate excitations. The actual results would then deviate from those ideal results. How much the deviation will be depends on how inaccurate the excitations are. An analysis of the sensitivity function has been given by Ma,[6] which should be helpful for estimating the deviation mentioned above. Mathematically, the sensitivity of a quantity A with respect to a parameter B, in a proportional form, may be defined as

$$S_B^A = \frac{\partial A/A}{\partial B/B} = \frac{B}{A}\cdot\frac{\partial A}{\partial B} \cong \frac{B}{A}\cdot\frac{\Delta A}{\Delta B}, \tag{2.73}$$

which gives the percentage change in A for a given percentage change of B. S_B^A can be first expressed as function of A, B, and some other parameters (if any) and then numerically calculated for a particular case. For simplicity, we can choose c_1 in the above example as B. Typical items we are interested in are percentage variations of the principal maximum, sidelobe level, excitation coefficients, positions of nulls and sidelobes, and directivity, due to a small percentage shift of the first null. Again, let the array with $n=7$ and $K^2=100$ be considered. The excitation coefficients according to (2.12) are given by the following:

$$\begin{aligned} I_0 &= I_6 = 1, \\ I_1 &= I_5 = c_1 + c_2 + c_3, \\ I_2 &= I_4 = 3 + c_1c_2 + c_2c_3 + c_1c_3, \\ I_3 &= 2(c_1 + c_2 + c_3) + c_1c_2c_3. \end{aligned} \tag{2.74}$$

Now suppose that the parameter c_1 is changed, for some reason, by a small amount Δc_1 so that it takes a new value

$$\overline{c_1} = c_1 + \Delta c_1, \tag{2.75}$$

and suppose that (2.65) still holds. That is, all the sidelobes of the new array after c_1 is changed still remain equal (of course, the level will be changed). Then, by (2.65), we have

$$\begin{aligned} \Delta c_2 &= 0.5359\Delta c_1, \\ \Delta c_3 &= 0.0718\Delta c_1, \\ \Delta y_1 &= -0.8038\Delta c_1, \\ \Delta y_2 &= -0.2680\Delta c_1, \end{aligned} \tag{2.76}$$

and

$$
\begin{aligned}
S_{c_1}^{c_2} &= \frac{0.5359c_1}{c_2} = -1.1819, \\
S_{c_1}^{c_3} &= \frac{0.0718c_1}{c_3} = -0.0377, \\
S_{c_1}^{y_1} &= \frac{-0.8039c_1}{y_1} = 2.0837, \\
S_{c_1}^{y_2} &= \frac{-0.2680c_1}{y_2} = -0.2073.
\end{aligned}
\tag{2.77}
$$

Here we can see that the position of the first sidelobe (y_1) is most sensitive and the location of the last null (c_3) is least sensitive to the change of c_1. In fact, they respectively increase (plus sign) and decrease (minus sign) by about 2% and 0.04% when c_1 is increased by 1%. Using (2.76) and (2.74) we can also compute the percentage change in I_i, $i=1,2,3,4,5$. The reader is asked to apply similar steps for evaluating the percentage change of other characteristics.

It should be noted that for arrays with a given number of elements, equal sidelobes, and $d \geqslant \lambda/2$, the value of c_1 is solely determined by a specified value of K^2 such as in (2.67). Once it is obtained, all the other c_i's are calculated according to (2.45) or (2.46), which in turn control the array polynomial such as that in (2.70) and other characteristics. Therefore, the current distribution in the array elements, when $d \geqslant \lambda/2$, is a function of K^2 only, and is independent of kd and θ_0. When the specification on K^2 is changed, it is equivalent to varying the parameter c_1. Then, $c_2, c_3, \ldots,$ will vary accordingly. If K^2 becomes larger and larger, so does c_1 to make the first-null beamwidth wider and wider, until the limit $c_1=2$ is reached, corresponding to $K^2=\infty$. Thus, the value of c_1 and the associated beamwidth increase monotonically with K^2, or decrease with increase in the sidelobe level. Their relationship is shown in Fig. 2.3 for a few number of n. On the other hand, the variation of D with respect to c_1 or K^2 is not monotonic, as can be seen from Fig. 2.4. Hence, we conclude that for each n, there exists a particular value of c_1 (or a particular sidelobe level) which will yield the maximum directivity. Of course, the maximum directivity here is true only when the constraint on equal sidelobes is held. Without this constraint, the directivity can be improved still further, the details of which will be studied in a later section.[7]

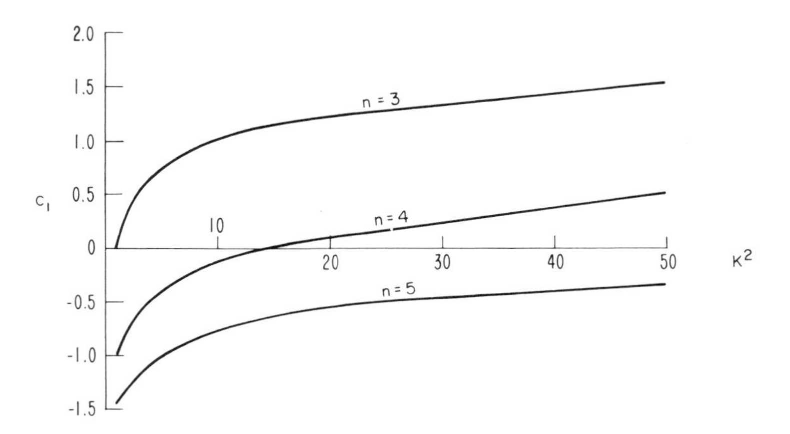

Fig. 2.3 First-null positions of arrays with equal sidelobes as a function of the main-lobe-to-sidelobe ratio.

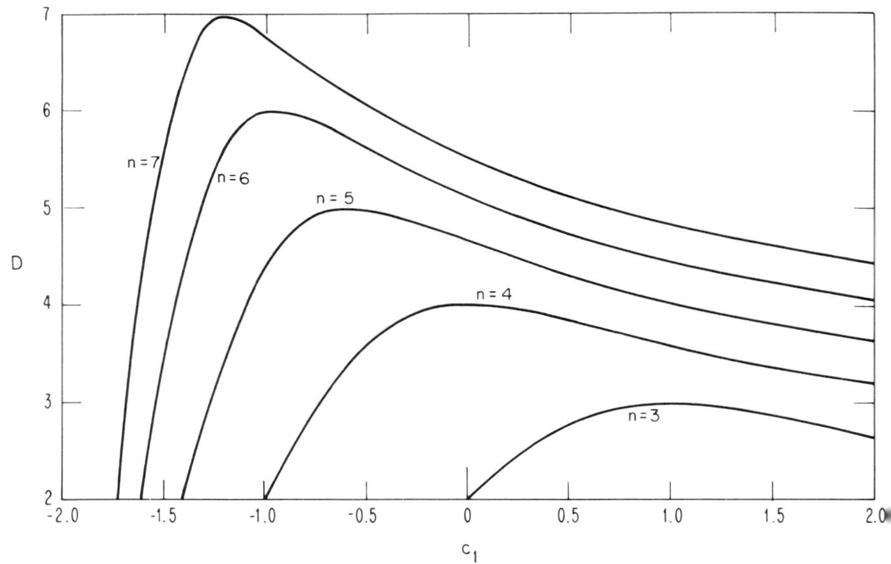

Fig. 2.4 Directivities of arrays with equal sidelobes as a function of the first-null position when $d=\lambda/2$ and $\theta_0=\pi/2$.

2.3 Optimization with Smaller Element Spacing

As we mentioned in the previous section, the results obtained there are considered optimum in Dolph's sense only when the element spacing is no less than one-half wavelength ($\theta_0 = \pi/2$). This section is intended to explain why that solution is no longer optimum when $d < \lambda/2$, and offers a method of optimization when such a case arises. Since there is a slight difference in the optimization procedure for the broadside and endfire arrays, we propose to discuss them separately.

A. Optimum Broadside Array: $u = kd\cos\theta$, $y = 2\cos u$. To keep the discussion simple and easy to understand, let us reexamine the numerical example given in the previous section where the condition $d = \lambda/2$ was used. Based on data obtained there, the entire pattern may be plotted as in Fig. 2.5(*a*). When the element spacing is greater than one-half wavelength, the visible range of u is larger, and more sidelobes of the same level will be brought into the visible range, to make the beamwidth narrower than that for $kd = \pi$. Note that these extra sidelobes are not independent by themselves in the sense that they are merely mirror images (with respect to $u = \pm\pi$) of those inside $0 \leqslant |u| \leqslant \pi$. Admittedly, the sidelobes inside $0 \leqslant |u| \leqslant \pi$ are not totally independent either since their positions are related by (2.45), but at least they as a whole can still move around depending on the specified value of K^2. Those extra sidelobes in $\pi \leqslant |u| \leqslant kd$, when $kd > \pi$, do not even have such limited freedom because they are completely fixed by those inside $0 \leqslant |u| \leqslant \pi$. The result for $d > \lambda/2$ is therefore still optimum in Dolph's sense.

When the element spacing is less than one-half wavelength, or $kd < \pi$, a portion of the sidelobe region shown in Fig. 2.5(*a*) will be invisible. Then, only the reduced visible portion of the pattern (in terms of u) will be expanded to occupy the same physical range of θ, $0 \leqslant \theta \leqslant \pi$. Under this condition, the pattern no longer containes the maximum possible number of sidelobes; the first-null beamwidth will be wider and the directivity will be lower than those for $kd = \pi$, although the sidelobe level remains unchanged. A sketch for this case is given in Fig. 2.5(*b*).

The optimization technique to be described is based on the idea of keeping the maximum possible number of sidelobes within the visible range of u even if $kd \leqslant \pi$. In order to achieve this, the following linear transformation of variables,

$$y' = A_1 y + A_2, \quad A_1 \neq 0, \tag{2.78}$$

is suggested where the constants A_1 and A_2 are to be determined by suitable conditions. Since the pattern of the broadside array is symmetric

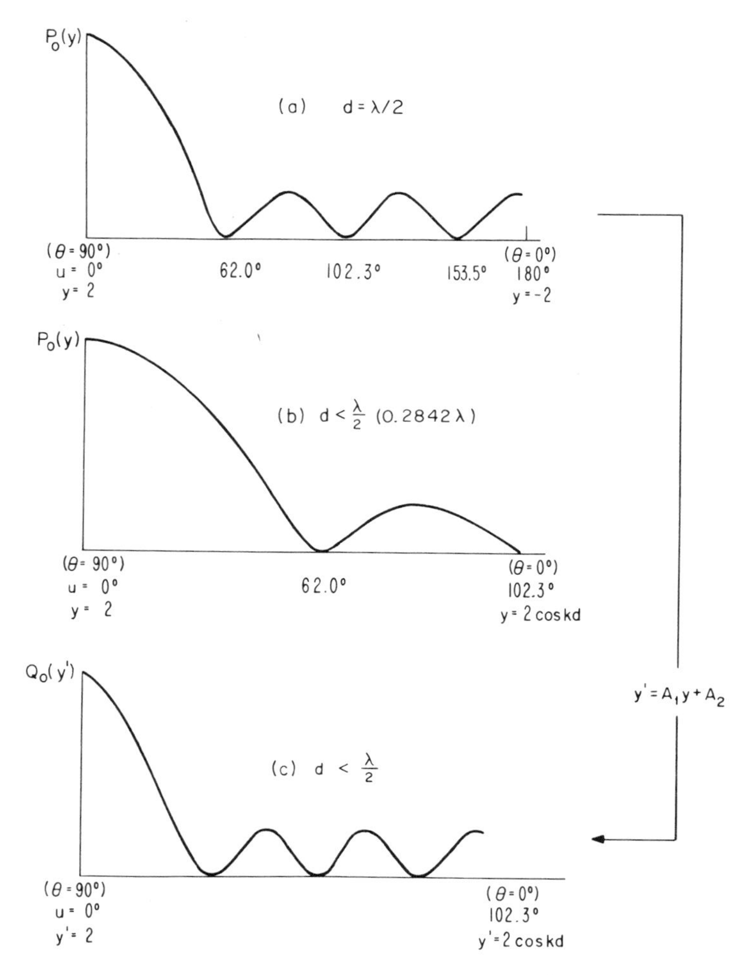

Fig. 2.5 Power patterns of a seven-element broadside array with equal sidelobes (-20 dB): (*a*) $kd=\pi$, (*b*) $kd<\pi$ without applying (2.78), (*c*) $kd<\pi$ with (2.78).

with respect to $u=0$ or $\theta=90°$, it is sufficient to consider the portion of the pattern in $0\leqslant\theta\leqslant 90°$. In this region, the visible range for $kd=\pi$ is $0\leqslant u\leqslant\pi$, or $-2\leqslant y\leqslant 2$, and that for $kd<\pi$ is $0\leqslant u\leqslant kd$, or $2\cos kd\leqslant y'\leqslant 2$. Therefore, the conditions to ensure the transformation of the maximum number of sidelobes in $-2\leqslant y\leqslant 2$, when $kd=\pi$, into $2\cos kd\leqslant y'\leqslant 2$, when $kd<\pi$ are

$$y'=2 \qquad \text{when } y=2,$$
$$y'=2\cos kd \qquad \text{when } y=-2. \tag{2.79}$$

The transformation between y and y' can be clearly illustrated by examining Figs. 2.5(*a*) and 2.5(*c*). From (2.78) and (2.79), we have

$$A_1=\tfrac{1}{2}(1-\cos kd) \qquad \text{and} \qquad A_2=1+\cos kd. \tag{2.80}$$

Note that for $kd=\pi$, $A_1=1$, $A_2=0$, and $y=y'$, an identical transformation. For $kd<\pi$ ($A_1\neq 1$ and $A_2\neq 0$), then

$$y=\frac{y'-A_2}{A_1}. \tag{2.81}$$

Substituting (2.81) into (2.64), we have the transformed power pattern

$$Q_0(y')=(A_1)^{-6}(y'+C_1)^2(y'+C_2)^2(y'+C_3)^2, \tag{2.82}$$

where

$$C_i=A_1c_i-A_2, \qquad i=1,2,3. \tag{2.83}$$

Based on (2.82), the new directivity and array polynomial will be, respectively,

$$D'=\frac{kdQ_0(2)}{\int_{y'_c}^{2}\left[Q_0(y')dy'/\sqrt{4-y'^2}\right]}, \qquad y'_c=2\cos kd \tag{2.84}$$

and

$$E'_0(z)=(A_1)^{-3}\prod_{i=1}^{3}(1+C_iz^{-1}+z^{-2}). \tag{2.85}$$

Equation (2.85) gives the new set of excitation coefficients required to keep the maximum number of sidelobes in the visible range even when $kd<\pi$.

As an example, let us consider $d=\lambda/4$ or $kd=\pi/2$. If we were to use the same excitation coefficients obtained in (2.70), the resulting array performance would have only one null in $0\leqslant\theta\leqslant 90°$, corresponding to $u=62°$ ($y=0.9382$), which now gives $\theta=46.5°$ (against $\theta=31.5°$ for $kd=\pi$). The other two nulls formerly at $u=102.3°$ and $153.5°$ would be shifted to the invisible range. By the same reason, only one sidelobe would occur at $y=0.3620$, $u=79.6°$, or $\theta=27.8°$. Although the sidelobe level remains unchanged, the first-null beamwidth and directivity would be $(\text{BW})_1 = 87°$ (against 40.2°) and $D = 3.3871$ (against 6.6560).

Now, using the transformation (2.78), we have

$$A_1 = \tfrac{1}{2}, \qquad A_2 = 1, \qquad y = 2(y' - 1),$$

$$C_1 = -1.4691, \qquad C_2 = -0.7873, \qquad C_3 = -0.1055,$$

$$Q_0(y') = 2^6 (y' - 1.4691)^2 (y' - 0.7873)^2 (y' - 0.1055)^2, \tag{2.86}$$

$$E'_0(z) = 8(1 - 2.3619z^{-1} + 4.3947z^{-2} - 4.8458z^{-3} + 4.3947z^{-4} - 2.3619z^{-5} + z^{-6}). \tag{2.87}$$

Principal maximum: $Q_0(2) = 95.2141$ (unchanged) at $\theta = 90°$.
Nulls: at $y' = 1.4691, 0.7873, 0.1055$;
$u = \pm 42.7°, \pm 66.8°, \pm 86.9°$;
$\theta = 15°, 42.1°, 61.7°, 118.3°, 137.9°, 165.0°$.
Sidelobes: at $y'_1 = 1.1810, y'_2 = 0.3936, y'_3 = 0$;
$u = \pm 53.8°, \pm 78.6°, \pm 90°$;
$\theta = 0°, 29.2°, 53.3°, 126.7°, 150.8°, 180°$;
$Q_0(y'_1) = Q_0(y'_a) = Q_0(y'_3) = 0.9521$ (unchanged).
First-null beamwidth: $(\text{BW})'_1 = 56.6°$.
Directivity: $D' = 4.7870$.

Three interesting points are noted. First, the number of physical nulls and sidelobes remains the same, although their positions are changed. Second, the first-null beamwidth and directivity, although poorer than those when $kd = \pi$, are better than those if no transformation is made. Third, the new excitation coefficients in (2.87) are dependent on kd, although they are still symmetric with respect to the central element. The only price we have to pay is the sign change (phase reversal) in the even-numbered excitation coefficients and the relatively larger ratio among these coefficients, as can be seen from (2.87).

At this point the reader may have experienced some confusion between y and y', and wonder why we did not use z' in (2.85). The reason is that, when $kd < \pi$, we synthesize the array by first obtaining a power pattern $P_0(y)$ as if $kd = \pi$, and then apply the transformation (2.78) to get $Q_0(y')$. Once this transformation is achieved, we actually can and should forget about the difference between y and y', and treat y' as y. Of course, the whole confusion would be totally eliminated if we could synthesize a $Q_0(y)$ directly when $kd < \pi$ without having to use the transformation between y

and y'. The truth is that we do not know how to do this with our approach of dealing with power patterns, or perhaps will have more complications if we do.

Although we have presented the optimization for arrays with equal sidelobes, when the element spacing is less than one-half wavelength, by a simple example with seven elements, the transformation (2.78) and the same procedure apply also to any other array with an odd number of elements. Specifically, whenever the power pattern of an array can be represented by (2.10), the transformation can be used such that the transformed power pattern and array polynomial will take, respectively, the following forms:

$$Q_0(y') = \frac{1}{A_1^{n-1}} \prod_{i=1}^{(n-1)/2} (y' + C_i)^2 \tag{2.88}$$

and

$$E'_0(z) = \frac{1}{A_1^{(n-1)/2}} \prod_{i=1}^{(n-1)/2} (1 + C_i z^{-1} + z^{-2}), \tag{2.89}$$

where

$$C_i = A_1 c_i - A_2, \qquad i = 1, 2, \ldots, \frac{n-1}{2}, \tag{2.90}$$

A_1 and A_2 are given in (2.80), and c_i's are the parameters in $P_0(y)$. The expression for directivity in (2.84) is also good for this general case.

The reader is asked to repeat the same with different values of d $(0 \leqslant d < \lambda/2)$ and plot D' versus d for $n=7$, $K^2=100$, and $\theta_0 = \pi/2$.

When the array has an even number of elements, we start with (2.11). In addition, if $kd < \pi$, the transformation (2.78) applies. The transformed power pattern will then be

$$Q_e(y') = \frac{1}{A_1^{n-1}} (y' - y'_c) \prod_{i=1}^{(n-2)/2} (y' + C_i)^2, \tag{2.91}$$

where

$$y'_c = A_2 - 2A_1 = 2\cos kd > -2 \qquad \text{when} \qquad kd < \pi. \tag{2.92}$$

The expression (2.91) still satisfies the necessary condition

$$Q_e(y') \geqslant 0 \tag{2.93}$$

in the visible range which is now

$$2\cos kd \leqslant y' \leqslant 2. \tag{2.94}$$

Comparing (2.91)–(2.93) with (2.7)–(2.8), we note that the lower end of the visible range has been changed from -2 to $2\cos kd$, and that the factor $y'-y'_c$ does not correspond to any of those listed in (2.8). Because of this, we are unable to determine an elementary array polynomial $E_e(z)$ such that $E_e(z)\overline{E_e(z)}$ will yield, with the aid of the relation $z+z^{-1}=y'$, the factor $y'-y'_c$. It means that (2.93) and (2.94) are necessary but not sufficient conditions. Therefore, the form (2.91) is not physically realizable —no array polynomial exists to give a power pattern such as that in (2.91). This leaves the problem of synthesizing broadside arrays yielding a better beamwidth-sidelobe relationship than that from the Dolph-Chebyshev array when the number of elements is even and the element spacing is less than one-half wavelength, mysteriously unsolved[5,8]. Although a general approach for trying to solve this kind of problem has been attempted by Pokrovskii[9,10], no specific numerical results have been obtained to prove that his formulation would lead to realizable arrays. The unfortunate conclusion just noted may be expected by observing the form of (2.87) where the new excitation coefficients for an array of odd number of elements after the transformation (2.78) is applied are still symmetric but change signs alternatively. The principal maximum which always occurs at $y'=2$ or $z=1$ for broadside arrays is $Q_0(2)=[E'(1)]^2>0$. If the same type of transformation were also successful when n is an even number, the final excitation coefficients would have the same pattern in sign changes and give a new array polynomial such as $E'_e(z)=1-C_1z^{-1}+C_2z^{-2}-C_2z^{-3}+C_1z^{-4}-z^{-5}$ $(n=6)$. Then, the array would have a null at $z=1$ (which is supposed to be the principal maximum). This fact confirms why the realization of such an array is impossible. It will be clear later that this same situation does not arise for the optimum endfire array whose principal maximum, although still at $\theta=0$, does not occur at $y'=2$ when the ordinary endfire condition is not imposed.

B. Optimum Endfire Array: $u=kd\cos\theta+\alpha$, $y=2\cos u$. Let us consider first the power pattern $P_0(y)$ given in (2.10) where all the sidelobes are supposed to be at an equal level. The entire pattern for $d=\lambda/2$ or $kd=\pi$ is plotted in Fig. 2.6(a) when the ordinary endfire condition $\alpha=-kd$ is used. Note that in this case there always is a large backlobe at $\theta=\pi$ having the same magnitude as the principal maximum at $\theta=0$. This type of bidirectional pattern is usually unwanted. If a smaller d is chosen, some portion of $P_0(y)$ on the left side of Fig. 2.6(a) will be invisible. It is clear that there exists an element spacing $d^*<\lambda/2$ for the ordinary endfire such that the whole visible range will end at $y^*=2\cos(-2kd^*)$ and that the final pattern will have a sidelobe at $\theta=\pi$ with the same level as the

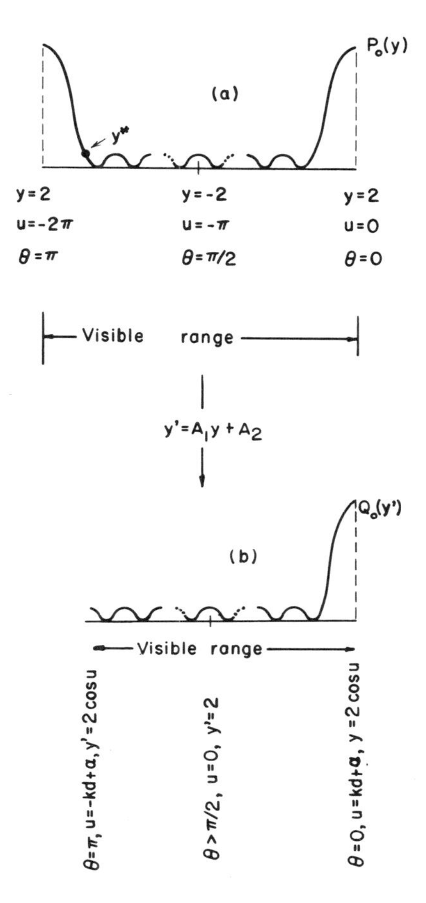

Fig. 2.6 Power patterns of an endfire array with equal sidelobes (n = odd): (a) ordinary endfire $kd=-\alpha=\pi$, (b) optimum endfire when $d<d^*$ and $\alpha \neq -kd$.

other sidelobes. In this case, no modification for improving the beamwidth or directivity is possible if the sidelobe level remains unchanged. This indeed is the case considered before by Rhodes.[11]

For $d<d^*$, some of the sidelobes will also be invisible; then only the visible portion will be expanded into the region $0 \leqslant \theta \leqslant \pi$, making the beamwidth wider than the case when $d=d^*$. In order to make improvements under this circumstance, an optimum endfire condition $\alpha \neq -kd$,

together with the transformation (2.78), is proposed such that the final pattern still contains much as it formerly did up to $y=y^*$. Since there are three unknowns [A_1, A_2 in (2.78), and α] involved, there should be three matching conditions to achieve the idea expressed above. They are

$$\begin{array}{lll} \text{(optimum endfire)} & \text{(original ordinary endfire)} & \\ y'=2\cos(kd+\alpha) & \text{when } y=2 \quad (\theta=0), & \\ y'=2\cos(-kd+\alpha) & \text{when } y=y^* \quad (\theta=\pi), & \quad (2.95) \\ y'=2 & \text{when } y=-2 \quad (\theta>\pi/2). & \end{array}$$

The first condition of (2.95) maps the location of the principal maximum, the second condition ensures that the maximum possible number of sidelobes at equal level be kept and that the large backlobe be avoided, and the third condition maps the sidelobe (for odd n) or the null (for even n) originally at $y=-2$ to $y'=2$. It is also this third condition which removes the difficulty of having an unrealizable power pattern encountered previously when the optimum broadside array with an even number of elements was discussed. The details on this point will be clear later.

The transformation (2.78) with the conditions given in (2.95) is shown in Fig. 2.6(*b*). From this figure we can see that, for $d<d^*$, the transformation always squeezes the whole pattern up to y^* into the visible range, and therefore helps to shift the pattern toward the main beam, making the beamwidth narrower and the directivity higher. Solving α, A_1, and A_2 using (2.95), we obtain

$$\tan\frac{\alpha}{2}=\cot^2\left(\frac{kd^*}{2}\right)\tan\left(\frac{kd}{2}\right),$$

$$A_1=-\sin^2\left(\frac{\alpha+kd}{2}\right)<0, \qquad (2.96)$$

$$A_2=2(1+A_1)=2\cos^2\left(\frac{\alpha+kd}{2}\right)>0,$$

where d^* is related to y^* by

$$y^*=2\cos(-2kd^*), \qquad (2.97)$$

and y^* can be easily determined by relations similar to those in (2.50), (2.59), and (2.61):

$$y^* = 2[1 - c_{(n+1)/4}] \qquad \text{for } n \text{ odd and } \frac{n-1}{2} \text{ odd,} \tag{2.98}$$

$$y^* = 2[1 + y_{(n-1)/4}] \qquad \text{for } n \text{ odd and } \frac{n-1}{2} \text{ even,} \tag{2.99}$$

or

$$y^* = 2 + y_1 - c_{(n-2)/2} \qquad \text{for } n \text{ even.} \tag{2.100}$$

It should be emphasized that d^* or y^* is fixed once the total number of elements n and the sidelobe level represented by K^2 are specified for arrays with equal sidelobes. The phase α and the transformation coefficients A_1 and A_2 required to optimize the endfire array can be calculated easily from (2.96) once a d ($<d^*$) is given. Since the value of kd^* is less than but near π, it can be concluded from the first equation of (2.96) that α should be in the first quadrant. Note that (2.96) also holds true for $d = d^*$, and in that case yields

$$\alpha^* = \pi - kd^* > 0, \tag{2.101}$$

$$A_1 = -1, \qquad \text{and} \qquad A_2 = 0. \tag{2.102}$$

From (2.101) it seems that there is an extra phase of π for α^* as compared with that required by the ordinary endfire condition. This extra phase of π is actually taken care of by the fact that $A_1 = -1$ and $A_2 = 0$. Of course, this apparent discrepancy can be removed if we try to map $y = -2$ onto $y' = -2$, replacing the third condition in (2.95). While this alternative matching condition seems more natural as far as the limiting case $d \to d^*$ is concerned, it will create a little mathematical difficulty later in dealing with the limiting case to the other end ($d \to 0$). In any case, these two approaches yield the same array depending on whether we like to express the final excitation coefficients with alternating signs, as will be clear later in an example, or to have the sign reversal absorbed in α if using the alternative transformation suggested above.

The optimization discussed above specifically for $P_0(y)$ (in Fig. 2.6) applies as well to $P_e(y)$ given in (2.11).

Substituting (2.81) into (2.10) and (2.11), we have the transformed power patterns

$$Q_0(y') = A_1^{-(n-1)} \prod_{i=1}^{(n-1)/2} (y' + C_i)^2 \qquad \text{for odd } n \tag{2.103}$$

and

$$Q_e(y') = A_1^{-(n-1)} (y' - 2) \prod_{i=1}^{(n-2)/2} (y' + C_i)^2$$

$$= -A_1^{-(n-1)} (2 - y') \prod_{i=1}^{(n-2)/2} (y' + C_i)^2 \qquad \text{for even } n, \tag{2.104}$$

where

$$C_i = A_1 c_i - A_2. \tag{2.105}$$

Note that the coefficient in the second form of (2.104) is positive since $A_1 < 0$ and $A_1^{-(n-1)} < 0$ for an even n. Note also that we now have a factor $(2 - y')$ in (2.104) rather than $(y' - 2\cos kd)$ in (2.91). In this case, we do fortunately have an elementary array polynomial $(1 - z^{-1})$ corresponding to $(2 - y')$. Therefore, the power patterns in (2.103) and (2.104) are both physically realizable. The corresponding array polynomials are, respectively,

$$E'_0(z) = A_1^{-(n-1)/2} \prod_{i=1}^{(n-1)/2} (1 + C_i z^{-1} + z^{-2}) \qquad \text{for odd } n \tag{2.106}$$

and

$$E'_e(z) = [-A_1^{-(n-1)}]^{1/2} (1 - z^{-1}) \prod_{i=1}^{(n-2)/2} (1 + C_i z^{-1} + z^{-2}) \quad \text{for even } n, \tag{2.107}$$

from which a new set of excitation coefficients required for synthesizing the optimum endfire array with equal sidelobes and $d < d^*$ can be derived.

The directivity D', the first-null beamwidth $(2\theta_1)$, and the half-power beamwidth $(2\theta_h)$ can be calculated, respectively, from

$$D' = \frac{2kdQ(y'_a)}{W'}, \tag{2.108}$$

$$-C_1 = 2\cos(kd\cos\theta_1 + \alpha), \tag{2.109}$$

and

$$y'_h = 2\cos(kd\cos\theta_h + \alpha), \tag{2.110}$$

where

$$W' = \int_{y'_a}^{y'_b} \frac{Q(y')\,dy'}{\sqrt{4-y'^2}}, \tag{2.111}$$

$$y'_b = 2\cos(-kd+\alpha), \tag{2.112}$$

$$y'_a = 2\cos(kd+\alpha), \tag{2.113}$$

and y'_h is determined by

$$Q(y'_h) = \tfrac{1}{2}Q(y'_a). \tag{2.114}$$

For the purpose of illustrating the theory thus presented, let us reconsider the example discussed in Section 2.2, where $n=7$ and $K^2=100$. The answer on c_i and y_l obtained in (2.68) still applies. If the ordinary endfire condition, $-\alpha = kd = \pi$, is used, the pattern will be bidirectional. To avoid this, the longest element spacing we can accept is d^*, which can be determined according to (2.98) and (2.97). That is,

$$y^* = 2(1-c_2) = 1.1492, \qquad 2kd^* = 305.1°, \qquad \text{or} \qquad d^* = 0.4238\lambda.$$

With $-\alpha^* = kd^* = 152.55°$, the array polynomial yielding this desired power pattern will be the same as that in (2.70), and the important characteristics are given below:

Principal maximum: at $y=2$, $u=0$, $\theta=0$, with $P_0(2)=95.2141$.
Nulls: at $y=0.9382$, -0.4254, -1.7890;
$u=-62.0°, -102.3°, -153.5°, -206.5°, -257.7°, -298.0°$;
$\theta=53.6°, 70.8°, 90.4°, 110.7°, 133.6°$, and $162.5°$.
Sidelobes: at $y_1=0.3620$, $y_2=-1.2128$, $y_3=-2$, $y_4=y^*=1.1492$;
$u=-79.6°, -127.3°, -180°, -232.7°, -280.4°, -305.1°$;
$\theta=61.3°, 80.5°, 100.4°, 121.7°, 147°, 180°$;
with $P_0(y_1)=P_0(y_2)=P_0(y_3)=P_0(y^*)=0.9521$.
First-null beamwidth: $(\text{BW})_1=107.2°$.
Half-power point: $y_h=1.8015$, $u_h=-25.7°$, $\theta_h=33.8°$,
$(\text{BW})_h=67.6°$.

Directivity: $y_b = y^* = 1.1492$, $y_a = 2$,

$$W = \int_{y_a}^{y_b} \frac{P_0(y)\,dy}{\sqrt{4-y^2}} = 46.7626,$$

$$D = \frac{2kd^* P_0(2)}{W} = 10.8423.$$

If $d = \lambda/4$ is assigned instead, and if we still use the ordinary endfire condition $-\alpha = kd = \pi/2$ without applying the transformation presented earlier in this section, the excitation coefficients would, of course, remain the same as those in (2.70). The entire visible range of u would be shrunk to $0 \geqslant u \geqslant -180°$, and as can be seen from the data listed above, the last three nulls and sidelobes would be shifted to the invisible range. As a consequence, the first null and half-power point in terms of θ would be changed, respectively, to $\theta_1 = 71.9°$ and $\theta_h = 44.4°$, making $(\text{BW})_1 = 143.8°$ and $(\text{BW})_h = 88.8°$, which are much wider than those for $d = d^*$. We also would have $D = 6.6560$, the same as that for the broadside array when $d = \lambda/2$ [see (1.48)].

Now, let us apply the transformation (2.78) together with the optimum endfire condition (2.96) to see how much improvement can be made for the case $d = \lambda/4$. Since $kd^* = 152.55°$ and $kd = 90°$, the required phase and the transformation coefficients, according to (2.96), will be $\alpha = 6.82°$, $A_1 = -0.5595$, and $A_2 = 0.8810$. We then have $u = 90° \cos\theta + 6.82°$,

$$Q_0(y') = 32.5986(y' - 0.3561)^2 (y' - 1.1190)^2 (y' - 1.8819)^2, \quad (2.115)$$

$$E'_0(z) = 5.7095(1 - 3.3570z^{-1} + 6.1745z^{-2} - 7.4640z^{-3} + 6.1745z^{-4} - 3.3570z^{-5} + z^{-6}). \quad (2.116)$$

Based on (2.115) and (2.116), the important characteristics of the new array are as follows:

Principal maximum: at $\theta = 0$, $u = 96.82°$, $y'_a = 2\cos 96.82° = -0.2378$, with $Q_0(y'_a) = 95.2141$.

Nulls: at $y' = 0.3561$, 1.1190, 1.8819;
$u = \pm 79.75°$, $\pm 56.0°$, $\pm 19.8°$;
$\theta = 35.9°$, $56.9°$, $81.7°$, $107.2°$, $134.3°$, $164.1°$.

Sidelobes: at $y_1' = 0.6785$, $y_2' = 1.5596$, $y_3' = 2.$, $y_4' = 0.2378$;
$u = \pm 70.2°$, $\pm 38.8°$, $0°$, $-83.18°$;
$\theta = 45.2°$, $69.2°$, $94.4°$, $120.4°$, $148.8°$, $180°$,
with $Q_0(y_1') = Q_0(y_2') = Q_0(y_3') = Q_0(y_4') = 0.9521$.
First-null beamwidth: $(\text{BW})_1 = 71.8°$.
Half-power point: $y_h' = -0.1268$, $u_h = 93.65°$, $\theta_h = 15.3°$,
$(\text{BW})_h = 30.6°$.
Directivity: $y_b' = 0.2378$, $y_a' = -0.2378$, $W' = 8.0561$, $D' = 37.1302$.

Note that $(\text{BW})_1$ and $(\text{BW})_h$ are much narrower than those when the ordinary endfire condition is used, and in fact are even better than those when $d = d^*$. Because of the improvement in beamwidths, the directivity is increased by 558% as compared with the case $-\alpha = kd = \pi/2$, and by 342% as compared with the case $-\alpha^* = kd^* = 152.55°$.

Optimum endfire arrays with an even number of elements can be handled in a similar fashion. Numerical results for $n = 3$ through $n = 7$ and for a sidelobe level of -20 dB are given in Figs. 2.7 and 2.8 as a function of d. For comparison, the corresponding directivities for the ordinary endfire array ($-\alpha = kd$) are also included in Fig. 2.7(a) as the dashed curves. The points marked with a star on these dashed curves are those when $d = d^*$.

From these figures we see that the smaller the d, the narrower will the beamwidth become, and the higher will the directivity be. The limiting characteristics for $n = 3$, 4, and 5 as $d \to 0$ are also included in these figures. They are obtained by taking appropriate limiting processes. Specifically, the equations in (2.96) should be replaced by the following as $d \to 0$:

$$\alpha \cong (kd)\cot^2\left(\frac{kd^*}{2}\right),$$

$$A_1 \cong -\left(x - \frac{x^3}{3!} + \frac{x^5}{5!} - \cdots\right)^2, \tag{2.117}$$

$$A_2 \cong 2\left(1 - \frac{x^2}{2!} + \frac{x^4}{4!} - \cdots\right)^2,$$

where

$$x = \frac{\alpha + kd}{2} \cong \left(\tfrac{1}{2}kd\right)\csc^2\left(\frac{kd^*}{2}\right).$$

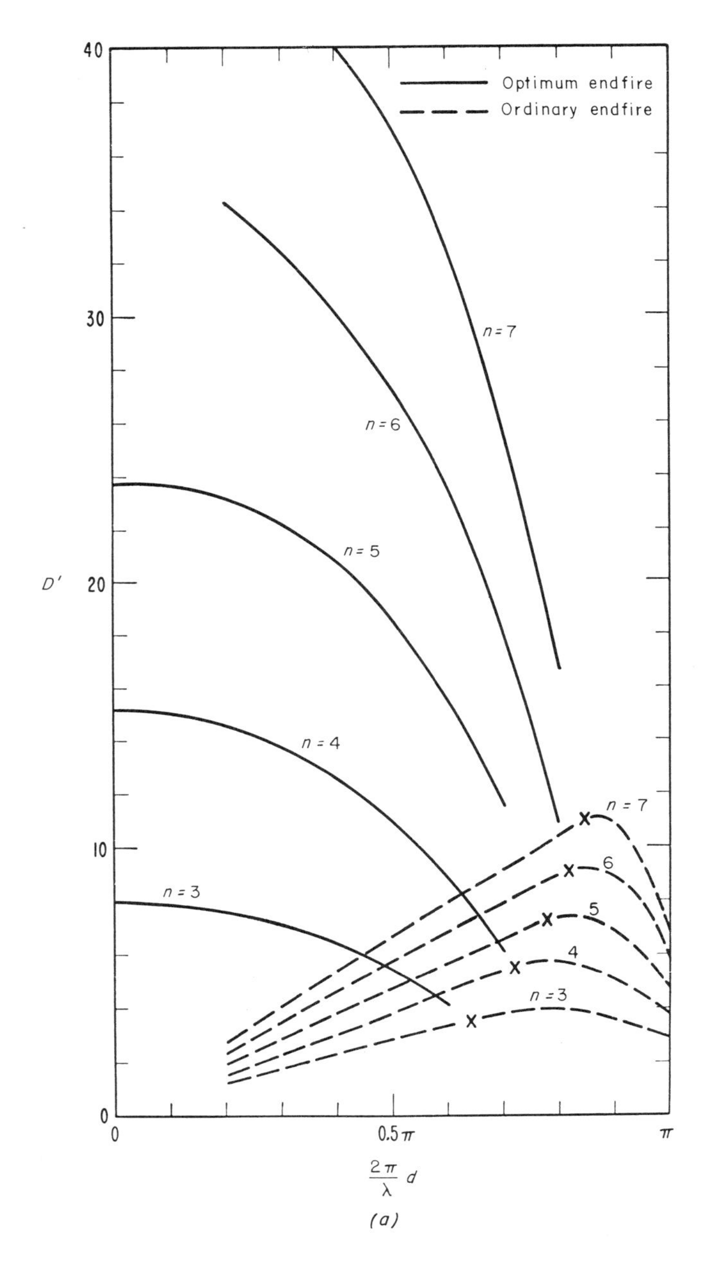

Fig. 2.7 Characteristics of optimum endfire arrays with equal sidelobe (−20 dB): (*a*) directivities, (*b*) beamwidths.

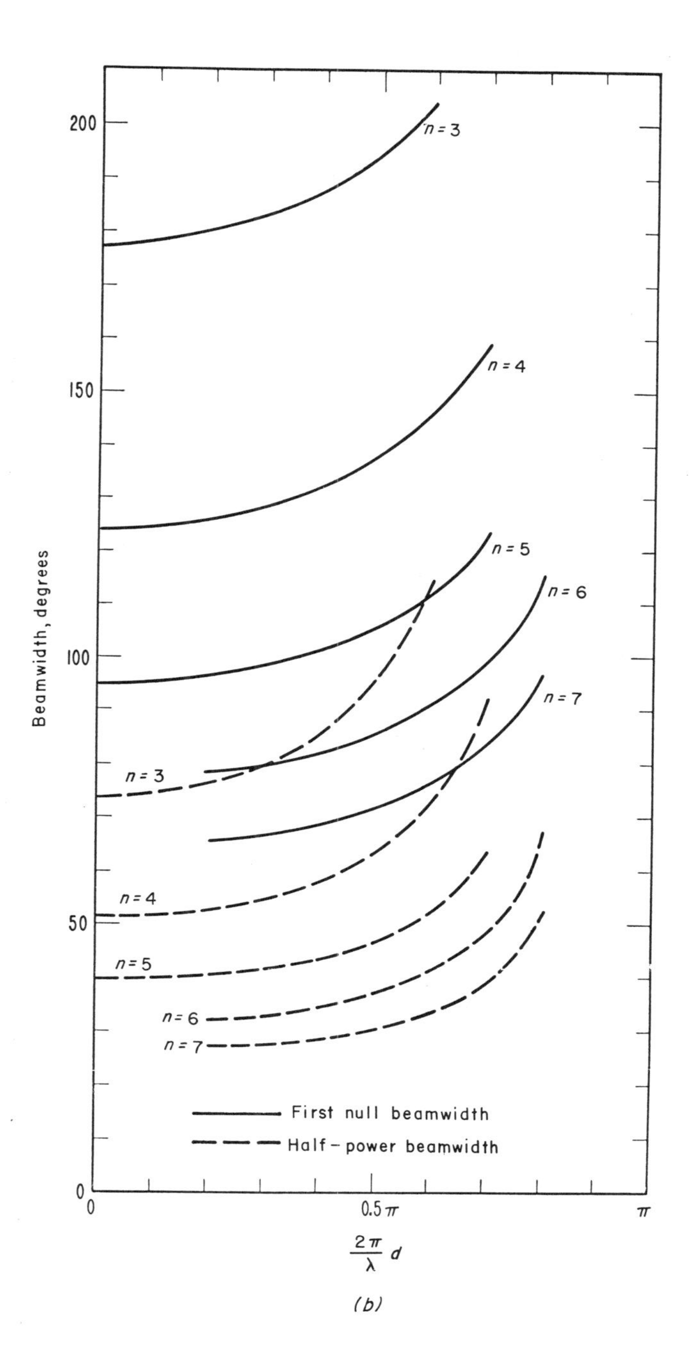
200
150
100
50
0
Beamwidth, degrees
n = 3
n = 4
n = 5
n = 6
n = 7
n = 3
n = 4
n = 5
n = 6
n = 7
First null beamwidth
Half-power beamwidth
0
0.5 π
π
2π/λ d

(b)

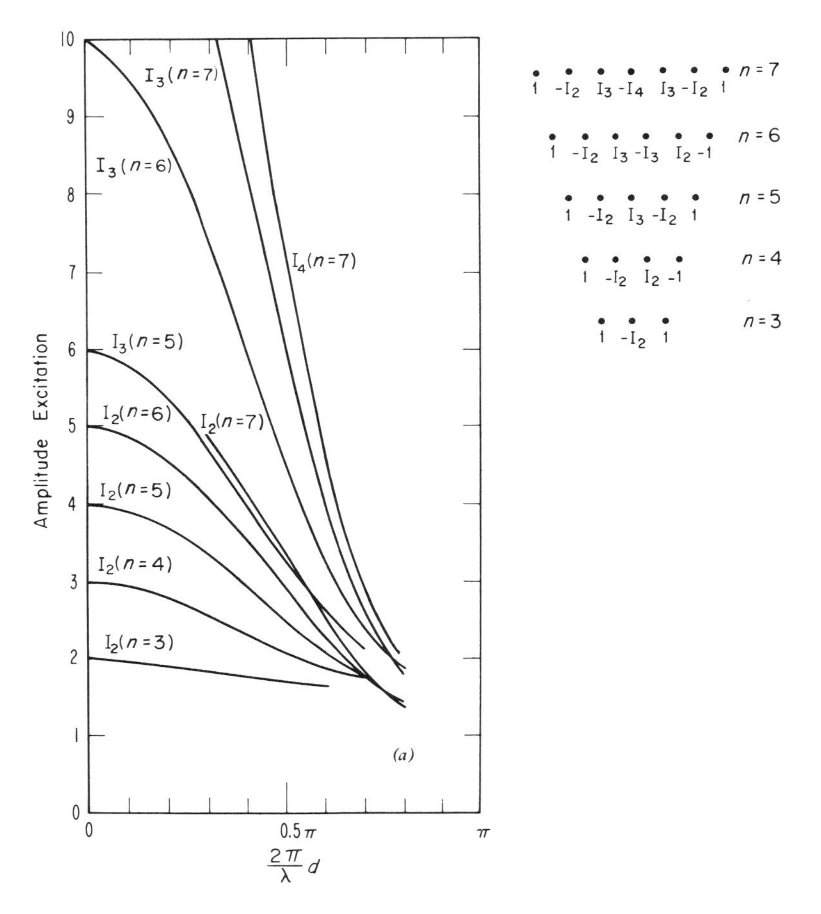

Fig. 2.8 (*a*) Excitation coefficients (relative to I_0) for optimum endfir arrays with equal sidelobes (−20 dB).

The number of terms to be carried for A_1 and A_2 in (2.117) is determine by n, the total number of elements in the array.

The limiting directivities as $d \to 0$ presented here are generally smalle than n^2 obtained by Uzkov,[12] who formulated the problem by means of a orthogonal transformation in vector space. The reason for this is again du to the constraint on equal sidelobes studied here.

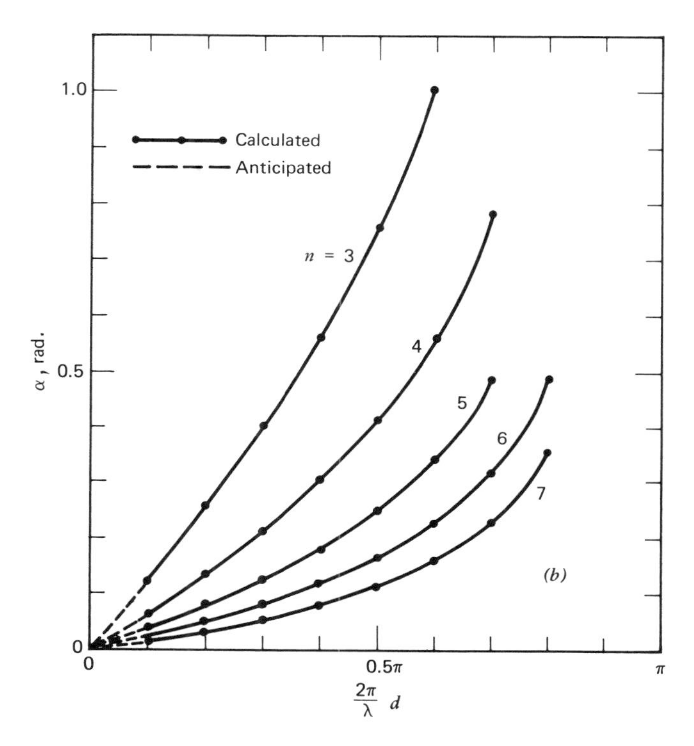

Fig. 2.8 (*b*) Phases required for optimum endfire arrays with equal sidelobes (−20 dB).

Although the optimized directivity increases when the element spacing is decreased [as demonstrated in Fig. 2.7(*a*)], this kind of superdirective array has long been considered impractical since it requires very large currents of opposite signs in closely spaced neighboring elements [see Fig. 2.8(*a*)], resulting in excessive heat loss and low radiation efficiency. For this reason, we define the main-beam radiation efficiency, η, as another criterion to evaluate the final performance of a superdirective array:

$$\eta = \frac{|E_{\max}|^2}{n \sum_{i=1}^{n} I_i^2} \times 100\%, \tag{2.118}$$

where

$$|E_{\max}|^2 = P(2) = Q(y'_a).$$

By Schwarz's inequality we see that η in (2.118) is always less than 100%. It can be 100% only when $I_i = 1$ (uniform array). Numerical results of η for the same set of cases considered in Fig. 2.7 are presented in Fig. 2.9. Clearly, we see that, when the element spacing becomes smaller, the excitations yielding superdirectivities also tend to produce lower main-beam radiation efficiency. A practical designer will choose a suitable spacing to compromise between the directivity and the main-beam radiation efficiency.

The problem of synthesizing optimum endfire arrays with equal sidelobes was first formulated by DuHamel[13] extending Riblet's method[5] of synthesizing broadside arrays. Their method was again based on Chebyshev polynomials and was valid only when n is odd. It was Pritchard[14] who, basing his work on Dolph's method,[2] developed the formulation also including the array when n is even. The material presented in this section was based on the unified approach proposed by Ma.[6,15] An extended table for characteristics of optimum endfire arrays with equal sidelobes when $n=3$ through $n=30$ with the sidelobe level varying from -10 through -50 dB has been published by Ma and Hyovalti[16] as an NBS monograph.

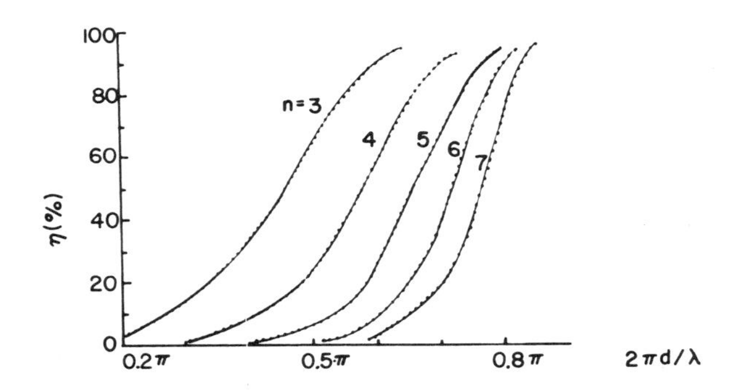

Fig. 2.9 Main-beam radiation efficiency for optimum endfire arrays with equal sidelobes (-20 dB).

2.4 Array Synthesis by Interpolation

In the previous two sections, we presented a simple kind of array synthesis with a specification on the sidelobe level or the beamwidth without considering the pattern details. In this section we will study a second kind of synthesis—to achieve a prescribed pattern shape within an acceptable tolerance on error. This problem is conventionally solved by expanding the prescribed pattern $F(u)$ into its Fourier series and then summing the first few terms for an approximation. One serious condition required by this method is that $d=\lambda/2$ or $kd=\pi$, since the entire visible range of u is then exactly 2π ($-\pi \leqslant u \leqslant \pi$ for broadside, and $-2\pi \leqslant u \leqslant 0$ for ordinary end-fire) in order to satisfy Dirichlet's conditions for Fourier expansion. As discussed by Jordan,[17] when $d<\lambda/2$, the visible interval for u is less than 2π, we then have to choose certain "extra" functions to fill in the remainder of the interval. Thus, there is an unlimited number of solutions available depending on what specific "fill-in" functions we have chosen. In case $d>\lambda/2$, the range of u is more than 2π. It is generally not possible to obtain a solution by Fourier method.

The objectives of this section are to take a refined review of the well-developed approximation[18,19] and interpolation[20] theories, and then to apply some results of this study to the synthesis of equally spaced linear arrays with arbitrary amplitude and phase distributions. Whenever possible, attention is always focused on devising a best power pattern $P(y)$ in the sense that the mean-square error or the maximum error committed by replacing the given pattern, prescribed either analytically or graphically, by the approximating power pattern is a minimum or that a prescribed maximum deviation is not exceeded. The methods, not limited by the element spacing, should help to obtain a best realizable power pattern, based on which the required excitations can then be determined, and also furnish the information as to how much mean-square error or maximum deviation has been introduced during the approximation process.

Given an arbitrary radiation pattern $f(\partial)$, we can reexpress it as $F(u)$ or $g(y)$, where $u=kd\cos\theta+\alpha$, $y=2\cos u$, if d and α are chosen or specified. The approximation problem is then to find a power pattern, whose possible expressions have been outlined in (1.129) and (1.136), satisfying the realization condition:

(a) for arrays with uniformly progressive phases (UPP),

$$P_{\mathrm{U}}(y)=\sum_{m=0}^{n-1} A_m y^m \geqslant 0, \qquad -2\leqslant y\leqslant 2 \tag{2.119}$$

or

(b) for arrays with nonuniformly progressive phases (NUPP),

$$P_{\text{NU}}(y)=\sum_{m=0}^{n-1} A'_m y^m+\left(\sum_{m=0}^{n-2} A''_m y^m\right)(4-y^2)^{1/2}\geqslant 0, \qquad -2\leqslant y\leqslant 2, \tag{2.120}$$

and of a type such that $P(y)$ and $g(y)$ agree as closely as possible under the given tolerance, with a minimum complexity. This complexity is measured in terms of the degree $(n-1)$ and the coefficients in $P(y)$, which in turn determine, respectively, the total number of elements and the excitation coefficients in the array elements.

With the substitution of $y=2\cos u$, (2.119) and (2.120) become, respectively,

$$S_e(u),$$

and $$-kd+\alpha\leqslant u\leqslant kd+\alpha \tag{2.121}$$

$$S_e(u)+S_0(u),$$

where $S_e(u)$ is an even while $S_0(u)$ is an odd function of u.

If the specified pattern is an even function of u, it certainly should be realized by an UPP array, $P_{\text{U}}(y)$. The task of replacing a given arbitrary function $g(y)$ by a polynomial $P_{\text{U}}(y)$ is guaranteed in principle by the well-known approximation theorem established by Weierstrass in 1885.[20] In terms of our notations, this approximation theorem says that if $g(y)$ is continuous in $a\leqslant y\leqslant b$, and if ϵ is a positive quantity, there exists a set of coefficients A_m and a positive integer n associated with the polynomial

$$P_{\text{U}}(y)=\sum_{m=0}^{n-1} A_m y^m, \tag{2.122}$$

such that

$$|g(y)-P_{\text{U}}(y)|<\epsilon \qquad \text{in} \qquad \alpha\leqslant y\leqslant b. \tag{2.123}$$

Unfortunately, this theorem is only an existence theorem. The way to determine A_m and the smallest n remains unanswered. Hence, the job of synthesizing equally spaced linear UPP arrays reduces to the one of devising a method or methods to determine n and A_m in (2.122) such that (2.123) and (2.119) will be satisfied.

A. Polynomial Interpolation. Since nearly all the standard formulas of interpolation are polynomial in form, it is natural to look into the possibility of accomplishing the task of finding an approximating $P_U(y)$ by using the knowledge of the interpolation theory. The only restriction is that $P_U(y)$ should satisfy the realization condition (2.119). Before presenting the details, it is worthwhile remarking that an interpolation formula does not always provide a general solution.[21] A solution may be obtained only when certain conditions on the given function and others are satisfied.

The general problem of interpolation involves representing a function with the aid of given values which this function takes at definite values of the independent variable. Thus, let $G(x)$ be a function whose values, $G(x_0), G(x_1), \ldots, G(x_{n-1})$, when the independent variable x assumes, respectively, $x_0, x_1, \ldots, x_{n-1}$ over a normalized interval $[-1,1]$, are known, and let $L(x)$ denote the approximating function so constructed that $L(x)$ takes the same values as $G(x)$ at $x_0, x_1, \ldots, x_{n-1}$. Then if $G(x)$ is replaced by $L(x)$ over the same interval $[-1,1]$, the process constitutes interpolation and the function $L(x)$ is a formula for interpolation. There are many interpolation formulas available;[20] the most important one, suitable for our purpose, is the Lagrange interpolation formula:

$$L(x) = \sum_{i=0}^{n-1} \frac{\pi(x)G(x_i)}{(x-x_i)\pi'(x_i)}, \tag{2.124}$$

where

$$\pi(x) = (x-x_0)(x-x_1)\cdots(x-x_{n-1}),$$
$$-1 \leqslant x_0, x_1, \ldots, x_{n-1} \leqslant 1, \tag{2.125}$$
$$\pi'(x_i) = \left.\frac{d\pi(x)}{dx}\right|_{x=x_i}.$$

It is noted that $L(x)$ given in (2.124) does not necessarily approach uniformly to $G(x)$ if G is an arbitrary function and x_i, $i=0,1,\ldots,(n-1)$ are arbitrarily chosen. In other words, whether $L(x)$ converges uniformly to $G(x)$ depends heavily on $G(x)$ and x_i. As shown by Buck[22] and Krylov,[23] if $G(x)$ is relatively smooth or absolutely continuous in $[-1,1]$, a proper choice of the sampling nodes x_i may yield a uniform convergence of $L(x)$ to $G(x)$ in $[-1,1]$.

The question concerning the accuracy of (2.123) can be answered by noting the remainder term, the derivation of which is similar to that of finding the remainder term in Taylor's expansion. That is,

$$\epsilon = \frac{G^n(\xi)}{n!}(x-x_0)(x-x_1)\cdots(x-x_{n-1}), \tag{2.126}$$

where ξ is some value of x between x_0 and x_{n-1}. The maximum possible error and the maximum possible mean-square error committed during the approximation process will be respectively,

$$\epsilon_{\max} \leqslant \frac{MN}{n!}, \tag{2.127}$$

$$\overline{\epsilon^2}_{\max} = \frac{1}{2}\frac{M^2}{(n!)^2}\int_{-1}^{1}\left[(x-x_0)(x-x_1)\cdots(x-x_{n-1})\right]^2 dx, \tag{2.128}$$

where

$$M = |G^n(x)|_{\max}, \tag{2.129}$$

$$N = |(x-x_0)(x-x_1)\cdots(x-x_{n-1})|_{\max}. \tag{2.130}$$

It is clear that $\epsilon_{\max}$ and $\overline{\epsilon^2}_{\max}$ depend not only on the behavior of the known function $G(x)$ but also on the number and the location of the sampling nodes $x_0, x_1, \ldots, x_{n-1}$. Nothing, of course, can be done to improve $G^n(\xi)$ since it depends totally on the specification. If the function $G(x)$ is well behaved and is given analytically by a rather simple form, the maximum value of its nth derivative, M, can always be found for each n. On the other hand, if $G(x)$ is either in such a form that it is not easy to determine M or specified in a graphical form, the best thing one then can do is to apply the technique of numerical differentiation. This procedure will introduce some extra error, the order of which is dependent on the general behavior of the specified graph and the details carried out when performing the numerical calculations. For estimating N in (2.130), we can, following and extending Krylov's proof,[23] choose the following two ways to make the task simple:

(a) Selecting sampling nodes x_i to coincide with the zeros of the Chebyshev polynomial $T_n(x) = \cos(n\cos^{-1}x)$, namely,

$$x_{n-1-i} = \cos\left(\frac{2i+1}{2n}\pi\right), \qquad i = 0, 1, 2, \ldots, (n-1), \tag{2.131}$$

we have

$$N=|(x-x_0)(x-x_1)\cdots(x-x_{n-1})|_{\max}=\frac{1}{2^{n-1}}|T_n(x)|_{\max}=\frac{1}{2^{n-1}}, \tag{2.132}$$

where the last step is obtained due to the fact

$$|T_n(x)|\leqslant 1 \qquad \text{for all } n \text{ in } -1\leqslant x\leqslant 1$$

with equality if and only if $x=\cos i\pi/n$, $i=0, 1, 2,\ldots, (n-1)$. Then the possible maximum deviation committed during the approximating process becomes

$$\epsilon_{\max}\leqslant\frac{MN}{n!}=\frac{M}{n!2^{n-1}}, \qquad -1\leqslant x\leqslant 1, \tag{2.133}$$

from which the maximum possible error can be estimated if the number of sampling nodes, n, is chosen beforehand or the minimum required number of n can be determined if an allowable $\epsilon_{\max}$ is specified. Other choices of sampling nodes might reduce the error at some points in the interval, but are likely to introduce more error at the other points.

The same choice of sampling points, as described above, can also be used when dealing with the mean-square error criterion. That is, starting with

$$\epsilon^2(x)=\left[\frac{G^n(\xi)(x-x_0)(x-x_1)\cdots(x-x_{n-1})}{n!}\right]^2$$

$$\leqslant\frac{M^2T_n^2(x)}{(n!2^{n-1})^2}\leqslant\frac{M^2}{(n!2^{n-1})^2}\cdot\frac{T_n^2(x)}{\sqrt{1-x^2}}, \quad -1\leqslant x\leqslant 1,$$

we obtain

$$\overline{\epsilon^2}\leqslant\tfrac{1}{2}\frac{M^2}{(n!2^{n-1})^2}\int_{-1}^{1}\frac{T_n^2(x)}{\sqrt{1-x^2}}\,dx$$

$$=\frac{\pi M^2}{(n!2^n)^2}, \tag{2.134}$$

where the fact that $T_n(x)$ is an orthogonal function with respect to the weighting function $(1-x^2)^{-1/2}$ has been considered in carrying out the last step. Hence, once again we can estimate the maximum mean-square error committed by replacing $G(x)$ by $L(x)$ if the number n is preassigned or the minimum required n can be determined if an allowable mean-square error is specified. Of course, there is no reason to believe that $\overline{\epsilon^2}$ obtained from (2.134) has been minimized.

(b) Selecting sampling nodes x_i to coincide with the zeros of the Legendre polynomial $p_n(x)$, we have

$$N=|(x-x_0)(x-x_1)\cdot(x-x_{n-1})|_{\max}=\frac{n!}{(2n-1)(2n-3)\cdots 1}|p_n(x)|_{\max}$$

$$=\frac{n!}{(2n-1)(2n-3)\cdots 1} \tag{2.135}$$

with the aid that

$$p_n(x)\leqslant 1, \qquad \text{for all } n, \qquad \text{in } -1\leqslant x\leqslant 1.$$

As $p_n(x)$ is a well-tabulated function, it should be easy to determine the sampling nodes once the order, n, is selected. With (2.135), (2.127) becomes

$$\epsilon_{\max}\leqslant\frac{MN}{n!}=\frac{M}{(2n-1)(2n-3)\cdots 1}, \qquad 1\leqslant x\leqslant 1. \tag{2.136}$$

A comparison of this with (2.133) reveals that selection (a) will yield a better approximation than selection (b), since $\epsilon_{\max}$ given in (2.133) is always less than that given in (2.136). Identifying all the sampling nodes as the zeros of $p_n(x)$, we can also estimate the mean-square error,

$$\overline{\epsilon^2}=\tfrac{1}{2}\int_{-1}^{1}\epsilon^2(x)\,dx\leqslant\tfrac{1}{2}\frac{M^2}{[(2n-1)(2n-3)\cdots 1]^2}\int_{-1}^{1}p_n^2(x)\,dx$$

$$=\frac{M^2}{(2n+1)[(2n-1)(2n-3)\cdots 1]^2}. \tag{2.137}$$

Although neither $\epsilon_{\max}$ nor $\overline{\epsilon_{max}^2}$ is minimized by coinciding sampling nodes with the zeros of Legendre polynomials, the method is still of analytical interest because it is easy to determine the sampling nodes and to estimate the errors committed in this approach of approximation.

Once a real polynomial $L(x)$ is determined by one of the methods thus described, it will be a solution as far as the approximation to $G(x)$ is concerned. In terms of array synthesis, $G(x)$ should be the given power pattern converted from $g(y)$ through an appropriate transformation of variables between x and y, and $L(x)$ will represent a realizable power pattern provided one more condition, namely, the realization condition, $L(x) \geqslant 0$ for all x in $[-1, 1]$, is also satisfied. $L(x)$ can then be transformed back, in terms of y, to become $P_{\mathrm{U}}(y)$ of the form (2.122). The deviation made during the approximation process is, therefore,

$$\epsilon(y) = g(y) - P_{\mathrm{U}}(y) = g(y) - \sum_{m=0}^{n-1} A_m y^m. \tag{2.138}$$

When the specified pattern is not an even function of u, the same procedure just described could still be applied for synthesizing an UPP array, with perhaps a large error. If, on the other hand, we use (2.120) with the intention of synthesizing the array as a NUPP array, the first summation terms in (2.120) are used to interpolate the given power pattern in the very same manner as that for UPP arrrys, thus determining A'_{m} with an error ϵ given by (2.138). The second summation terms are then used to interpolate the function ϵ to determine the coefficients A''_m with a final amount of error:

$$\epsilon' = \epsilon - \sum_{m=0}^{n-2} A''_m y^m (4-y^2)^{1/2} = g(y) - P_{NU}(y). \tag{2.139}$$

Note that the factor $\sum_{m=0}^{n-2} A''_m y^m (4-y^2)^{1/2}$ serves as an extra correction to ϵ and that the final error ϵ' is the result of a sort of "second approximation"; it can be made smaller than ϵ if a right second approximation is applied. This implies that, with one more degree of freedom to control the phase distribution, the synthesis of NUPP arrays seems likely to yield more accurate results than the UPP arrays if the given pattern is not an even function of u. With $P_{\mathrm{NU}}(y)$ completely obtained in this manner and if the condition $P_{\mathrm{NU}}(y) \geqslant 0$ in $-2 \leqslant y \leqslant 2$ is also satisfied, $P_{\mathrm{NU}}(y)$ will then represent a realizable power pattern.

Once a $P_{\mathrm{U}}(y)$ or $P_{\mathrm{NU}}(y)$, whatever the case may be, is determined, we can then extract a corresponding array polynomial $E_{\mathrm{U}}(z)$ according to (2.8) and (2.9), or $E_{\mathrm{NU}}(z)$ from (1.137) and (1.138), giving the required excitation coefficients.

B. Trigonometric Interpolation. The two interpolation methods discussed above are distinguished by the facts that the sampling nodes x_i are nonequally spaced and that the approximating function always results in a polynomial. When the values of a function at equidistant nodes are given, the method of trigonometric interpolation[24] is found to be exceedingly well suited. The numerical procedure for obtaining the harmonic components of a function given at equidistant intervals is simple and straightforward, and at the same time well convergent. Since the expressions given in (2.121) for the power pattern are preferable when dealing with transcendental functions, we assume that $F(u_i)$, the functional values of the given function $F(u)$ at the following nodes,

$$u_i = \frac{i\pi}{n-1}, \qquad i = 0, 1, \ldots, (n-1), \tag{2.140}$$

are known. Note that the range of u, $[0, \pi]$, has been divided into $(n-1)$ equal intervals and that the original range of u, $[-kd+\alpha, kd+\alpha]$, can be transformed to $[0, \pi]$ through a simple translation of coordinates.

Since every given function can be rewritten as the sum of its even and odd components,

$$F(u) = F_e(u) + F_0(u), \tag{2.141}$$

where

$$\begin{aligned} F_e(u) &= \tfrac{1}{2}[F(u) + F(-u)], \\ F_0(u) &= \tfrac{1}{2}[F(u) - F(-u)], \end{aligned} \tag{2.142}$$

Lanczos[24] formulated in his book that an approximating function $S(u)$ to interpolate the given data $F(u_i)$ can be obtained as

$$S(u) = S_e(u) + S_0(u), \tag{2.143}$$

with the even component given by

$$S_e(u) = \tfrac{1}{2}a_0 + a_1 \cos u + \cdots + \tfrac{1}{2}a_{n-1} \cos(n-1)u \tag{2.144}$$

and the odd component by

$$S_0(u) = b_1 \sin u + b_2 \sin 2u + \cdots + b_{n-2} \sin(n-2)u, \tag{2.145}$$

where

$$a_k = \frac{2}{n-1} \sum_{i=0}^{n-1}{}' F_e(u_i) \cos\frac{ki\pi}{n-1}, \qquad k=0, 1,\ldots,(n-1), \quad (2.146)$$

$$b_k = \frac{2}{n-1} \sum_{i=1}^{n-2} F_0(u_i) \sin\frac{ki\pi}{n-1}, \qquad k=1, 2,\ldots,(n-2). \quad (2.147)$$

Note that the coefficients a_k and b_k here should not be confused with those used in Chapter 1. The prime attached to the summation sign in (2.146) signifies that the two end terms of the sum are taken with half weight in order to be consistent with the factor $\frac{1}{2}$ appearing in the two end terms in (2.144). The success of expressing a_k and b_k in terms of summation instead of definite integral (as the ordinary Fourier method) makes the calculation extremely simple. Equation (2.143) is an analytical expression in the form of a trigonometric polynomial of the lowest order to fit the given data exactly. The accuracy for values of u between the nodes depends on the given function. The power of trigonometric interpolation lies in the fact that, with increasing n, the approximating function $S(u)$ approximates $F(u)$ with ever-diminishing oscillations. For every given function of bounded variation, the trigonometric interpolation converges unlimitedly to $F(u)$ at every point in the given range as the number of data nodes increases. This behavior of the trigonometric interpolation is in marked contrast to that of equidistant data by powers such as the polynomial interpolation. While we can always find a polynomial of $(n-1)$th order which will also exactly fit n equidistant data in a finite range, the error oscillations between the data nodes do not have the tendency in general to diminish in amplitudes as n increases. In fact, near the end of the range, the error oscillations may increase indefinitely, thus giving an arbitrarily large error everywhere except at the data nodes. The trigonometric kind of interpolation is entirely free of this peculiar difficulty. It is thus superior to the ordinary polynomial interpolation for data which are given equidistantly.

Both $S_e(u)$ and $S_0(u)$ will be present in $S(u)$ if the given $F(u)$ is an arbitrary function. When $F(u)$ is given as an even function, only $S_e(u)$ will appear. The function $S(u)$ obtained by using the above technique will represent a realizable power pattern if it also satisfies the realization condition. This realizable $S(u)$ can then be converted to $P_{\mathrm{U}}(y)$ or $P_{\mathrm{NU}}(y)$ from which an array polynomial $E_{\mathrm{U}}(z)$ or $E_{\mathrm{NU}}(z)$ is obtainable.

Two examples are given in order to illustrate the methods described in this section.[6,25] The first is to synthesize the pattern $[J_1(v)/v]^2$, $0 \leqslant v \leqslant 6$, known as the power pattern from a uniformly illuminated circular aperture,[26,27] with $n=5$. Here J_1 is the first-order Bessel function. Note that this example is of practical interest and that it cannot be done by the conventional Fourier approach. In the second example, the pattern with a Gaussian distribution such as $e^{-(x-1)^2}$, $-1 \leqslant x \leqslant 1$, is to be synthesized with $n=4$. Since the analytic expression of the pattern in the second example is simple, the maximum value of its nth derivative may be calculated to predict the approximation errors beforehand. Only the cases $P_{\mathrm{U}}(y)$ or $S_e(u)$ are considered since the first is an even function and the second is symmetric with respect to $x=1$ (the principal maximum).

Example 1. Desired pattern: $g(v)=[J_1(v)/v]^2$, $0 \leqslant v \leqslant 6$, with $n=5$. The appropriate transformation of independent variables for the polynomial interpolation, when $kd=\pi$ and $\alpha=0$ is $v=1.5(2-y)=3(1-x)$, where $x=y/2=\cos u$, $-1 \leqslant x \leqslant 1$.

(a) Approximation by coinciding sampling nodes with the zeros of $T_5(x)$: In this case, we have

$$x_0=-0.9511, \qquad x_1=-0.5878, \qquad x_2=0,$$

$$x_3=0.5878, \qquad x_4=0.9511; \quad G(x_0)=0.00268, \quad G(x_1)=0.00375,$$

$$G(x_2)=0.0128, \quad G(x_3)=0.1683, \quad G(x_4)=0.2486.$$

The approximating polynomial, according to (2.124), is

$$L(x)=-0.15552x^4-0.02079x^3+0.26820x^2+0.14508x+0.01280.$$

The $L(x)$ obtained above does not satisfy the realization condition $L(x) \geqslant 0$ in $-1 \leqslant x \leqslant 1$. This fact may be expected because the desired pattern has a null approximately at $v=3.832$ or $x=-0.277$. Therefore, $L(x)$ may have a minimum with negative value near $x=-0.277$. This is confirmed by noting that $x=-0.289$ is a root of $(d/dx)L(x)=0$. The minimum value of $L(x)$ at this point is $L(-0.289)=-0.00696$. Thus, we must use a modified approximating polynomial,

$$L_m(x)=L(x)+0.00696$$

$$=-0.15552(x+0.2890)^2(x-1.48716)(x+1.04284), \qquad (2.148)$$

to synthesize the desired pattern with an additional error of 0.00696. It is seen that $L_m(x) \geqslant 0$ in $-1 \leqslant x \leqslant 1$. Based on (2.148), the realizable power pattern becomes

$$P_U(y) = -0.00972(y+0.5780)^2(y-2.97432)(y+2.08568). \quad (2.149)$$

The array polynomial required to yield this power pattern is

$$E_U(z) = 0.09859 E_1(z) E_2(z) E_3(z), \quad (2.150)$$

where $E_1(z)$, corresponding to the factor $(y+0.5780)^2$, is uniquely given by

$$E_1(z) = 1 + 0.5780 z^{-1} + z^{-2}, \quad (2.151)$$

$E_2(z)$, corresponding to $(y-2.97432)$, is given by

$$E_2(z) = \frac{1}{1.60870}(1-2.58792z^{-1}) \quad \text{or} \quad \frac{1}{0.62161}(1-0.3864z^{-1}), \quad (2.152)$$

and $E_3(z)$, corresponding to $(y+2.08568)$, is

$$E_3(z) = \frac{1}{1.15698}(1+1.33867z^{-1}) \quad \text{or} \quad \frac{1}{0.86430}(1+0.74701z^{-1}). \quad (2.153)$$

Since neither $E_2(z)$ nor $E_3(z)$ is unique, the final answer for $E_U(z)$ in (2.150) may have four different solutions:

$$0.05297(1-0.67125z^{-1}-3.18644z^{-2}-3.25166z^{-3}-3.46437z^{-4}),$$

$$0.07091(1-1.30251z^{-1}-2.02014z^{-2}-2.99790z^{-3}-1.93320z^{-4}),$$

$$0.13708(1+1.53027z^{-1}+1.03315z^{-2}+0.65329z^{-3}-0.51726z^{-4}),$$

or

$$0.18351(1+0.93861z^{-1}+0.91979z^{-2}+0.19377z^{-3}-0.28864z^{-4}).$$

Any one of the above will yield the power pattern (2.149), from which other important radiation characteristics may be easily calculated. The synthesized pattern in terms of v is plotted as curve b in Fig. 2.10 to compare with the desired pattern, curve a.

Note that the choice of the relation $v = 1.5(2-y) = 3(1-x)$ in this example is based on $kd = \pi$ and $\alpha = 0$, because under this condition the visible range for y is $-2 \leqslant y \leqslant 2$, enough to cover the desired interval $0 \leqslant v \leqslant 6$. If $kd = 2\pi/3$ and $\alpha = 0$ are specified instead, the visible range will

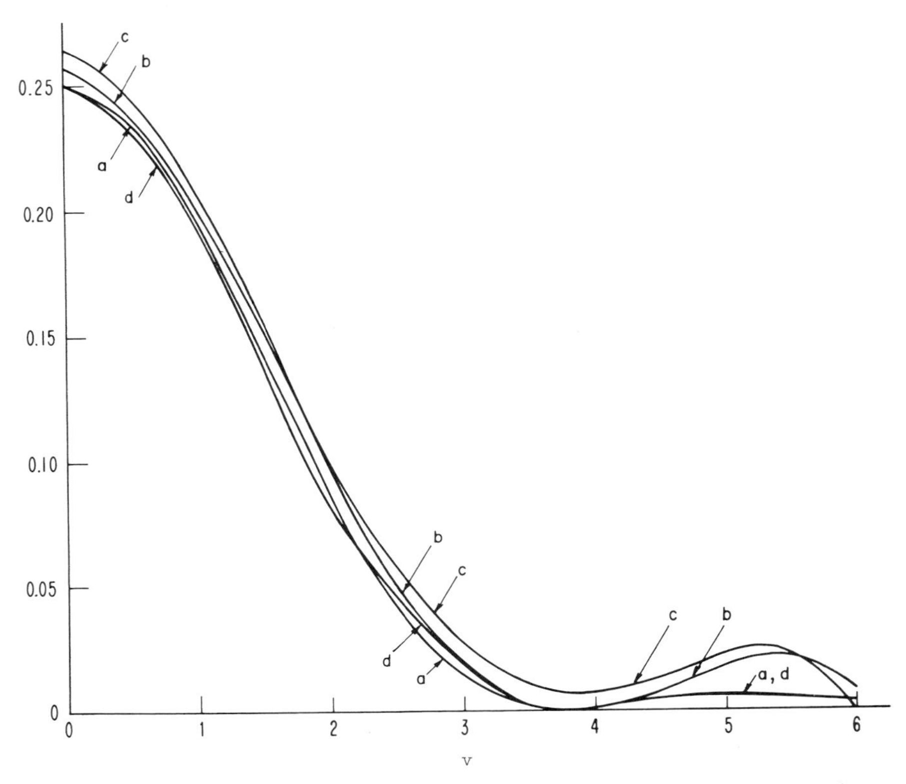

Fig. 2.10 $g(v)=[J_1(v)/v]^2$ as the specified pattern (curve a); Curves b, c, and d represent, respectively, the synthesized power patterns obtained by choosing sampling nodes as the zeros of $T_5(x), P_5(x)$, and by trigonometric interpolation.

be $-1 \leqslant y \leqslant 2$. The appropriate relationship among variables should then be

$$v=2(2-y)=3(1-x), \qquad x=\tfrac{1}{3}(2y-1), \tag{2.154}$$

such that the interval for x is still $-1 \leqslant x \leqslant 1$, required by (2.131). Although the expression for $L(x)$ or $L_m(x)$ obtained in (2.148) remains the same under this new condition, the corresponding $P_{\text{U}}(y)$ will be different from that given in (2.149) in view of (2.154). The new $P_{\text{U}}(y)$ will naturally result in different array excitations. It is clear then that the applicability of the polynomial interpolation is indeed flexible and does not necessarily demand $d=\lambda/2$ as required by the Fourier expansion method mentioned at the beginning of this section.

(b) Approximation by coinciding sampling nodes with the zeros of $p_5(x)$: In this case, we have

$$x_0=-0.9062, \quad x_1=-0.5384, \quad x_2=0,$$

$$x_3=0.5384, \quad x_4=0.9062; \quad G(x_0)=0.00317, \quad G(x_1)=0.00318,$$

$$G(x_2)=0.0128, \quad G(x_3)=0.1516, \quad G(x_4)=0.2449.$$

Substituting these in Lagrange's interpolation formula (2.124), we obtain

$$L(x)=-0.16442x^4-0.00837x^3+0.27045x^2+0.14028x+0.01280. \tag{2.155}$$

Again, the above $L(x)$ does not satisfy the realization condition. It happens that $L(-1)=-0.01308$ is the largest negative value in the interval $-1\leqslant x\leqslant 1$. Therefore, we may use the following as the modified realizable approximating polynomial:

$$\begin{aligned} L_m(x) &= L(x)+0.01308 \\ &= -0.16442(x+1)(x-1.488)(x^2+0.5389x+0.1061) \\ &\geqslant 0, \qquad -1\leqslant x\leqslant 1. \end{aligned} \tag{2.156}$$

Once $L_m(x)$ is determined, the remaining task of finding $P_{\mathrm{U}}(y)$, through an appropriate transformation between x and y dependent on kd and α, and $E_{\mathrm{U}}(z)$ is similar to that for case (a). The reader is asked to carry out the details. Equation (2.156) is plotted with $v=3(1-x)$ as curve c in Fig. 2.10 for comparison.

(c) Trigonometric interpolation: In this case, we require that the entire visible range for u be $[0,\pi]$. This can be met if we choose $kd=\pi$, $\alpha=0$, and $u=\pi v/6$. The sampling nodes for $n=5$ are

$$u_0=0, \quad u_1=\pi/4, \quad u_2=\pi/2, \quad u_3=3\pi/4, \quad u_4=\pi;$$

$$v_0=0, \quad v_1=1.5, \quad v_2=3.0, \quad v_3=4.5, \quad v_4=6.0.$$

The functional values at these points are

$$F_e(u_0)=0.25000, \qquad F_e(u_1)=0.13835, \qquad F_e(u_2)=0.01280,$$

$$F_e(u_3)=0.00264, \qquad F_e(u_4)=0.00212.$$

Substituting these into (2.146), we obtain

$$a_0=0.13993, \qquad a_1=0.10995, \qquad a_2=0.05663,$$

$$a_3=0.01398, \qquad a_4=-0.00106.$$

The approximating trigonometric polynomial is then

$$S_e(u)=0.06996+0.10995\cos u+0.05663\cos 2u+0.01398\cos 3u$$
$$-0.00052\cos 4u, \tag{2.157}$$

which happens to satisfy the realization condition. Therefore, no modification in this case is necessary. Based on (2.157), the realizable power pattern is

$$P_U(y)=-0.0002657(y^2+1.6905y+0.7149)(y+14.7692)(y-42.7769)$$
$$\geqslant 0, \qquad \text{in } -2\leqslant y\leqslant 2. \tag{2.158}$$

The determination of an array polynomial $E_U(z)$ for (2.158) is again omitted. The realized power pattern in terms of v is presented as curve d in Fig. 2.10.

Example 2. Desired pattern: $g(y)=\exp[-(y-2)^2/4]$, $-2\leqslant y\leqslant 2$, with $n=4$. The specified pattern may be expressed as $G(x)=\exp[-(x-1)^2]$ in $-1\leqslant x\leqslant 1$ when we choose $x=y/2$. Since there is no null in the specified pattern, we may expect that the approximating polynomial obtained directly from Lagrange's formula will also satisfy the realization condition.

(a) Approximation by choosing the zeros of $T_4(x)$ as the sampling nodes: Here we have

$$x_0=-0.924, \qquad x_1=-0.383, \qquad x_2=0.383, \qquad x_3=0.924;$$

$$G(x_0)=0.0247, \qquad G(x_1)=0.1481, \qquad G(x_2)=0.6840, \qquad G(x_3)=0.9958.$$

Lagrange's formula for interpolation yields

$$L(x) = -0.24628x^3 + 0.13343x^2 + 0.73586x + 0.39643$$

$$= -0.24628(x-2.2170)(x^2+1.67520x+0.72604) \geqslant 0, \text{ in } -1 \leqslant x \leqslant 1,$$

or

$$P_U(y) = -0.03078(y-4.4340)(y^2+3.33504y+2.9042) \geqslant 0$$

$$\text{in} \quad -2 \leqslant y \leqslant 2. \tag{2.159}$$

The required array polynomial giving the excitation coefficients is

$$E_U(z) = 0.1755E_1(z)E_2(z), \tag{2.160}$$

where

$$E_1(z) = \frac{1}{2.0484}(1-4.1957z^{-1}) \quad \text{or} \quad \frac{1}{0.4882}(1-0.2383z^{-1}), \tag{2.161}$$

$$E_2(z) = \frac{1}{1.3013}(1+2.1064z^{-1}+1.6931z^{-2}) \quad \text{or} \quad \frac{1}{0.7685}(1+1.2440z^{-1}+0.5906z^{-2}). \tag{2.162}$$

Again, the final solution for $E_U(z)$ is not unique. The given pattern $g(y)$ and the synthesized pattern $P_U(y)$ are respectively plotted as curves a and b in Fig. 2.11. Since the analytic expression of the given pattern is relatively simple, we can calculate the fourth derivative of $G(x)$ rather easily:

$$G^4(x) = (16x^4 - 64x^3 + 48x^2 + 32x - 20)e^{-(x-1)^2},$$

yielding

$$M = |G^4(x)|_{\max} = G^4(x)|_{x=1} = 12.$$

Then, according to (2.133), the estimated worst maximum deviation committed by replacing $G(x)$ by $L(x)$ should be 0.0625. The actual maximum deviation between the two patterns occurs at $x \cong 0$ with

$$\epsilon(0) = L(0) - G(0) = 0.0284,$$

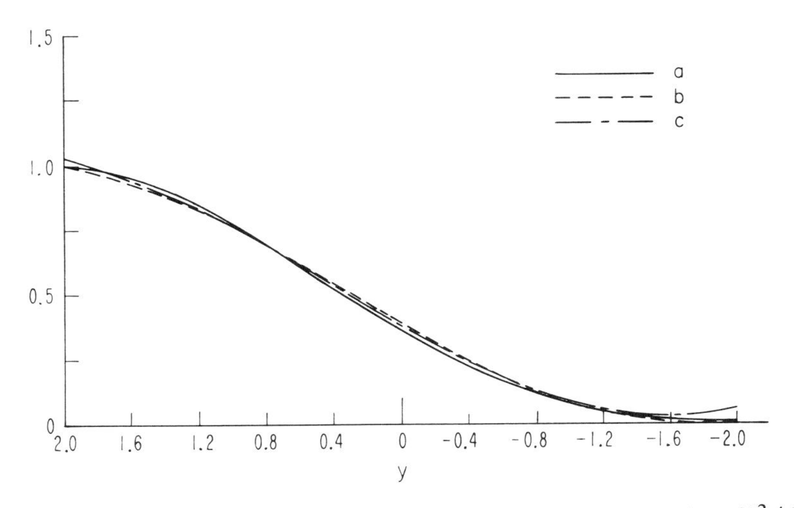

Fig. 2.11 Curve a shows the specified pattern $g(y)=\exp[-(y-2)^2/4)]$; Curves b and c are, respectively, the synthesized power patterns when sampling nodes are chosen as the zeros of $T_4(x)$ and $p_4(x)$.

verifying that the estimated error will never be exceeded by using the approximation described.

(b) Approximation by choosing the zeros of $p_4(x)$ as the sampling nodes: Here we have

$$x_0=-0.861, \qquad x_1=-0.340, \qquad x_2=0.340, \qquad x_3=0.861;$$

$$G(x_0)=0.0314, \quad G(x_1)=0.1662, \quad G(x_2)=0.6471, \quad G(x_3)=0.9807;$$

$$\begin{aligned} L(x) &= -0.2492x^3+0.1588x^2+0.7360x+0.3883 \\ &= -0.2492(x-2.2543)(x^2+1.6168x+0.6913) \geqslant 0, \end{aligned}$$

$$\text{in} \quad -1 \leqslant x \leqslant 1. \tag{2.163}$$

The realizable power pattern in terms of y becomes

$$\begin{aligned} P_U(y) &= -0.0311(y-4.5086)(y^2+3.2336y+2.7651) \\ &\geqslant 0, \quad \text{in} \quad -2 \leqslant y \leqslant 2, \end{aligned} \tag{2.164}$$

from which we obtain the final array polynomial

$$E_{\mathrm{U}}(z)=0.1764E_1(z)E_2(z),$$

where

$$E_1(z)=\frac{1}{2.0675}(1-4.2747z^{-1}) \quad \text{or} \quad \frac{1}{0.4837}(1-0.2339z^{-1}), \tag{2.165}$$

$$E_2(z)=\frac{1}{1.3521}(1+2.0902z^{-1}+1.8281z^{-2}) \quad \text{or}$$

$$\frac{1}{0.7396}(1+1.1434z^{-1}+0.5470z^{-2}). \tag{2.166}$$

The synthesized power pattern in (2.164) is plotted as curve c in Fig. 2.11. According to (2.136) the estimated maximum deviation in the process of approximation should be 0.1143. The actual maximum error between $G(x)$ and $L(x)$ is, however, $\epsilon_{\max}=|\epsilon(0)|=0.0203$, which is also under the estimated value.

(c) Trigonometric interpolation: Here we have, when $kd=\pi$ and $\alpha=0$ are again chosen:

$$u_0=0, \qquad u_1=\pi/3, \qquad u_2=2\pi/3, \qquad u_3=\pi;$$

$$F_e(u_0)=1, \qquad F_e(u_1)=0.779, \qquad F_e(u_2)=0.105, \qquad F_e(u_3)=0.018;$$

$$a_0=0.9287, \qquad a_1=0.5520, \qquad a_2=0.0447, \qquad a_3=-0.1220.$$

The approximating trigonometric polynomial becomes, therefore,

$$S_e(u)=0.4643+0.5520\cos u+0.0447\cos 2u-0.0610\cos 3u,$$

or

$$P_{\mathrm{U}}(y)=0.0305y^3+0.0223y^2+0.3675y+0.4197$$

$$=-0.0305(y-4.2893)(y^2+3.5571y+3.2081), \tag{2.167}$$

which does satisfy the realization condition in $-2\leqslant y\leqslant 2$. With (2.167), the required array polynomial can readily be obtained by the same procedure. The reader is asked to find the final result for this. The synthesized power pattern (2.167) is not plotted in Fig. 2.11 since it is so close to the desired pattern. The maximum deviation between $g(y)=\exp[-(y-2)^2/4]$ and $P_{\mathrm{U}}(y)$ occurs at $y=0$ with $\epsilon_{\max}=|\epsilon(0)|=0.0522$.

When comparing the results obtained by these three methods, we note that they all give very accurate approximation.

2.5 Approximation by Bernstein Polynomials

An approximation method using polynomial or trigonometric interpolations has been described in Section 2.4 to synthesize a given power pattern. While the accuracy of approximation shown there by two numerical examples is very good, there is one distinct disadvantage with any one of the interpolation methods, namely, the resultant approximating polynomial may not satisfy the realization condition. In that case, a modified realizable polynomial has to be devised by adding a small positive quantity to that derived from the interpolation formula. This addition of course will introduce an extra error between the synthesized and desired patterns.

In this section, we present a different method of approximation, which automatically guarantees the fulfillment of the realization condition, so that no modification is necessary once the approximation is carried out. This new method is primarily due to Bernstein, who was the first one to give a constructive proof to Weierstrass' approximation theorem.

For a real function $f(x)$ defined in the interval $0 \leqslant x \leqslant 1$, we can construct a polynomial in x of degree $\leqslant n$,

$$B_n^f(x) = \sum_{i=0}^{n} C_i^n f\left(\frac{i}{n}\right) x^i (1-x)^{n-i}, \tag{2.168}$$

where

$$C_i^n = \frac{n!}{i!(n-i)!}. \tag{2.169}$$

The expression given in (2.168) is called the Bernstein polynomial of order n of the function $f(x)$.[28] The following theorems concerning the general properties of $B_n^f(x)$ are taken from the books by Lorentz[28] and Achieser.[18]

Theorem 1. For a function $f(x)$ bounded in $0 \leqslant x \leqslant 1$, the relation

$$\lim_{n \to \infty} B_n^f(x) = f(x) \tag{2.170}$$

holds true at each point of continuity of f, and it holds uniformly in $0 \leqslant x \leqslant 1$ if $f(x)$ is also continuous in this interval.

Theorem 2. Bernstein polynomials are linear with respect to the function $f(x)$ in the following sense:

$$B_n^f(x) = a_1 B_n^{f_1}(x) + a_2 B_n^{f_2}(x), \tag{2.171}$$

if

$$f(x) = a_1 f_1(x) + a_2 f_2(x), \tag{2.172}$$

where a_1 and a_2 are arbitrary constants.

Theorem 3. If the given function $f(x)$ is bounded, $B_n^f(x)$ will also be bounded with the same upper and lower bounds. That is,

$$m \leqslant B_n^f(x) \leqslant M \quad \text{in} \quad 0 \leqslant x \leqslant 1 \quad \text{for all } n \tag{2.173}$$

if

$$m \leqslant f(x) \leqslant M \quad \text{in} \quad 0 \leqslant x \leqslant 1. \tag{2.174}$$

The actual positions of these extrema may not coincide with those of $f(x)$ for a finite n.

The accuracy of approximation of a function $f(x)$ in $0 \leqslant x \leqslant 1$ by $B_n^f(x)$ can be simply expressed in terms of the modulus of continuity $\omega(\delta)$ which is defined as [19]

$$\omega(\delta) = \max |f(x_1) - f(x_2)|, \tag{2.175}$$

for all x_1 and x_2 in [0, 1], and $|x_1 - x_2| \leqslant \delta$. Clearly $\omega(\delta)$ decreases to zero with δ if $f(x)$ is continuous.

Theorem 4. If $f(x)$ is continuous and $\omega(\delta)$ is the modulus of continuity of $f(x)$ defined in (2.175), then

$$|f(x) - B_n^f(x)| \leqslant \tfrac{5}{4}\omega(n^{-1/2}). \tag{2.176}$$

Theorem 5. If $\omega_1(\delta)$ is the modulus of continuity of the first derivative of $f(x)$, then

$$|f(x) - B_n^f(x)| \leqslant \tfrac{3}{4} n^{-1/2} \omega_1(n^{-1/2}). \tag{2.177}$$

It should be noted that the above relations giving the deviation between $f(x)$ and $B_n^f(x)$ hold true for all continuous $f(x)$ or $f'(x)$, requiring nothing about the continuity of derivatives of higher orders. For a well-behaved given function of which the higher-order derivatives do exist and are continuous, the estimation of accuracy from (2.177) is very conservative and, in that case, a much better degree of approximation should be expected. Nevertheless, it is convenient to be able to estimate the worst possible error involved in an approximation before the details are derived.

In terms of the pattern synthesis discussed here, the function $f(x)$ should represent the given power pattern which is normally specified in a finite interval and is bounded with the upper and lower bounds respectively representing the principal maximum and nulls. Then, the Bernstein polynomial $B_n^f(x)$ obtained from (2.168) will be an approximating polynomial with the degree of accuracy depending on n and $f(x)$ to be estimated by (2.176) or (2.177). This approximation polynomial will, in view of (2.173) and (2.174), always satisfy the realization condition $B_n^f(x) \geqslant 0$ in $0 \leqslant x \leqslant 1$ and, therefore, represent a realizable power pattern.[25]

Before applying the theorems outlined above to the synthesis of linear arrays, we must make a change of variables so that the whole visible range of the given power pattern in terms of y will be transformed into the interval $0 \leqslant x \leqslant 1$ required by (2.168). Generally, the transformation

$$x = A_1 y + A_2 \tag{2.178}$$

will suffice. The exact values of A_1 and A_2 depend on kd and α. In particular, if the visible range for the given power pattern $g(y)$ is $-2 \leqslant y \leqslant 2$, Eq. (2.178) will become

$$x = \tfrac{1}{4}(y+2). \tag{2.179}$$

Designating the transformed pattern by $G(x)$, the associated Bernstein polynomial will be

$$B_n^G(x) = \sum_{i=0}^{n} C_i^n G\left(\frac{i}{n}\right) x^i (1-x)^{n-i}. \tag{2.180}$$

The number of terms used under the summation sign, which determines the number of elements in the synthesized array, can be estimated by (2.176) or (2.177) when an allowable maximum error is also specified. From (2.180), we see that $B_n^G(x)$ can even be derived when the values of $G(x)$ at $x = i/n$, $i = 0, 1, \ldots, n$ are given. After obtaining $B_n^G(x)$, we then transform it back as a function of y, from which an array polynomial giving the synthesized excitation coefficients can be determined in a usual manner.

Let us consider the same examples as given in Section 2.4 to demonstrate the application of Bernstein polynomials. For the first example where $[J_1(v)/v]^2$ was the specified power pattern, the transformation (2.179) may be used if the visible range of y is indeed $[-2,2]$. Then, we have

$$G(x)=\left\{\frac{J_1[6(1-x)]}{6(1-x)}\right\}^2, \qquad G(0)=0.00212, \qquad G(\tfrac{1}{4})=0.00264,$$

$$G(\tfrac{1}{2})=0.01280, \qquad G(\tfrac{3}{4})=0.13835, \qquad G(1)=0.25000;$$

$$\begin{aligned}B_4^G(x)&=G(0)(1-x)^4+4G(\tfrac{1}{4})x(1-x)^3+6G(\tfrac{1}{2})x^2(1-x)^2\\&\qquad+4G(\tfrac{3}{4})x^3(1-x)+G(1)x^4\\&=-0.23502x^4+0.42294x^3+0.05790x^2+0.00206x+0.00212.\end{aligned} \tag{2.181}$$

Here we do not need to verify if the realization condition is satisfied with $B_4^G(x)$ in $0 \leqslant x \leqslant 1$, because we know from the theorem that it must have been.

Substituting (2.179) into (2.181), we obtain

$$\begin{aligned}P_U(y)&=-\frac{0.23502}{256}(y^4+0.8016y^3-23.1320y^2-70.7085y-60.7853)\\&=-\frac{0.23502}{256}(y+2.8288)(y-5.7232)(y^2+3.6960y+3.7556)\\&\geqslant 0 \quad \text{in} \quad -2 \leqslant y \leqslant 2.\end{aligned} \tag{2.182}$$

The required array polynomial $E_U(z)$ is

$$E_U(z)=0.0303E_1(z)E_2(z)E_3(z), \tag{2.183}$$

where

$$E_1(z)=\frac{1}{0.6435}(1+0.4141z^{-1}) \quad \text{or} \quad \frac{1}{1.5540}(1+2.4147z^{-1}),$$

$$E_2(z)=\frac{1}{0.4248}(1-0.1804z^{-1}) \quad \text{or} \quad \frac{1}{2.3543}(1-5.5428z^{-1}),$$

$$E_3(z)=\frac{1}{0.6111}(1+1.0047z^{-1}+0.3734z^{-2}) \quad \text{or}$$

$$\frac{1}{1.6365}(1+2.6913z^{-1}+2.6780z^{-2}).$$

The error involved in this approximation can also be evaluated. Since $n=4$, $f(v)=[J_1(v)/v]^2$, $v=6(1-x)$, we have

$$G'(x)=\frac{dv}{dx}f'(v)=\frac{12J_1(v)[2J_1(v)-vJ_0(v)]}{v^3},$$

v	x	$G'(x)$
0	1	0
0.5	5.5/6	0.3558
1.0	5.0/6	0.6068
1.5	4.5/6	0.7753
2.0	4.0/6	0.6104
2.5	3.5/6	0.4257
3.0	3.0/6	0.2197

From the short table given above, it is clear that the modulus of continuity of the first derivative of $G(x)$ for $n=4$ is

$$\omega_1(\tfrac{1}{2})=G'(4.5/6)-G'(1)=0.7753,$$

and the maximum error estimated by (2.177) should be 0.2907. The actual maximum error between $G(x)$ and $B_4^G(x)$ occurs at $x=0.52$ with $\epsilon_{max}=0.0435$. This analysis of error reveals that the estimation by (2.177) is indeed very conservative, because the given pattern does possess continuous higher derivatives.

The result of (2.181) is given in Table 2.1, which may be compared with those shown in Fig. 2.10 synthesized by the interpolation methods. Generally speaking, the approximation accuracy is poorer when a Bernstein polynomial of a low order is used. The advantage of this method lies on the fact that the accuracy can be uniformly improved by increasing n without having to worry about the fulfillment of the realization condition. For the purpose of demonstrating this point, the results by Bernstein polynomials with $n=5$ and 6 are also shown in Table 2.1. From there we can see that $B_n^G(x)$ does approach $G(x)$ uniformly at each point of continuity when n is increased.

The results for the second example considered in Section 2.4 where $G(x)=\exp[-4(1-x)^2]$, $x=\frac{1}{4}(y+2)$, are listed in Table 2.2. The uniform convergence of $B_n^G(x)$ to $G(x)$ when n is increased and that all $B_n^G(x)$ remain non-negative in the entire range $0\leqslant x\leqslant 1$ are again noted. It can be

Table 2.1 Approximation of $G(x)=\{J_1[6(1-x)]/6(1-x)\}^2$ by Bernstein Polynomial of Various Orders.

x	$G(x)$	$B_4^G(x)$	$B_5^G(x)$	$B_6^G(x)$
1.0	0.250	0.250	0.250	0.250
0.9	0.2283	0.2049	0.2076	0.2099
0.8	0.1724	0.1611	0.1622	0.1632
0.7	0.1044	0.1206	0.1187	0.1173
0.6	0.0469	0.0851	0.0849	0.0772
0.5	0.0128	0.0558	0.0503	0.0461
0.4	0.0007	0.0333	0.0280	0.0242
0.3	0.0011	0.0175	0.0137	0.0111
0.2	0.0039	0.0078	0.0060	0.0049

Table 2.2 Approximation of $G(x)=\exp[-4(1-x)^2]$ by Bernstein Polynomials of Various Orders.

x	$G(x)$	$B_3^G(x)$	$B_4^G(x)$	$B_5^G(x)$
1.0	1.0	1.0	1.0	1.0
0.9	0.961	0.889	0.902	0.910
0.8	0.852	0.775	0.788	0.797
0.7	0.698	0.658	0.666	0.671
0.6	0.527	0.543	0.542	0.541
0.5	0.368	0.431	0.423	0.416
0.4	0.237	0.326	0.311	0.300
0.3	0.141	0.229	0.212	0.200
0.2	0.077	0.144	0.129	0.119
0.1	0.039	0.073	0.063	0.042
0.0	0.018	0.018	0.018	0.018

concluded that the approximation of a given power pattern can always be achieved by Bernstein polynomials and that the degree of accuracy can also be unlimitedly improved by increasing n, although the rate of improvement may be somewhat slow.

2.6 Inverse Z-Transform Method

In Section 1.4 we showed many advantages of applying the finite Z-transform theory to the analysis of linear arrays where the amplitude excitations are not equal. That approach enabled us to sum the array polynomial of n terms into a finite ratio form from which the analysis then proceeded. In this section the reverse problem, synthesizing linear arrays by applying the inverse Z-transform theory, is studied.[29]

According to what we learned before, the power pattern of an equally spaced linear array with uniformly progressive phases can be written as

$$|E(z,z^{-1})|^2 = E(z)\,\overline{E(z)} = \left(\sum_{i=0}^{n-1} I_i z^{-i}\right)\left(\sum_{i=0}^{n-1} I_i z^{i}\right), \tag{2.184}$$

where I_i representing the amplitude excitations in the array are all real. From the analysis viewpoint, with I_i first assigned, $|E|^2$ can be calculated in a straightforward manner. From the synthesis viewpoint, a desired pattern expressed in terms of the observation angle θ is usually specified. We should (a) try to obtain $|E|^2$ as a function of z and z^{-1} from the specified pattern, (b) extract $E(z)$ and $\overline{E(z)}$ from it with the relation $E(z)\,\overline{E(z)} = |E|^2$ kept in mind throughout, and then (c) determine an approximate set of I_i such that $\sum_{i=0}^{n-1} I_i z^{-i}$ will approximate the $E(z)$ so extracted within a tolerable limit of error.

Generally, the first step involves no approximation since it merely is a transformation of variables. The second step is also exact if $|E|^2$ happens to be separable into two component parts which are, respectively, functions of z and z^{-1} alone. We then designate one component as $E(z)$ and the other as $\overline{E(z)}$. In this section we assume this condition is a prerequisite. The approximation which is usually involved in the third step is discussed in the following.

If the factor $E(z)$ so extracted is an analytic function and is regular at the origin,[30] we can expand $E(z)$ as

$$E(z) = c_0 + c_1 z^{-1} + c_2 z^{-2} + \cdots, \tag{2.185}$$

where the coefficients c_p can be determined by one of the following methods:[31]

$$c_p = \frac{1}{2\pi j}\int_\Gamma E(z)z^{p-1}dz$$

$$= \text{sum of residues of } E(z)z^{p-1}, \qquad p=0,1,2,\ldots, \tag{2.186}$$

or

$$c_0 = \lim_{z\to\infty} E(z),$$

$$c_1 = \frac{-1}{1!}\lim_{z\to\infty} z^2 \frac{\partial E(z)}{\partial z},$$

$$c_2 = \frac{1}{2!}\lim_{z\to\infty} z^2 \frac{\partial}{\partial z}\left[z^2 \frac{\partial E(z)}{\partial z}\right], \tag{2.187}$$

$$\cdots$$

Equation (2.186) is known as the residue method where Γ is a closed contour enclosing the singularities of $E(z)$, if any, but not enclosing infinity with a counterclockwise direction of integration. Relations given in (2.187) are obtained from the power series method.

Strictly speaking, (2.185) is the infinite-series representation of $E(z)$. Physically, it means that an array of infinite number of elements with c_p as the corresponding excitations should be constructed to produce the power pattern $|E|^2$ without committing an error. Practically, we can only allow a finite number of elements in the array, say, n. The synthesis task is then accomplished by designating c_p up to $p=n-1$ as the required excitations I_i with an error due to the remainder terms,

$$R(z) = c_n z^{-n} + c_{n-1}z^{-(n-1)} + \cdots. \tag{2.188}$$

In terms of the power pattern, we have

$$|E_s|^2 = \text{synthesized power pattern}$$

$$= [E(z) - R(z)]\left[\overline{E(z)} - \overline{R(z)}\right]$$

$$= |E|^2 + |R|^2 - R(z)\,\overline{E(z)} - \overline{R(z)}\,E(z). \tag{2.189}$$

If the series expansion in (2.185) converges well, consideration of the first term of (2.188) alone may be sufficient to give a reasonable estimation of accuracy. That is,

$$|E_s|^2 \cong |E|^2 + c_n^2 - c_n\left[z^n E(z) + z^{-n}\overline{E(z)}\right]$$

or

$$\text{error} = c_n^2 - c_n\left[z^n E(z) + z^{-n}\overline{E(z)}\right]. \tag{2.190}$$

As an example, suppose a "Gaussian" pattern such as

$$f(\theta) = \exp\ \left\{-4[1-\cos(kd\cos\theta+\alpha)]^2\right\}$$

is the desired power pattern. Remembering the notations used in Chapter 1,

$$z = e^{-ju}, \qquad u = kd\cos\theta + \alpha,$$

we can transform the given power pattern into

$$\begin{aligned} |E|^2 &= \exp\left\{-4[1-\tfrac{1}{2}(z+z^{-1})]^2\right\} \\ &= \exp[-(z^2-4z+6-4z^{-1}+z^{-2})], \end{aligned} \tag{2.191}$$

which can readily be separated as

$$|E|^2 = \exp[-(z^2-4z+3)]\exp[-(z^{-2}-4z^{-1}+3)],$$

yielding

$$\begin{aligned} \overline{E(z)} &= \exp[-(z^2-4z+3)]. \\ E(z) &= \exp[-(z^{-2}-4z^{-1}+3)]. \end{aligned} \tag{2.192}$$

With $E(z)$ extracted from $|E|^2$, we then have, according to (2.187),

$$c_0 = e^{-3}, \qquad c_1 = 4e^{-3}, \qquad c_2 = 7e^{-3}, \qquad c_3 = \tfrac{20}{3}e^{-3}, \qquad c_4 = \tfrac{19}{6}e^{-3}.$$

If we choose $n=4$, the synthesized array may approximately be represented by

$$E_s = e^{-3}(1 + 4z^{-1} + 7z^{-2} + \tfrac{20}{3}z^{-3}), \tag{2.193}$$

and the difference between $|E_s|^2$ and $|E|^2$ is given by

$$\text{error} \cong \left(\tfrac{19}{6}e^{-3}\right)^2 - \tfrac{19}{6}e^{-3}\left[z^4E(z) + z^{-4}\overline{E(z)}\right]$$

$$= 0.0248 - 0.1577\left[z^4E(z) + z^{-4}\overline{E(z)}\right], \qquad (2.194)$$

where the second term can be calculated with the help of (2.192) as a function of θ when kd and α are known.

2.7 Application of Haar's Theorem

All the synthesis methods described so far in this chapter were applicable only to linear arrays with equal element spacing. Now we are ready to study a new technique which is useful for synthesizing a nonuniformly but symmetrically spaced array including, of course, that with equal spacings as the special case. The technique is based on the application of Haar's theorem.[18]

A set of real functions, $g_i(u)$, $i = 0, 1 \ldots N$, is said to form a Chebyshev system with respect to an interval $[a,b]$ if the following conditions are satisfied:

1. $g_i(u)$, $i = 0, 1, \ldots, N$, are bounded and continuous in $[a,b]$,
2. $g_i(u)$ are linearly independent real functions of the real variable u, and
3. no linear combination of the form

$$G(c_i, u) = \sum_{i=0}^{N} c_i g_i(u), \qquad (2.195)$$

which is not identically zero due to condition 2, has more than N distinct real zeros in $[a,b]$, where the c_i's in (2.195) are real coefficients, not all zero.

According to Haar's theorem,[18] for any specified real function $f(u)$, also bounded and continuous in $[a,b]$, the function $G(c_i, u)$ given in (2.195) can be uniquely determined to approximate $f(u)$ best in the Chebyshev sense. The best approximation in the Chebyshev sense is characterized by the fact that the deviation function,

$$e(u) = f(u) - G(c_i, u), \qquad (2.196)$$

attains its maximal magnitude at no less than $N+2$ consecutive points i
$[a, b]$, the sign of $e(u)$ at these points being alternately plus and minus, an
that $\max|e(u)|$ in $a \leqslant u \leqslant b$ is minimized.

The above theorem is very general. The $g_i(u)$'s and $f(u)$ can represen
any physical system with u as the general variable.[32] In order to appl
Haar's theorem to the synthesis of a nonuniformly but symmetricall
spaced linear array such as that shown in Fig. 1.22, whose pattern functio
may be written as

$$E(u) = I_0 + 2\sum_{i=1}^{N} I_i \cos u_i$$

$$= I_0 + 2\sum_{i=1}^{N} I_i \cos b_i u, \qquad (2.197$$

it is clear that $g_i(u)$ in our application should take the form of $\cos b_i$
including the constant term where $b_0 = 0$, and that $f(u)$ will represent
desired field pattern. In fact, by comparing (2.197) with (2.195), we see
one-to-one correspondence:

$$g_i(u) \to \cos b_i u,$$

$$G(c_i, u) \to E(u)$$

$$c_i \to I_0 \quad \text{or} \quad 2I_i \quad (i \neq 0).$$

In (2.197), I_i is the amplitude excitation of the ith pair of elements,

$$u_i = \frac{2\pi}{\lambda} d_i(\cos\theta - \cos\theta_0) = b_i u,$$

$$b_i = 2d_i/\lambda, \qquad (2.198$$

$$u = \pi(\cos\theta - \cos\theta_0),$$

and θ_0 is the position of the main beam measured from the array axis. Th
latest definition of u given in the last equation of (2.198) is chosen pure
for the convenience of presentation here, and should not be confused wit
the previous definition.

We shall concentrate the discussion on the case of an odd total numb
of elements as is explicitly expressed by (2.197). The case with an eve
number of elements can be presented in a parallel manner.

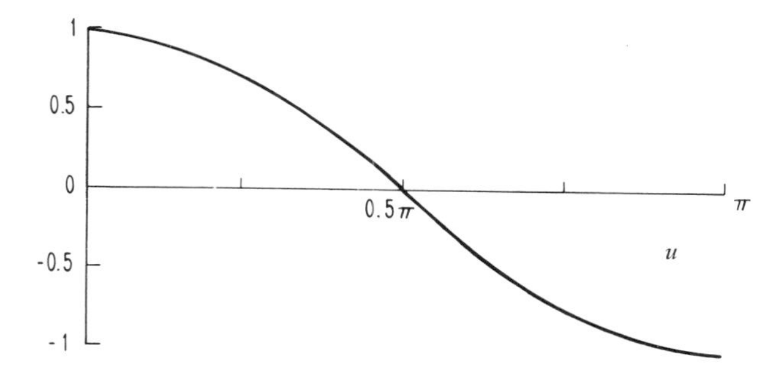

Fig. 2.12 A sketch of cos u in $0 \leqslant u \leqslant \pi$.

Obviously, the set of real functions $\cos b_i u$ with $0=b_0<b_1<\cdots<b_N$ are inearly independent, continuous, and bounded in a finite interval $[a,b]$ vhich becomes, respectively, $[0,\pi]$ and $[0,2\pi]$ when $\theta_0=\pi/2$ (broadside) nd $\theta_0=0$ (ordinary endfire). These intervals are obtained after consider- ng the fact that $\cos b_i u$, $i=1,2,\ldots,N$ are all even functions of u. It is clear hat the first two conditions required for a Chebyshev system have been atisfied. The third condition can also be met by an appropriate choice of $_i$. Let us consider a simple case where $N=1$ to illustrate this point. The rray for this simple case has only three elements (one pair plus one at the rray center), and the field is given by $E_1(u)=I_0+2I_1\cos b_1 u$. If we choose $_1=\lambda/2$ $(b_1=1)$ and $\theta_0=\pi/2$, the part of $\cos u$ in the visible range $\leqslant u \leqslant \pi$ is shown in Fig. 2.12. It is obvious that E_1 cannot have more than ne real zero in $0\leqslant u\leqslant\pi$ no matter what I_0 and I_1 may be. All the onditions required for a Chebyshev system are thus satisfied. This point annot be easily seen, however, for a larger N. In that case, the only thing e can do is to choose the largest spacing d_N in such a way,

$$d_N \leqslant \begin{Bmatrix} N\lambda/2 \\ N\lambda/4 \end{Bmatrix} \quad \text{or} \quad b_N \leqslant \begin{Bmatrix} N \\ N/2 \end{Bmatrix} \quad \text{for} \quad \theta_0 = \begin{Bmatrix} \pi/2 \\ 0 \end{Bmatrix}, \tag{2.199}$$

hat the most oscillating term alone, $\cos b_N u$, does not have more than N eal zeros in $0\leqslant u\leqslant\pi$ for $\theta_0=\pi/2$, or in $0\leqslant u\leqslant 2\pi$ for $\theta_0=0$, and that the equired third condition may "likely" be fulfilled. Of course, the condition 2.199) is only a necessary one because whether $\cos b_i u$ will form a hebyshev system depends also heavily on $b_1,b_2,\ldots,b_{N-1}$. Since there is no imple way, so far, to establish a clear criterion for the general case, we can nly restrict ourselves to (2.199) and choose other b_i from a practical onsideration, and hope for the best. Actually, we can also learn something about the appropriate choice of b_i from the numerical process to be presented.

Supposing all the conditions for a Chebyshev system are now satisfied we can then determine uniquely, according to Haar's theorem, the best $E(u)$ in the form of (2.197) such that

$$\max|e(u)|=\max|f(u)-E(u)| \qquad \text{in} \qquad a\leqslant u\leqslant b, \tag{2.200}$$

is minimized.

Now we are ready to show how to obtain the $E(u)$ satisfying (2.200) upon a specification of $f(u)$. Since there are $N+1$ unknowns of I_i to be determined, the simplest way of accomplishing this is to choose a set of b_i under the restriction of (2.199) and interpolate the given $f(u)$ at a set of $N+1$ points, $u=U_j$, $j=1,2,\ldots,(N+1)$, which may or may not be uniformly distributed in $[a,b]$, such that an initial set of I_i, denoted by I_i^0 can be determined. Mathematically, the above statement means that, upon taking

$$E^0(u)=I_0^0+2\sum_{i=1}^{N} I_i^0\cos b_i u \tag{2.201}$$

with a choice of b_i and then setting

$$E^0(U_j)=f(U_j), \qquad a<U_1<U_2<\cdots<U_{N+1}<b, \tag{2.202}$$

we can write the solution for I_i^0 in matrix form as follows:

$$\begin{bmatrix} I_0^0 \\ I_1^0 \\ \vdots \\ I_N^0 \end{bmatrix}=\begin{bmatrix} 1 & 2\cos b_1U_1 & 2\cos b_2U_1 & \cdots & 2\cos b_NU_1 \\ 1 & 2\cos b_1U_2 & 2\cos b_2U_2 & \cdots & 2\cos b_NU_2 \\ \cdots & \cdots & \cdots & \cdots & \cdots \\ 1 & 2\cos b_1U_{N+1} & 2\cos b_2U_{N+1} & \cdots & 2\cos b_NU_{N+1} \end{bmatrix}$$

$$\times\begin{bmatrix} f(U_1) \\ f(U_2) \\ \vdots \\ f(U_{N+1}) \end{bmatrix} \tag{2.203}$$

Here U_j's are employed as the interpolating points. The existence of the inverse matrix is guaranteed by the conditions imposed on a Chebyshev system. The $N+2$ extrema points of the initial deviation function,

$$e^0(u) = f(u) - E^0(u), \tag{2.204}$$

are then found at u_i^0, $i=0,1,\ldots,(N+1)$ with $a \leqslant u_0^0 < u_1^0 \cdots < u_{N+1}^0 = b$ by setting $de^0(u)/du = 0$. The magnitudes of $e^0(u)$ at these extrema are, of course, not equal in general since this is only an initial arbitrary interpolation. The next step is to assume that

$$E^1(u) = I_0^1 + 2\sum_{i=1}^{N} I_i^1 \cos b_i u, \tag{2.205}$$

with I_i^1 as the new coefficients to be determined by equalizing

$$e^1(u) = f(u) - E^1(u) \tag{2.206}$$

at u_i^0, $i=0,1\ldots(N+1)$ such that the following will be true:

$$e^1(u_i^0) = (-1)^i \epsilon_1. \tag{2.207}$$

Equation (2.207) consists of a system of $N+2$ equations which are just enough to solve for I_i^1 and ϵ_1. Here ϵ_1 may be either positive or negative depending on whether $e^1(u_0^0) > 0$ or $e^1(u_0^0) < 0$. The set of coefficients I_i^1 so obtained, though satisfying (2.207), does not necessarily ensure that $e^1(u)$ varies with equal ripples since $e^1(u)$ now has a new set of extrema points, u_i^1, $a \leqslant u_0^1 < u_1^1 < \cdots < u_{N+1}^1 = b$. From these points we shall again try to equalize the deviation function until a final set of extrema points, u_i^k, are obtained such that the values of $e^k(u) = f(u) - E^k(u)$ at these points are equal in magnitude to a certain accuracy but with signs alternately plus and minus. The superscript k signifies the kth iteration after an initial starting. The iterative process is proved to be convergent so long as the conditions for a Chebyshev system have been satisfied.[32,33]

Once a solution is obtained, it not only provides precise information concerning the amplitude excitation, element positions, and the minimum possible absolute deviation between the synthesized and desired patterns (with respect to the chosen set of b_i), but also is unique and optimum in the Chebyshev sense.

Throughout this section a broadside ($\theta_0 = \pi/2$, $[a,b] = [0,\pi]$) Gaussian pattern, $f(u) = e^{-Au^2}$ with A as a positive real number, is chosen as our desired pattern. This choice is justified by the following reasons: (a) $f(u)$ is

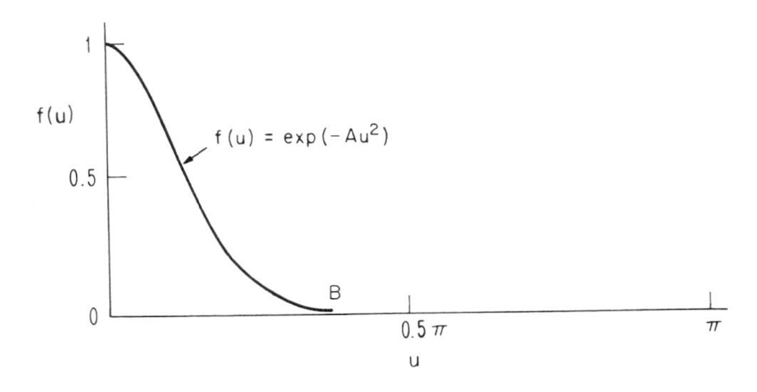

Fig. 2.13 A possible desired pattern.

Table 2.3 Synthesized Results when $f(u)=e^{-2u^2}$ and $N=3$ (seven element (a) $b_2=2.0, b_3=3.0$; (b) $b_2=1.8, b_3=3.3$.

b_1	I_0	I_1	I_2	I_3	Maximum Deviation (ϵ)	$(BW)_h$ deg.	Sidelobe dB	Directivi
				(a)				
1.5	0.2971	0.2310	−0.0062	0.1030	0.0473	17.24	−26.08	6.42
1.4	0.2848	0.2016	0.0345	0.0968	0.0494	17.33	−25.68	6.39
1.3	0.2698	0.1842	0.0641	0.0910	0.0516	17.41	−25.28	6.36
1.2	0.2515	0.1751	0.0867	0.0854	0.0540	17.50	−24.86	6.32
1.1	0.2285	0.1726	0.1049	0.0799	0.0567	17.61	−24.42	6.28
1.0	0.1995	0.1759	0.1203	0.0740	0.0599	17.73	−23.91	6.24
0.9	0.1608	0.1864	0.1333	0.0681	0.0635	17.87	−23.37	6.19
0.8	0.1061	0.2069	0.1437	0.0628	0.0670	17.01	−22.87	6.14
0.7	0.0259	0.2418	0.1520	0.0580	0.0704	18.14	−22.41	6.10
0.6	−0.0981	0.2997	0.1587	0.0538	0.0736	18.26	−22.00	6.05
				(b)				
1.4	0.3098	0.1174	0.1263	0.0861	0.0304	16.57	−30.08	6.46
1.2	0.2915	0.0963	0.1578	0.0846	0.0310	16.60	−29.90	6.51
1.0	0.2613	0.0970	0.1732	0.0834	0.0316	16.62	−29.74	6.54
0.8	0.2059	0.1166	0.1819	0.0824	0.0320	16.63	−29.61	6.56
0.6	0.0864	0.1716	0.1873	0.0817	0.0324	16.65	−29.52	6.58
0.4	−0.2548	0.3393	0.1905	0.0812	0.0326	16.66	−29.45	6.59

a very well-behaved function, (b) the parameter A can be varied to adjust the desired beamwidth, and (c) since $f(u)$ practically vanishes after a point B (see Fig. 2.13), all the sidelobes of the synthesized pattern $E^k(u)$ will be approximately at the same level represented by $20\log|\epsilon_k/E^k(0)|$.

Numerical results for $A=2$, $N=3$, and two different combinations of b_2 and b_3 are given below. For completeness, the associated half-power beamwidth and directivity are also included. The initial interpolating points U_j are chosen uniformly inside the interval $[0,\pi]$. That is, $U_j = j\pi/(N+2), j=1,2,\ldots,(N+1)$.

Table 2.3(a) gives results for the case of $N=3$ (seven elements), $f(u) = e^{-2u^2}$, $b_3=3.0$, and $b_2=2.0$ with various b_1, after three iterations. Immediate conclusions for this case are (a) the radiation characteristics are rather insensitive to the change of b_1 as long as $b_2=2.0$ and $b_3=3.0$, (b) the case with $b_1=1.0$ corresponding to the equally spaced array does not yield the maximum directivity or the lowest sidelobe level, (c) since I_0 changes sign from $b_1=0.6$ to $b_1=0.7$, it vanishes for $0.6<b_1<0.7$ and, therefore, the central element can be omitted. This indicates that essentially the same result can be achieved with only six elements in the array if we choose $b_1 \cong 0.65$, and (d) the same performance can even be expected from an array of five elements if b_1 somewhere between 1.4 and 1.5 is selected to make I_2 vanish.

Included in Table 2.3(b) are the results for the same array but with $b_2=1.8$ and $b_3=3.3$. It shows that the maximum directivity occurs approximately at the same time when $I_0=0$ for $0.4<b_1<0.6$. This implies that a favorable solution can be achieved with a fewer total number of elements for a given specified array length if the positions of the inner elements are wisely assigned. It also reveals that an increase of the number of elements from six to seven by allowing $I_0 \neq 0$ does not necessarily mean an improvement in directivity if the elements are poorly spaced. Note that in this case ($b_3=3.3$) the condition (2.199) has definitely been violated. Even though the entire basis functions, $\cos b_i u$, $i=0$, 1, 2, and 3, have not formed a Chebyshev system, the iterative process still works. This should not be surprising, however, because the actual $f(u)=e^{-Au^2}$ and $\cos b_i u$ used here behave much better than the minimum requirements of continuity and boundedness in Haar's theorem. It is apparent that the third condition for a Chebyshev system may be relaxed to a certain degree for a "better class" of $f(u)$ and $g_i(u)$. The precise information on this is unfortunately not yet clear.

Cases with many other combinations of b_i and various A's have been considered with the same method described. Their results, too numerous to be included here, have been reported elsewhere.[34]

The technique of applying Haar's theorem can also be generalized for arrays consistng of directive elements whose element pattern is represented by $W(u)$. For example, $W(u)$ for a half-wavelength thin linear dipole with the element pattern approximately given by $\cos[(\pi/2)\cos\theta]/\sin\theta$, when a sinusoidal current distribution is assumed, will be

$$W(u)=\frac{\cos(u/2)}{\sqrt{1-u^2/\pi^2}}\geqslant 0, \qquad 0\leqslant u\leqslant\pi. \tag{2.208}$$

If $W(u)$ is bounded, continuous, and non-negative everywhere inside the visible range as it usually is, the product of $W(u)$ and $f(u)$ [e.g., $f(u) = e^{-Au^2}$], which can be considered as a new desired pattern, is also bounded and continuous. If the field pattern $E(u)$, synthesized previously for arrays of isotropic elements, has satisfied the condition of having no more than N real zeros, $W(u)E(u)$ may also satisfy the same condition because $W(u)\geqslant 0$. Under that case, the set of the following,

$$W(u),, W(u)\cos b_1 u, \ldots, W(u)\cos b_N u,$$

will also form a Chebyshev system. The application of Haar's theorem then yields that

$$\begin{aligned}\max|e'(u)| &= \max|W(u)f(u)-W(u)E(u)| \\ &= \max W(u)|f(u)-E(u)|, \qquad 0\leqslant u\leqslant\pi,\end{aligned} \tag{2.209}$$

is a minimum. Note that the expression shown in (2.209) may be considered as a weighted maximum deviation, and that $W(u)E(u)$ now represents the synthesized optimum pattern for an array with half-wavelength thin dipoles as its elements.

2.8 Perturbation Method

In the previous section a method was presented to synthesize a linear array whose field pattern approximates a desired pattern in the minimax sense. Since the Gaussian type pattern, $f(u)=e^{-Au^2}$, was particularly chosen as the desired pattern, the sidelobes in the synthesized pattern were also approximately at the same level. Both the amplitude excitations and element spacings were allowed to vary.

Now we are showing another method of synthesis based on a perturbation technique. The idea is to derive a desired pattern from a known one through small changes of element spacings or/and amplitude excitations. Although the method to be described is valid when any known pattern is chosen to start with, we will concentrate only on that of a uniform array for the purpose of demonstrating the method. Since the overall pattern of a uniform array is reasonable except perhaps that the level of its first sidelobe is rather high, our objective here is to apply the perturbation technique to reduce the level of the first few sidelobes by sacrificing those far away from the main beam. Ultimately we will again try to equalize all the sidelobes approximately.

Two kinds of perturbation are possible: One is to perturb the element spacing only[35] and the other is to change the amplitude excitation only. We will discuss the former first in view of the practical preference of equal amplitude excitation. Since, for many occasions before, we formulated specifically the case when the total number of elements in the array is odd, we now present the material by considering the array with an even number of elements in order to get a balanced picture although the method applies equally well to both cases.

For a nonuniformly but symmetrically spaced array of $2N$ elements as shown in Fig. 2.14, the broadside field pattern may be written as

$$E(u) = 2 \sum_{i=1}^{N} I_i \cos b_i u, \tag{2.210}$$

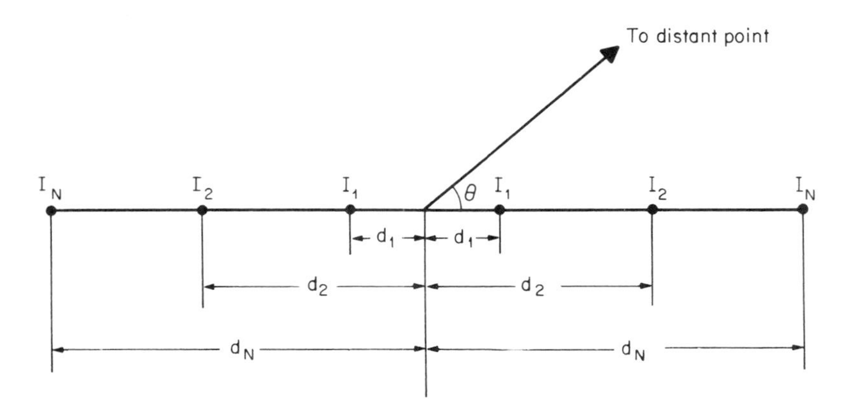

Fig. 2.14 A linear array of $2N$ nonuniformly but symmetrically spaced elements.

where

$$b_i = 2d_i/\lambda, \qquad u = \pi\cos\theta, \tag{2.211}$$

with d_i the distance from the ith pair of elements to the array center where no element is present.

In general, the pattern function, $E(u)$ in (2.210), consists of a main lobe at $u=0$ and several sidelobes at $u_j = \pi\cos\theta_j$ of level ϵ_j, $j=1,2,\ldots,m$. The last sidelobe can be made to be directed along the array axis ($\theta=0$) if we put $u_m = \pi$. Mathematically, the above picture can be represented by the following:

$$E(u_j) = \epsilon_j, \qquad j=1,2,\ldots,m, \tag{2.212}$$

$$\left.\frac{\partial E(u)}{\partial u}\right|_{u=u_j} = 0, \qquad j=1,2,\ldots,(m-1). \tag{2.213}$$

Note that, in (2.213), $j \neq m$, since the derivative at the last sidelobe is not necessarily equal to zero in the general case. It is clear that (2.212) and (2.213) consist of $(2m-1)$ equations with $(N+m-1)$ unknowns $(b_1, b_2, \ldots, b_N;\ u_1, u_2, \ldots, u_{m-1})$. A solution exists in the range $0 \leqslant u \leqslant \pi$ if $m \leqslant N$. That is, the total number of sidelobes including that at $u_m = \pi$ in the field pattern can at most be equal to N, half of the number of elements in the array. The solution will be unique if $m=N$.

A. For a uniformly excited array, we can set, without loss of generality, $I_i = \frac{1}{2}$. Equation (2.210) thus becomes

$$E(u) = \sum_{i=1}^{N} \cos b_i u. \tag{2.214}$$

To make all the sidelobes at the same level ϵ, the right-hand side of (2.212) should be $(-1)^j\epsilon$. That is,

$$E(u_j) = (-1)^j \epsilon. \tag{2.215}$$

Of course, it will be difficult to solve (2.213) and (2.215) directly for b_i for a multielement array. That is the reason a perturbation technique is employed here.

With an initial choice of b_i, say b_i^0, we can calculate the pattern $E^0(u)$ giving the positions of the sidelobes u_j^0 and their levels ϵ_j^0. We thus have

$$\epsilon_j^0 = \sum_{i=1}^{N} \cos b_i^0 u_j^0, \qquad j=1,2\ldots N;$$
$$\sum_{i=1}^{N} b_i^0 \sin b_i^0 u_j^0 = 0, \qquad j=1,2,\ldots,(N-1). \tag{2.216}$$

Note that if the uniform array with a half-wavelength element spacing is chosen as the starting point, $b_i^0 = \frac{1}{2}(2i-1)$, u_j^0 and ϵ_j^0 in (2.216) are all known.

Now, we wish to perturb the initial spacings by a small change so that the new set of spacings b_i^1 is

$$b_i^1 = b_i^0 + \Delta b_i. \tag{2.217}$$

As a consequence, the positions of the sidelobes (except perhaps the last one at $u_N^0 = \pi$) and their levels will be changed to

$$u_j^1 = u_j^0 + \Delta u_j \qquad \text{and} \qquad \epsilon_j^1 = \epsilon_j^0 + \Delta\epsilon_j. \tag{2.218}$$

Since the new u_j^1 and ϵ_j^1 should also satisfy (2.212) and (2.213), we have

$$\epsilon_j^1 = \sum_{i=1}^{N} \cos\left(b_i^0 + \Delta b_i\right)\left(u_j^0 + \Delta u_j\right)$$
$$= \sum_{i=1}^{N} \left[\cos\left(b_i^0 u_j^0\right)\cos\varphi - \sin\left(b_i^0 u_j^0\right)\sin\varphi\right], \tag{2.219}$$

$$\sum_{i=1}^{N} \left[\left(b_i^0 + \Delta b_i\right)\sin\left(b_i^0 + \Delta b_i\right)\left(u_j^0 + \Delta u_j\right)\right]$$
$$= \sum_{i=1}^{N} \left(b_i^0 + \Delta b_i\right)\left[\sin\left(b_i^0 u_j^0\right)\cos\varphi + \cos\left(b_i^0 u_j^0\right)\sin\varphi\right] = 0, \tag{2.220}$$

where

$$\varphi = b_i^0 \Delta u_j + u_j^0 \Delta b_i + \Delta b_i \Delta u_j. \tag{2.221}$$

If Δb_i's are indeed very small, so will Δu_j be. The following approximations may then be applied:

$$\cos\varphi \cong 1, \qquad \sin\varphi \cong \varphi \cong b_i^0 \Delta u_j + u_j^0 \Delta b_i. \tag{2.222}$$

Equations (2.219) and (2.220) become, respectively,

$$\Delta\epsilon_j = -u_j^0 \sum_{i=1}^{N} \Delta b_i \sin\left(b_i^0 u_j^0\right), \tag{2.223}$$

$$\Delta u_j \sum_{i=1}^{N} \left(b_i^0\right)^2 \cos\left(b_i^0 u_j^0\right) + u_j^0 \sum_{i=1}^{N} b_i^0 \Delta b_i \cos\left(b_i^0 u_j^0\right)$$

$$+ \sum_{i=1}^{N} \Delta b_i \sin\left(b_i^0 u_j^0\right) = 0. \tag{2.224}$$

In the above, (2.216) has been used, and the second-order terms such as $(\Delta b_i)^2$ and $\Delta b_i \Delta u_j$ have been neglected.

With b_i^0, u_j^0, and ϵ_j^0 known and ϵ_j^1 set in any desired manner, (2.223) and (2.224) are nothing but ordinary algebraic simultaneous equations. We can solve for Δb_i's first from (2.223) and then substitute them into (2.224) to determine Δu_j's. With $b_i^1 = b_i^0 + \Delta b_i$ as the new set of element spacings, the perturbed pattern $E^1(u)$ should yield approximately the sidelobes to the desired levels ϵ_j^1. In view of the approximation involved, the results from (2.223) and (2.224) will be more accurate if smaller $\Delta\epsilon_j$'s are demanded at a time. The process can be continued until a pattern with equal sidelobes (or satisfying some other criterion) is achieved. Examples on this for arrays of seven and eight elements are respectively shown as curves (a) and (b) in Fig. 2.15. The final required spacings after four perturbations are also indicated there.

B. When a set of element spacings, b_i, is given and we wish to perturb the amplitude excitations to improve the pattern shape, the problem can be formulated in a completely parallel manner. Instead of using (2.217) we will have

$$I_i^1 = I_i^0 + \Delta I_i, \qquad i = 1, 2 \ldots N, \tag{2.224}$$

where I_i^0's are the initial amplitude excitations chosen to start the perturbation process. One natural (though not necessary) choice is $I_i^0 = 1$. Substituting (2.224) and (2.218) into (2.212) and applying the approximations

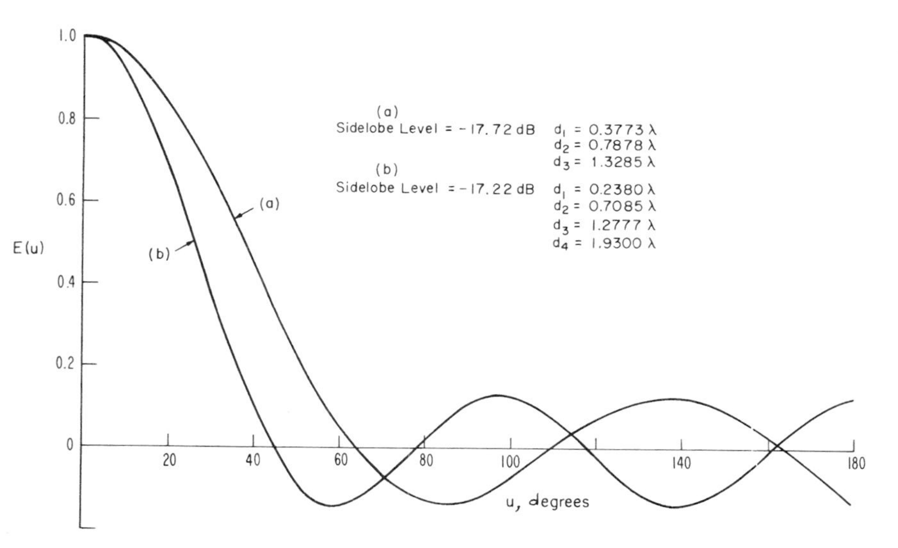

Fig. 2.15 Field patterns (normalized) of uniformly excited arrays synthesized by spacing perturbation: (*a*) $n=7$, (*b*) $n=8$.

similar to those in (2.222), we obtain the final governing equations for determining ΔI_i's:

$$\frac{\Delta\epsilon_j}{2} = \sum_{i=1}^{N} \Delta I_i \cos\left(b_i u_j^0\right), \qquad j=1,2\ldots N. \tag{2.225}$$

As is expected, the solutions of the final amplitude excitations for the equally spaced case with $d=\lambda/2$ $[b_i=\frac{1}{2}(2i-1)]$, when the same criterion of equalizing all the sidelobes is employed, should be identical to those discussed in Section 2.2. Since the method is not limited to the case with $b_i=\frac{1}{2}(2i-1)$, we illustrate it by giving a numerical example for $N=3$ (six elements), $b_1=0.6$, $b_2=1.4$, and $b_3=2.5$. The perturbed pattern function is plotted in Fig. 2.16 where the final required amplitude excitations are also given.

In view of the fact that the solution so obtained is not unique and depends heavily on the amount $\Delta\epsilon_j$ we demand in each perturbation step, we list all the intermediate results in Table 2.4 for this simple example in order to gain an insight of the entire process.

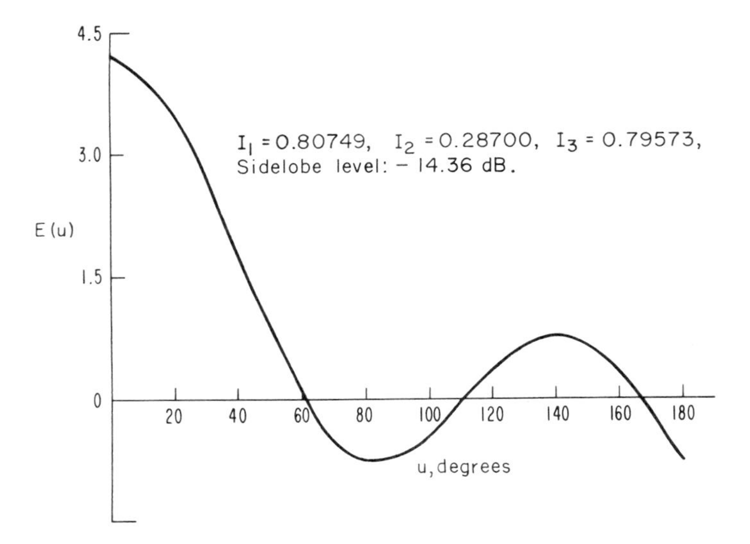

Fig. 2.16 Field pattern of a six-element nonuniformly spaced array where $b_1=0.6, b_2=1.4$, and $b_3=2.5$, synthesized by amplitude perturbation.

From Table 2.4 we see that four perturbations are involved to bring all the sidelobes to approximately same level (-14.36 dB). In each step, the actual results are quite close to what are expected. In fact, one perturbation could be saved should we take larger $\Delta\epsilon_j$'s than those in the above example. We then, of course, run the risk of overcorrection and commit larger errors in each step.

Should we demand to lower $|\epsilon_1|$ and $|\epsilon_3|$ only and not allow ϵ_2 to increase, the result for the final sidelobe level would be better by requiring perhaps more perturbation steps. It is here we can control the overall pattern shape to make the solution very flexible.

Note that the approach presented in this section is formulated in terms of u. It certainly is also applicable to the endfire array, although the broadside array is chosen for discussion. In our method, we always require solving a set of simultaneous algebraic equations. This process, although straightforward and effective, can be burdensome if the array has a larger number of elements. For this reason Harrington[36] devised a different perturbational procedure for the same problem as case A presented above.

Table 2.4 Example Solution Obtained by Perturbation ($N=3$, $b_1=0.6$, $b_2=1.4$, $b_3=2.5$, $I_i^0=1$).

Starting point	Pattern function	$E^0(u)=2(\cos 0.6u+\cos 1.4u+\cos 2.5u)$
	Sidelobe positions	$u_1^0=88.4°$, $u_2^0=142.9°$, $u_3^0=180°$.
	Sidelobe values	$\epsilon_1^0=-1.4170$, $\epsilon_2^0=0.2737$, $\epsilon_3^0=-1.2360$
First perturbation	Demanded new sidelobes	$\epsilon_1^1=-1.2170$, $\epsilon_2^1=0.4737$, $\epsilon_3^1=-1.0360$
	Solutions from (2.225)	$\Delta I_1=-0.121488$, $I_2=-0.202136$, $I_3=-0.080568$
	New pattern function	$E^1(u)=2(0.878512\cos 0.6u+0.797864\cos 1.4u+0.919432\cos 2.5u)$
	New sidelobe positions	$u_1^1=86.6°$, $u_2^1=143.1°$, $u_3^1=180°$
	Actual sidelobe values	$\epsilon_1^1=-1.2242$, $\epsilon_2^1=0.4688$, $\epsilon_3^1=-1.0360$
Second perturbation	Demanded new sidelobes	$\epsilon_1^2=-1.0742$, $\epsilon_2^2=0.6188$, $\epsilon_3^2=-0.8860$
	Solutions from (2.225)	$\Delta I_1=-0.089192$, $I_2=-0.153526$, $I_3=-0.062547$
	New pattern function	$E^2(u)=2(0.789320\cos 0.6u+0.644338\cos 1.4u+0.856885\cos 2.5u)$
	New sidelobe positions	$u_1^2=85.1°$, $u_2^2=142.0°$, $u_3^2=180°$
	Actual sidelobe values	$\epsilon_1^2=-1.0764$, $\epsilon_2^2=0.6194$, $\epsilon_3^2=-0.8860$
Third perturbation	Demanded new sidelobes	$\epsilon_1^3=-0.9264$, $\epsilon_2^3=0.7694$, $\epsilon_3^3=-0.8400$
	Solutions from (2.225)	$\Delta I_1=0.021449$, $I_2=-0.095883$, $I_3=-0.017635$
	New pattern function	$E^3(u)=2(0.810769\cos 0.6u+0.548454\cos 1.4u+0.839250\cos 2.5u)$
	New sidelobe positions	$u_1^3=84.1°$, $u_2^3=141.3°$, $u_3^3=180°$
	Actual sidelobe values	$\epsilon_1^3=-0.9282$, $\epsilon_2^3=0.7700$, $\epsilon_3^3=0.8400$
Fourth perturbation	Demanded new sidelobes	$\epsilon_1^4=-0.8000$, $\epsilon_2^4=0.8000$, $\epsilon_3^4=-0.8000$
	Solutions from (2.225)	$\Delta I_1=-0.003277$, $I_2=-0.061448$, $I_3=-0.043523$
	New pattern function	$E^4(u)=2(0.807492\cos 0.6u+0.487006\cos 1.4u+0.795727\cos 2.5u)$
	New sidelobe positions	$u_1^4=83.8°$, $u_2^4=140.8°$, $u_3^4=180°$
	Actual sidelobe values	$\epsilon_1^4=-0.8003$, $\epsilon_2^4=0.8000$, $\epsilon_3^4=0.8000$

His method also starts from the known uniform array whose normalized pattern function when $n=2N$ and $d=\lambda/2$ is

$$E_u(u)=\frac{1}{N}\sum_{i=1}^{N}\cos b_i^0 u, \tag{2.226}$$

where

$$b_i^0=\tfrac{1}{2}(2i-1).$$

Now, if the spacing b_i^0 is changed to $b_i^1=b_i^0+\Delta b_i$ by a small increment Δb_i, the pattern function of the perturbed array becomes

$$\begin{aligned}E(u)&=\frac{1}{N}\sum_{i=1}^{N}\cos\left(b_i^0+\Delta b_i\right)u\\&=\frac{1}{N}\sum_{i=1}^{N}\left[\cos\left(b_i^0 u\right)\cos\left(\Delta b_i u\right)-\sin\left(b_i^0 u\right)\sin\left(\Delta b_i u\right)\right]\\&\cong E_u(u)-\frac{u}{N}\sum_{i=1}^{N}\Delta b_i\sin\left(b_i^0 u\right),\end{aligned}$$

or

$$\frac{1}{N}\sum_{i=1}^{N}\Delta b_i\sin\left(b_i^0 u\right)\cong\frac{E_u(u)-E(u)}{u}, \tag{2.227}$$

from which we obtain

$$\Delta b_i\cong\frac{2N}{\pi}\int_0^{\pi}\frac{E_u(u)-E(u)}{u}\sin\left(b_i^0 u\right)du. \tag{2.228}$$

Hence, if a desired pattern $E(u)$ is assumed, fractional changes in element spacing from uniform spacing can be calculated directly from (2.228). For example, if we wish to reduce the first sidelobe of $E_u(u)$ to the level of the second as shown in Fig. 2.17(*a*), where $N=3$, the desired $E(u)$ would appear as the dashed curve. Then $(E_u-E)/u$, represented by Fig. 2.17(*b*), can be inserted in (2.228) for calculating the required fractional changes in spacings. If an analytic expression of $(E_u-E)/u$ is not known, a numerical integration would be involved. To avoid this, Harrington[36] suggested

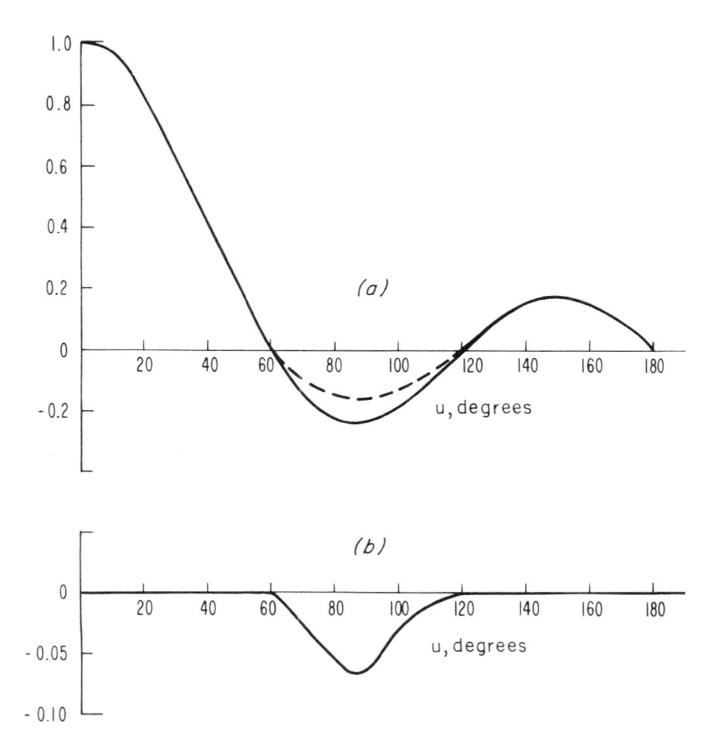

Fig. 2.17 (*a*) Normalized field pattern of a six-element uniform array E_u (solid curve), desired pattern E (dashed), (*b*) the function $(E_u - E)/u$.

another approximation procedure. It essentially expands $(E_u - E)/u$ into a series of impulse functions as follows:

$$\frac{E_u - E}{u} \cong \frac{1}{u} \sum_{j=1}^{K} a_j \delta(u - u_j), \tag{2.229}$$

where u_j and a_j denote, respectively, the position and strength of the jth impulse, and K, the number of impulses applied, is equal to the number of sidelobes reduced. To have an effective modification of the sidelobe level, u_j is usually taken to coincide with the position of the jth sidelobe in E_u. For a large array, u_j is approximately equal to [also see (1.29)]

$$u_j \cong \frac{\pi}{n}(2j+1), \qquad j = 1, 2, \ldots,$$

where n is the total number of elements.

Substitution of (2.229) into (2.228) yields

$$\Delta b_i = \frac{2N}{\pi} \sum_{j=1}^{K} a_j \frac{\sin b_i^0 u_j}{u_j}. \tag{2.230}$$

Unfortunately, the impulse strength a_j needed in (2.230) can only be determined by trial in order to make Δb_i small as required, and at the same time to ensure the condition of (2.229). For details guiding the choice of a_j and related numerical examples, the reader is referred to Harrington's original paper.[36] With his method, the sidelobe level can be reduced to about $2/n$ times the level of the main beam, while the beamwidth can be maintained approximately the same as that of a uniform array.

Since only finite terms are involved in the left-hand side of (2.227), the answer obtained by this procedure yields an approximation minimizing the mean-square error between $\Sigma \Delta b_i \sin(b_i^0 u_j)$ and $(E_u - E)/u$.

Clearly, the method described above can also be applied to case B where the element spacing is fixed and amplitude excitations are allowed to vary for improving the pattern shape. The reader is asked to formulate the necessary details.

2.9 Quadratic Form Approach—Maximization of Directivity

In the previous sections we were concerned with a few synthesis approaches whereby either the sidelobe level in a synthesized pattern is minimized or a specific desired pattern is approximated with different criteria. Array directivity was not duly considered in all of these approaches. Now we are presenting a different synthesis aimed at maximizing the array directivity without considering the detailed pattern shape. More precisely, the purpose is to determine an optimum set of excitations (amplitude and phase) to yield the maximum directivity with respect to a prespecified number of elements and spacings. The success of this synthesis is based on the fact that the array directivity can be expressed as a ratio of two quadratic forms[37] so that the method developed for eigenvalue problems[38] can be applied here. Arrays of both equal and nonequal spacings are considered.

In order to cover the general case, let us go back to an arbitrarily spaced and excited linear array of n elements shown in Fig. 1.2, whose field pattern was given in (1.4). We repeat it here for the convenience of discussion:

$$E = f(\theta,\varphi) \sum_{i=1}^{n} I_i e^{j\alpha_i} \exp(jkz_i \cos\theta). \tag{2.231}$$

The directivity of this arbitrary linear array, according to (1.9), can be written as

$$D = \frac{4\pi |E(\theta_0,\varphi_0)|^2}{\int_0^{2\pi}\int_0^{\pi} |E|^2 \sin\theta \, d\theta \, d\varphi}, \tag{2.232}$$

where (θ_0, φ_0) is the position of the main beam determined by I_i and α_i.

If we define two $n \times 1$ column matrices $[I]$ and $[e]$,

$$[I] = \begin{bmatrix} I_1 e^{j\alpha_1} \\ I_2 e^{j\alpha_2} \\ \vdots \\ I_n e^{j\alpha_n} \end{bmatrix}, \tag{2.233}$$

$$[e] = \begin{bmatrix} \exp(-jkz_1 \cos\theta_0) \\ \exp(-jkz_2 \cos\theta_0) \\ \vdots \\ \exp(-jkz_n \cos\theta_0) \end{bmatrix}, \tag{2.234}$$

we can readily express (2.232) as

$$D = \frac{[I]^+ [A][I]}{[I]^+ [B][I]}, \tag{2.235}$$

where $[I]^+$ is the conjugate transpose (adjoint) of $[I]$; $[A]$ is an $n \times n$ square matrix

$$[A] = [e][e]^+, \tag{2.236}$$

whose typical element a_{lm} is given by

$$a_{lm} = \exp[jk(z_m - z_l)\cos\theta_0]; \tag{2.237}$$

and the typical element b_{lm} in the square matrix [B] (also $n \times n$) is

$$b_{lm} = \frac{1}{4\pi}\int_0^{2\pi}\int_0^{\pi} f^2 \exp[jk(z_m - z_l)\cos\theta]\sin\theta\, d\theta\, d\varphi. \tag{2.238}$$

In (2.235) we have made the normalization that $f(\theta_0, \varphi_0) = 1$.

According to a theorem known in matrix theory for eigenvalue problems,[38] whenever a function is expressed as a ratio of two quadratic forms such as D in (2.235), where [A] and [B] are both Hermitian and [B] is positive definite, we will have the following:

1. the eigenvalues ($p_1 \leqslant p_2 \leqslant \cdots \leqslant p_n$) or the roots of the eigenequation

$$\det\{[A] - p[B]\} = 0, \tag{2.239}$$

are real, where det means "the determinant of";

2. p_1 and p_n represent respectively the lower and upper bounds for the value of D; that is

$$p_1 \leqslant D \leqslant p_n; \tag{2.240}$$

3. the left equality in (2.240) is possible when [I] satisfies

$$[A][I] = p_1[B][I]; \tag{2.241}$$

4. the right equality in (2.240) is attainable when [I] satisfies

$$[A][I] = p_n[B][I]. \tag{2.242}$$

As far as the optimization of (2.235) is concerned here, the problem reduces to the verification of [A] and [B] to see whether they satisfy the required conditions, the evaluation of p_n representing the maximum obtainable directivity, and the determination of [I] from (2.242) giving the required excitations. Fortunately, all these steps can be easily performed or calculated in our application.

From (2.237) we see that the main-diagonal elements of $[A]$ when $l=m$ are real, and that the off-diagonal elements when $l \neq m$ are complex conjugate, $a_{lm}=\bar{a}_{ml}$. Therefore, $[A]$ is Hermitian. The matrix $[B]$ whose element is given in (2.238) can be proven to be Hermitian by the same consideration. To prove the positive definiteness of $[B]$, we note a well-known theorem[39] stating that the Hermitian matrix $[B]$ is to be positive definite if the associated quadratic form, $[I]^+[B][I]$, is positive for any real or complex $[I]$. Since

$$[I]^+[B][I]=\frac{1}{4\pi}\int_0^{2\pi}\int_0^{\pi}|E|^2\sin\theta\,d\theta\,d\varphi \tag{2.243}$$

represents the total power radiated by the array, it has to be positive from the physical consideration. Therefore, the foregoing optimization theorem applies. Furthermore, because of the special form of $[A]$ in (2.236), it can be shown[37,40] that all eigenvalues of (2.239) are zero except the largest one, p_n, which is given explicitly by

$$p_n=D_{\max}=[e]^+[B]^{-1}[e]>0, \tag{2.244}$$

and that the required optimum excitation matrix found from (2.242) is

$$[I]_{\text{opt}}=[B]^{-1}[e], \tag{2.245}$$

where $[B]^{-1}$ is the inverse of $[B]$.

Equations (2.244) and (2.245) constitute the complete solution to the directivity optimization problem. Note that this does not require the solution of simultaneous equations. Once the element positions z_i are given, b_{lm} can be calculated from (2.238), which is independent of α_i and θ_0. The matrix $[e]$ can be easily obtained if θ_0 is also known. Then, finding the inverse of $[B]$ is the only rather complicated operation to get the final answer. Substituting (2.245) into (2.244) and noting the fact that $[B]$ is symmetric in application, we have

$$p_n=D_{\max}=[I]_{\text{opt}}^+[B][I]_{\text{opt}}, \tag{2.246}$$

which is the denominator of (2.235). Of course, under this condition, the numerator of (2.235) should be $(D_{\max})^2$.

As an example, let us consider the equally spaced $(d=\lambda/2)$ broadside array of n isotropic elements. In this case, $\theta_0 = \pi/2$, $f(\theta,\varphi)=1$, $z_i=(i-1)d = (i-1)\lambda/2$;

$$[e] = \begin{bmatrix} 1 \\ 1 \\ \vdots \\ 1 \end{bmatrix}, \tag{2.247}$$

$$[A] = \begin{bmatrix} 1 & 1 & \cdots & 1 \\ 1 & 1 & \cdots & 1 \\ \vdots & & & \\ 1 & 1 & \cdots & 1 \end{bmatrix}, \tag{2.248}$$

$$b_{lm} = \frac{\sin k(z_m - z_l)}{k(z_m - z_l)} = \begin{cases} 0, & l \neq m \\ 1, & l = m \end{cases}, \tag{2.249}$$

and (2.239) reduces to

$$\begin{bmatrix} (1-p) & 1 & \cdots & 1 \\ 1 & (1-p) & \cdots & 1 \\ \vdots & & \ddots & \vdots \\ 1 & 1 & \cdots & (1-p) \end{bmatrix} = (-1)^n p^{n-1}(p-n) = 0. \tag{2.250}$$

It is clear that the only nonzero root of (2.250) is $p_n = n$, which can also be obtained by substituting (2.247) and (2.249) into (2.244). Since both $[B]$ and $[B]^{-1}$ are now unit matrices, the answer for the required excitation matrix, according to (2.245), is

$$[I]_{\text{opt}} = [e], \qquad (\text{or} \quad I_i = 1, \qquad \alpha_i = 0), \tag{2.251}$$

Table 2.5 Optimum Directivities of an Array with $\theta_0=\pi/2$, $f(\theta,\varphi)=1$, $z_i=(i-1)d$, and $n=3$.

d/λ	D_{max}	$I_1=I_3$	I_2
0.2	2.3404	2.7565	−3.1726
0.3	2.4658	1.4794	−0.4929
0.4	2.6737	1.0923	0.4891
0.5	3.0000	1.0000	1.0000
0.6	3.4800	1.0728	1.3345
0.7	4.0397	1.2496	1.5404
0.8	4.2514	1.3670	1.5173
0.9	3.7255	1.2344	1.2566
1.0	3.0000	1.0000	1.0000

a well-known result checking with that learned from Chapter 1.

Strictly speaking, the directivity obtained above is optimum only with respect to the particular element spacing used. When d is not a multiple of $\lambda/2$, the off-diagonal elements of $[B]$ will not always vanish, the final solution will be different even though the other condition such as $\theta_0=\pi/2$ remains unchanged. For the purpose of illustration, numerical results of D_{max} and $[I]_{opt}$ for a simple array of three isotropic elements are presented in Table 2.5, as a function of d. It is seen that only when the spacing is a multiple of $\lambda/2$, the directivity reaches its maximum with $I_i=1$. Optimum directivities of equally spaced broadside arrays with a larger number of elements and with simple dipoles can be found elsewhere.[7] As another example, let us determine the maximum directivity obtainable from an equally spaced endfire array of five isotropic elements with $d=\lambda/4$. In this case, $\theta_0=0$, $f(\theta,\varphi)=1$, $z_i=(i-1)\lambda/4$. We then have

$$[e]=\begin{bmatrix} 1 \\ -j \\ -1 \\ j \\ 1 \end{bmatrix}, \qquad [A]=\begin{bmatrix} 1 & j & -1 & -j & 1 \\ -j & 1 & j & -1 & -j \\ -1 & -j & 1 & j & -1 \\ j & -1 & -j & 1 & j \\ 1 & j & -1 & -j & 1 \end{bmatrix}, \tag{2.252}$$

$$b_{lm} = \frac{\sin k(z_m - z_l)}{k(z_m - z_l)}, \qquad l,m = 1,2,\ldots,5,$$

$$[B] = \begin{bmatrix} 1 & 0.6366 & 0 & -0.2111 & 0 \\ 0.6366 & 1 & 0.6366 & 0 & -0.2122 \\ 0 & 0.6366 & 1 & 0.6366 & 0 \\ -0.2122 & 0 & 0.6366 & 1 & 0.6366 \\ 0 & -0.2122 & 0 & 0.6366 & 1 \end{bmatrix},$$

$$[B]^{-1} = \begin{bmatrix} 11.9641 & -24.3152 & 29.0256 & -21.2782 & 8.3863 \\ -24.3152 & 55.5026 & -68.3902 & 51.9247 & -21.2782 \\ 29.0256 & -68.3902 & 88.0769 & -68.3902 & 29.0256 \\ -21.2782 & 51.9247 & -68.3902 & 55.5026 & -24.3152 \\ 8.3863 & -21.2782 & 29.0256 & -24.3152 & 11.9641 \end{bmatrix}$$

$$D_{\max} = [e]^+ [B]^{-1} [e] = 19.8342, \tag{2.253}$$

$$[I]_{\text{opt}} = [B]^{-1}[e] = \begin{bmatrix} 9.1906 \exp(j160.7^\circ) \\ 23.0758 \exp(-j8.9^\circ) \\ 30.0273 \exp(j180^\circ) \\ 23.0758 \exp(j8.9^\circ) \\ 9.1906 \exp(j199.3^\circ) \end{bmatrix}, \tag{2.254}$$

or relatively,

$$I_1 : I_2 : I_3 : I_4 : I_5 = 1 : 2.5108 : 3.2672 : 2.5108 : 1,$$

$$\alpha_1 = 0, \quad \alpha_2 = -169.6^\circ, \quad \alpha_3 = 19.3^\circ, \quad \alpha_4 = -151.8^\circ, \quad \alpha_5 = 38.6^\circ.$$

The normalized pattern based on the excitations in (2.254) is given in Fig 2.18. The directivity determined in (2.253) can be compared with thos

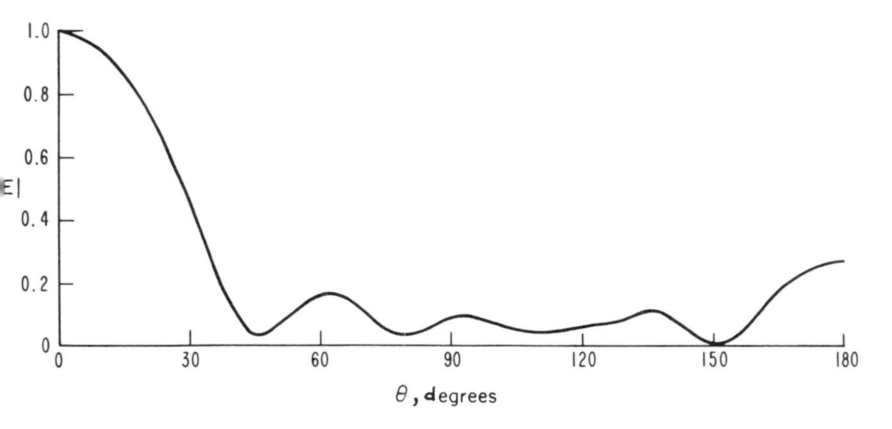

ig. 2.18 Normalized pattern of the five-element endfire array with $d=\lambda/4$ yielding maximum directivity.

btained before in Chapter 1, Sections 2.2 and 2.3. Table 2.6 is prepared or this purpose. Before concluding this section we should note that the nethod outlined here also applies to nonuniformly spaced arrays,[41,42] lthough arrays with equal spacings are specifically chosen in the above wo examples for illustrative purpose.

.10 Synthesis of Monopulse Arrays

n Section 1.7 we formulated the problem of monopulse arrays and nalyzed the characteristics associated with the sum and difference modes. Now we are ready to study the reverse problem—synthesis of a monopulse rray. Since all the techniques presented so far in this chapter also apply to he sum pattern in a monopulse array, we will concentrate on the lifference pattern only. Specifically, three methods will be examined. The irst is to extend the application of Haar's theorem, presented in Section .7, to the synthesis of a difference pattern which will approximate a lesired pattern in the minimax sense. The second method is to employ the ame perturbation technique discussed in Section 2.8 to reduce the side-obe level in a difference pattern, also from a known monopulse array. Jltimately, all the sidelobes will be approximately equalized. The third pproach is similar to that given in Section 2.9, namely, the determination of an optimum set of excitations required to maximize the difference lirectivity for a given squint angle, element spacing, and number of lements in the array.

Table 2.6[a] A Comparison of Directivities by Different Approaches for a Five-Element, Equally Spaced, Endfire Array with $d=\lambda/4$.

	Directivity	$\frac{I_2}{I_1}=\frac{I_4}{I_1}$	$\frac{I_3}{I_1}$	α_1	α_2	α_3	α_4	α_5
Ordinary uniform endfire (Section 1.2)	5.00	1.0000	1.0000	0°	−90°	−180°	90°	0°
Improved uniform endfire with $\delta=35°$ (Section 1.3)	8.77	1.0000	1.0000	0°	−125°	110°	−15°	−140°
Endfire with NUPP phases (Section 1.6) curve *I* in Fig. 1.28	9.81	1.0000	1.0000	0°	−145°	100°	−15°	−159.5°
Ordinary endfire with equal *s*. 1. (−20 dB)	4.68	1.6080	1.9341	0°	−90°	−180°	90°	0°
Optimized endfire with equal *s*. 1. (−20 dB) (Section 2.3)	18.47	2.5033	3.2867	0°	−165.5°	29.1°	−136.4°	58.2°
Endfire with maximum directivity (Section 2.9)	19.83	2.5108	3.2672	0°	−169.6°	19.3°	−151.8°	38.6°

[a] For this table, $I_5=I_1$, is implied. While the characteristics for the ordinary endfire array with equal sidelobes (−20 dB) were not specifically worked out in this book, they can easily be calculated by the method described in Section 2.2.

A. Application of Haar's Theorem. If the same notation as that in Section 2.8 is also adopted here, the difference pattern of a symmetrically spaced and excited monopulse array of $2N+1$ elements may be written as

$$\Delta(u)=4\sum_{i=1}^{N} I_i \sin(b_i u_s)\sin(b_i u), \tag{2.255}$$

where

$$b_i=2d_i/\lambda, \qquad u=\pi\cos\theta, \qquad u_s=\pi\cos\theta_s, \tag{2.256}$$

with d_i the distance of the ith pair of elements to the array center, and θ_s defined in Fig. 1.29(a).

In general, the difference function given in (2.255) has a maximum at $u=u_1$ (previously called u_m) and several sidelobes at $u_2, u_3,\ldots,u_p$ in the visible range $0<u\leqslant\pi$. The sidelobe levels at u_j, $j=2, 3,\ldots,p$ may not be satisfactory if I_i and b_i are arbitrary. To reduce all the sidelobes to an approximately same level, we again apply Haar's theorem as discussed in Section 2.7. Since $\Delta(u)$ is an odd function and for the same reasons outlined before, we choose this time $f(u)=ue^{-Au^2}$ as the desired difference pattern, where the positive parameter A may be used to control the pattern shape. Because the basis function is now sin $(b_i u)$, it is clear that the first two conditions required by a Chebyshev system have been satisfied. The third condition that no linear combination of $\sin(b_i u)$ should have more than N-1 distinct real zeros in a finite interval $[a,b]$ depends again on the choice of b_i.

In the case $b_i=i$ (equal spacing with $d=\lambda/2$), the difference function in (2.255) becomes

$$\begin{aligned}\Delta(u)&=4\sum_{i=1}^{N} I_i\sin(iu_s)\sin(iu)\\ &=4\sin u\left(I'_{N-1}\cos^{N-1}u+I'_{N-2}\cos^{N-2}u+\cdots+I'_0\right).\end{aligned} \tag{2.257}$$

Under this condition $\Delta(u)$ will always have two zeros at $u=0$ and $u=\pi$ because of the presence of $\sin u$. The condition on the maximum allowable number of real distinct zeros of $\Delta(u)$, no matter what I_i and u_s are, will be satisfied if we choose an open interval $(a,b)=(0,\pi)$. Once the conditions are established, the actual process of determining a unique set of I_i to approximately equalize all the sidelobes of $\Delta(u)$ follows what was outlined in Section 2.7. For illustration, a simple example for $N=3$, $u_s=15.69°$ ($\theta_s=85°$), and $A=1$ is given here. First, we match $\Delta(u)$ in (2.257) to

$f(u) = ue^{-u^2}$ at three arbitrary points U_j $[0 < U_1 < U_2 < U_3 < \pi]$ to obtain an initial set of I_i (called I_i^0), $i = 1$, 2, and 3. That is, I_i^0 satisfies the following

$$4 \sum_{i=1}^{3} I_i^0 \sin(i \times 15.69°) \sin(iU_j) = f(U_j), \quad j = 1, 2, \text{ and } 3.$$

The initial deviation function, $e^0(u) = f(u) - \Delta^0(u)$, vanishes at $u = 0$, U_1, U_2 and U_3; and $de^0(u)/du = 0$ at u_0^0, u_1^0, u_2^0, and u_3^0 where $0 < u_0^0 < U_1 < u_1^0 < U_2 < u_2^0 < U_3 < u_3^0 < \pi$. The magnitude of $e^0(u)$ at u_j^0 are, of course, not equal initially. The next step is to assume

$$\Delta^1(u) = 4 \sum_{i=1}^{3} I_i^1 \sin(i \times 15.69°) \sin(iu)$$

with I_i^1 as the new set of coefficients to be determined by equalizing $e^1(u) = f(u) - \Delta^1(u)$ at u_j^0 so that the following will be true:

$$e^1(u_j^0) = (-1)^j \epsilon_1, \qquad j = 0, 1, 2, \text{ and } 3. \tag{2.258}$$

Equation (2.258) consists of a system of four equations (in general, $N+1$) which are just enough to solve for I_i^1 and ϵ_1. The process can be repeated until the sidelobes are practically equalized. For this particular example after three iterations, we obtain the final solution:

$$I_1 = 0.2029, \qquad I_2 = 0.0995, \qquad I_3 = 0.0314; \qquad \epsilon_3 = 0.0210;$$

$$u_0 = 0.3687, \qquad u_1 = 1.1187, \qquad u_2 = 1.9020, \qquad u_3 = 2.7226;$$

$$\Delta(u_m) = \Delta(u)_{\max} = 0.4282 \text{ at } u_m = 0.7604 \text{ (or } 43.56°).$$

In the above, we have omitted the burdensome superscript which should be used to identify the number of iterations.

This procedure for determining I_i by use of Haar's theorem can be easily generalized for any N. As long as the element spacing does not exceed $\lambda/2$, which happens to be the most practical case, all the conditions required by a Chebyshev system are satisfied. The solution so obtained is therefore unique and optimum with respect to the chosen set of b_i.

In the case b_i are arbitrary corresponding to nonuniform spacings, it is rather difficult to verify at the beginning whether the condition on the maximum allowable number of real distinct zeros of $\Delta(u)$ can be satisfied in the interval $(0, \pi)$. The only way of satisfying it is perhaps by choosing all the b_i small, say $b_i < i$, which would then not be very practical. When $b_i > i$ for some i so that the condition is violated, the iterative procedure described above would not always converge, depending largely on what the initial matching points U_j are used. Even if it does converge, we still have no knowledge about the uniqueness of the final answer.

B. Application of Perturbation Method. As was done in Section 2.8, the problem can be divided into two classes, namely, either the element spacing or the amplitude excitation alone is perturbed at a time. We again start from a known difference function satisfying

$$\Delta(u_j) = 4 \sum_{i=1}^{N} I_i^0 \sin(b_i^0 u_s) \sin(b_i^0 u_j^0) = \epsilon_j^0,$$

$$j = 1, 2, \ldots, p, \tag{2.259}$$

$$\left. \frac{\partial \Delta(u)}{\partial \Delta} \right|_{u=u_j^0} = 4 \sum_{i=1}^{N} I_i^0 b_i^0 \sin(b_i^0 u_s) \cos(b_i^0 u_j^0) = 0,$$

where u_1^0 is the known position of the beam maximum whose value is ϵ_1^0, and $u_2^0, u_3^0, \ldots, u_p^0$ are those of sidelobes with respective levels $\epsilon_2^0, \epsilon_3^0, \ldots, \epsilon_p^0$.

B.1 When the element spacing alone is to be varied while the amplitude excitation is kept constant in order to control the sidelobe level, we can insert $I_i^0 = 1$, $b_i^1 = b_1^0 + \Delta b_i$, and $u_j^1 = u_j^0 + \Delta u_j$ into (2.259) to solve for Δb_i and Δu_j. Then we have a system of $2p$ equations with $p+N$ unknowns ($\Delta b_1, \Delta b_2, \ldots, \Delta b_N, \Delta u_1, \Delta u_2, \ldots, \Delta u_p$). To have a unique solution, we require $p = N$. Fortunately, this condition can be easily met if we choose $b_i^0 = i$. After a few perturbations, the final b_N may be greater than N. In this case, the situation $p > N$ will occur. We then select the first N extrema points to work with and suppress the remaining into the invisible range by a simple rescaling process. This extra process can best be understood with a later example.

After making an approximation similar to that in (2.222), we obtain the final set of simultaneous equations in terms of the perturbed variables:

$$\Delta\epsilon_j = 4 \sum_{i=1}^{N} \left[\Delta b_i \left(u_s \cos b_i^0 u_s \sin b_i^0 u_j^0 + u_j^0 \sin b_i^0 u_s \cos b_i^0 u_j^0 \right) \right.$$

$$\left. + \Delta u_j \left(b_i^0 \sin b_i^0 u_s \cos b_i^0 u_j^0 \right) \right], \tag{2.260}$$

$$0=\sum_{i=1}^{N}\left\{\Delta b_i\left[\left(\cos b_i^0u_j^0-b_i^0u_j^0\sin b_i^0u_j^0\right)\sin b_i^0u_s\right.\right.$$

$$\left.+b_i^0u_s\cos b_i^0u_s\cos b_i^0u_j^0\right]$$

$$\left.-\Delta u_j\left(b_i^0\right)^2\sin b_i^0u_s\sin b_i^0u_j^0\right\}. \tag{2.261}$$

With b_i^0 and u_j^0 known and setting $\epsilon_j^1=\epsilon_j^0+\Delta\epsilon_j$ in any desired level through small steps, we can solve for Δb_i to determine the required new element spacings. Eventually, all the sidelobes can approximately be equalized by this procedure.

Let us consider again an example with $N=3$, $b_i^0=i$, and $u_s=15.69°$. After three iterations, the difference function becomes

$$\Delta(u)=4(0.1822\sin 0.6699u+0.5292\sin 2.0379u$$

$$+0.8295\sin 3.5749u). \tag{2.262}$$

The levels of Δ(u) at the first three extrema are

$$\Delta(28.85°)=5.2830,\qquad \text{the main lobe,}$$

$$\Delta(80.87°)=-1.9895,\qquad \text{the first sidelobe,}$$

$$\Delta(124.2°)=1.9978,\qquad \text{the second sidelobe,}$$

and the required spacings are

$$b_1=0.6699,\qquad b_2=2.0379,\qquad b_3=3.5749,$$

or

$$d_1=0.3350\lambda,\qquad d_2=1.0190\lambda,\qquad d_3=1.7875\lambda.$$

Since b_3 is now substantially larger than 3.0, resulting in another extremum near $u=180°$ (see curve a in Fig. 2.19), and $|\Delta(180°)|>1.9900$, we must suppress a small portion of (2.262) into the invisible range. This can be done by determining a u^* so that $124.2°=u_3<u^*<180°$, and that $\Delta(u^*)=-1.9900$. Solving this equation, we get $u^*=158.2°$. Multiplying b_i obtained above by $158.2/180=0.8789$, we will have the modified difference function,

$$\Delta_m(u)=4(0.1822\sin 0.5888u+0.5292\sin 1.7911u$$

$$+0.8295\sin 3.1419u), \tag{2.263}$$

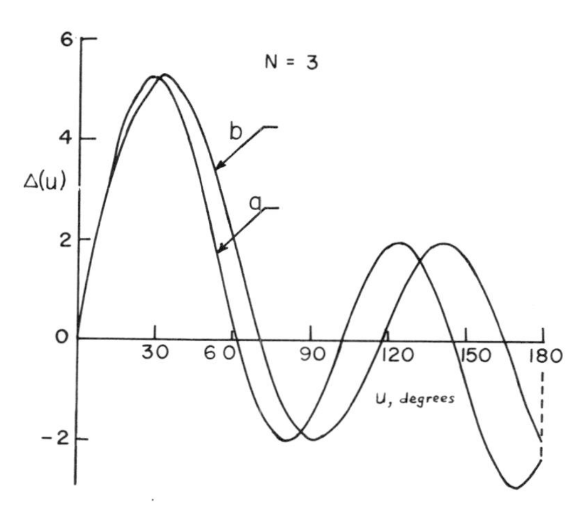

Fig. 2.19 Synthesized difference patterns of a uniformly excited seven-element monopulse array with $u_s = 15.69°$: (*a*) $d_1 = 0.3350\lambda, d_2 = 1.0190\lambda, d_3 = 1.7875\lambda$, (*b*) $d_1 = 0.2944\lambda, d_2 = 0.8955\lambda, d_3 = 1.5709\lambda$.

all the sidelobes of which are now nearly equal in level, as can be seen from curve b in Fig. 2.19. The final spacings will be

$$d_1 = 0.2944\lambda, \quad d_2 = 0.8955\lambda, \quad \text{and} \quad d_3 = 1.5709\lambda.$$

B.2 When the amplitude excitation is allowed to vary while $b_i = i$ is kept throughout, the perturbation can be performed in a similar manner. The final set of equations in terms of the perturbed variables will then be

$$\Delta\epsilon_j = 4 \sum_{i=1}^{N} \left[\Delta I_i \sin(iu_s) \sin(iu_j^0) + \Delta u_j i I_i^0 \sin(iu_s) \cos(iu_j^0) \right],$$

$$0 = \sum_{i=1}^{N} i \sin(iu_s) \left[\Delta I_i \cos(iu_j^0) - \Delta u_j i I_i^0 \sin(iu_j^0) \right]. \tag{2.264}$$

For $N = 3, b_i = i, u_s = 15.69°$, and $I_i^0 = 1$, the required amplitude excitations obtained from (2.264) are, after two iterations,

$$I_1 = 1.1783, \quad I_2 = 0.9115, \quad \text{and} \quad I_3 = 1.0027.$$

The resulting important characteristics are:

$$\Delta(35.11°)=5.3514, \quad \text{the main lobe,}$$

$$\Delta(98.56°)=-1.9450, \quad \text{the first sidelobe,}$$

$$\Delta(152.2°)=1.9450, \quad \text{the second sidelobe,}$$

and

$$\Delta(u)=0 \quad \text{at} \quad u=0°, 75.6°, 124.8°, \quad \text{and} \quad 180°.$$

Clearly, the case that the number of extrema points, p, may be greater than N can never happen here, since $\Delta(u)$ always vanishes at $u=180°$.

c. Maximization of Difference Directivity. The difference directivity of a monopulse array is defined as

$$D_d=\frac{4\pi\Delta^2(u_m)}{\int_0^{2\pi}\int_0^{\pi}\Delta^2(u)\sin\theta\, d\theta\, d\varphi}, \tag{2.265}$$

where u_m is the positon of the beam maximum in the difference pattern. Substituting (2.255) into (2.265), we obtain

$$D_d=\frac{[I']^T[R][I']}{[I']^T[Q][I']}, \tag{2.266}$$

where

$$[I']=\begin{bmatrix} I_1\sin b_1u_s \\ I_2\sin b_2u_s \\ \vdots \\ I_N\sin b_Nu_s \end{bmatrix}, \tag{2.267}$$

$[I']^T$ is the transpose of $[I']$, and the typical elements in $[R]$ and $[Q]$ are, respectively,

$$R_{ij}=\sin b_iu_m\sin b_ju_m, \tag{2.268}$$

$$Q_{ij}=\frac{1}{2}\left[\frac{\sin(b_i-b_j)\pi}{(b_i-b_j)\pi}-\frac{\sin(b_i+b_j)\pi}{(b_i+b_j)\pi}\right], \tag{2.269}$$

$$i,j=1,2,\ldots,N. \tag{2.270}$$

From (2.268) we also have

$$[R]=[V][V]^T, \tag{2.271}$$

where

$$[V]=\begin{bmatrix} \sin b_1 u_m \\ \sin b_2 u_m \\ \vdots \\ \sin b_N u_m \end{bmatrix}. \tag{2.272}$$

Since (2.266) is expressed as a ratio of two quadratic forms, $[R]$ and $[Q]$ are real matrices (special form of Hermitian matrices), and $[Q]$ is positive definite, we again can apply the theorem presented in Section 2.9 to conclude that the maximum value of D_d should be given by the largest root of $\det\|[R]-p[Q]\|=0$. Also, because of the special form of $[R]$, the eigenequation $\det\|[R]-p[Q]\|=0$ degenerates, and the only nonzero root is the largest one (representing the maximum difference directivity) which is explicitly given by

$$\begin{aligned}(D_d)_{\max}&=[V]^T[Q]^{-1}[V]\\ &=\frac{1}{|Q|}\sum_{i=1}^{N}\sum_{j=1}^{N} q_{ij}\sin b_i u_m \sin b_j u_m,\end{aligned} \tag{2.273}$$

where q_{ij} is the cofactor of Q_{ji} in $[Q]$, and $|Q|$ is the determinant of $[Q]$.

The excitation required to yield the maximum difference directivity is

$$[I']_{\text{opt}}=[Q]^{-1}[V], \tag{2.274}$$

or

$$[I]_{\text{opt}} = \frac{1}{|Q|} \begin{bmatrix} \dfrac{\sum_{i=1}^{N} q_{i1} \sin b_i u_m}{\sin b_1 u_s} \\ \dfrac{\sum_{i=1}^{N} q_{i2} \sin b_i u_m}{\sin b_2 u_s} \\ \vdots \\ \dfrac{\sum_{i=1}^{N} q_{iN} \sin b_i u_m}{\sin b_N u_s} \end{bmatrix}. \tag{2.275}$$

Three important points should be noted here. First, $(D_d)_{\max}$ obtained in (2.273), although a function of u_m, b_i, and N, is independent of u_s. Second, the difference directivity so determined is optimum only with respect to the particular values of b_i and N used. Third, since u_m is denoted as the position of the beam maximum of the difference pattern, it should satisfy the following:

$$\left[\frac{d}{du}\Delta(u)\right]_{u=u_m} = 0$$

or

$$\sum_{i=1}^{N} I_i' b_i \cos(b_i u_m) = 0, \tag{2.276}$$

from which we can determine u_m when b_i and N are specified. The smallest positive solution of (2.276) should be used in (2.273) and (2.274) to calculate $(D_d)_{\max}$ and $[I']_{\text{opt}}$. The final $[I]_{\text{opt}}$ can then be determined if u_s is also given.

For arrays of equal spacing with $d = \lambda/2\,(b_i = i)$, the results are very simple. In this case, we have

$$[Q] = \frac{1}{2}\begin{bmatrix} 1 & 0 & 0 & \cdots & 0 \\ 0 & 1 & 0 & \cdots & 0 \\ \cdots & \cdots & \cdots & \cdots & \cdots \\ 0 & 0 & 0 & \cdots & 1 \end{bmatrix}, [Q]^{-1} = \begin{bmatrix} 2 & 0 & 0 & \cdots & 0 \\ 0 & 2 & 0 & \cdots & 0 \\ \cdots & \cdots & \cdots & \cdots & \cdots \\ 0 & 0 & 0 & \cdots & 2 \end{bmatrix}$$

$$(D_d)_{\max}=2\sum_{i=1}^{N}\sin^2(iu_m)=N-\frac{\cos(N+1)u_m\sin Nu_m}{\sin u_m}, \tag{2.277}$$

$$[I']_{\text{opt}}=2\begin{bmatrix}\sin u_m\\ \sin 2u_m\\ \vdots\\ \sin Nu_m\end{bmatrix}, \tag{2.278}$$

and the value of u_m should, in view of (2.276), satisfy

$$\sum_{i=1}^{N} i\sin(iu_m)\cos(iu_m)=\tfrac{1}{2}\sum_{i=1}^{N} i\sin(2iu_m)=0 \tag{2.279}$$

Note that the second expression in (2.277) can be derived with the help of the finite Z-transform theory and Table 1.1 discussed in Chapter 1.

Table 2.7 gives $(D_d)_{\max}$ and u_m as a function of N when $b_i=i$. For other element spacings (also equal spacing), the calculated $(D_d)_{\max}$ and u_m for $N=2$ and 3 are presented in Fig. 2.20. The method can also be applied for arrays with nonuniform spacings. The reader is asked to work out a simple example on this as an exercise.

Table 2.7 Maximum Directivities and Main Beam Positions of Difference Patterns for a Monopulse Array of $2N+1$ Elements.

N	u_m (deg)	$(D_d)_{\max}$
2	52.24	3.1249
3	37.04	4.3158
4	28.73	5.5195
5	23.47	6.7342
6	19.84	7.9411
7	17.19	9.1541
8	15.16	10.3701
9	13.56	11.5815
10	12.27	12.7985

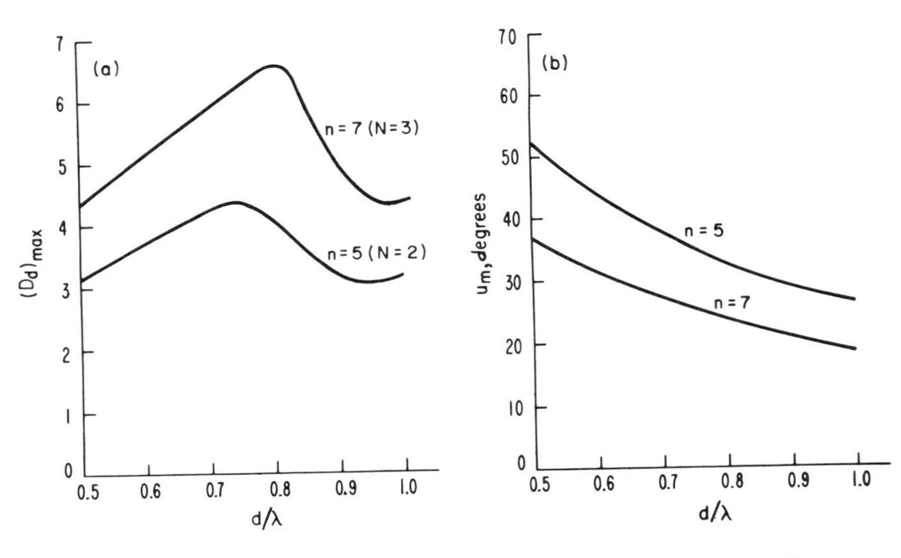

Fig. 2.20 Difference characteristics of an equally spaced monopulse array of $2N+1$ elements: (*a*) maximum directivity, (*b*) main beam position.

2.11 Linear Array—Passive Network Analogy

In the previous sections various theories and techniques were presented for the synthesis of linear arrays with different criteria. Because of the surprising similarity in mathematics involved, we now wish to note the analogy between two different fields, linear arrays and passive networks. Since a significant body of knowledge in the area of passive network synthesis has been developed in the past,[43] the material presented in this section may be helpful for finding new approaches in array synthesis.

For many occasions in the first two chapters, we have dealt with the field or power patterns associated with different kinds of arrays in the following forms:

$$E_1(u) = \sum_{i=0}^{n-1} I_i e^{jiu}, \tag{2.280}$$

$$E_2(u) = I_0 + 2\sum_{i=1}^{N} I_i \cos b_i u, \tag{2.281}$$

$$P_U(y) = \sum_{m=0}^{n-1} A_m y^m \geqslant 0, \tag{2.282}$$

$$P_{NU}(y) = \sum_{m=0}^{n-1} A_m' y^m + \left(\sum_{m=0}^{n-2} A_m'' y^m \right) (4-y^2)^{1/2} \geqslant 0, \tag{2.283}$$

and

$$p_N(x) = \frac{Q(x)}{(1+x^2)^{n-1}} \geqslant 0, \tag{2.284}$$

where the meaning of the notations has been explained before.

In the theory of passive network synthesis, a desired impulse response of a network system is usually specified. The designer is faced with the problem of translating the given time response specification, also under a specified tolerance on accuracy, into a form that is amenable to a determination of the network structure and the element values. One way of solving this problem is known as the time-domain systhesis. Before doing so, it is recalled that the Laplace transform of a realizable impulse response (transfer function) can be expressed as a ratio of two polynomials in frequency with real coefficients and that from a consideration of the system stability, the roots of the denominator polynomial of the transfer function must all have nonpositive real parts. These requirements imply that the impulse response of a realizable network should be of the typical form

$$h(t) = \sum_{i=0}^{M} a_i e^{s_i t}, \tag{2.285}$$

where each a_i and s_i may be real, complex, or pure imaginary, but $h(t)$ itself is a real function of t, and s_i must have a negative real part if it is complex. In writing (2.285), it has been assumed that the degree of denominator polynomial of

$$H(s) = \mathfrak{L}[h(t)] = \frac{H_1(s)}{H_2(s)} \tag{2.286}$$

is at least one degree greater than that of the numerator polynomial, as it should be for a passive network. Now, the network synthesis problem can be stated as follows: Assuming $f(t)$ is a known function of time representing a desired impulse response of the network system, we must find a function $h(t)$ of the form (2.285) which will approximate the given $f(t)$, under a specified tolerance on accuracy, such that $H(s)$ can immediately be expressed as a ratio of two polynomials without further approximation and satisfies the stability condition. From this $H(s)$, a network structure and the element values are then realized by one of the standard procedures.[43]

A comparison of (2.285) with (2.280) and (2.281), and of (2.286) with (2.282) and (2.284) reveals that the mathematical forms are very similar although the symbols represent different physical entities in these two different fields. It is natural that we should take advantage of this similarity to deal with array problems by using the existing knowledge from network problems, or vice versa, just as the familiar network-potential analogy[44] has been employed to solve a class of network problems by using some results from the classical potential problems. Since tremendous results pertinent to the network synthesis have been obtained,[45-50] some new techniques for array synthesis may be developed through a serious study of the analogy.[51,52]

2.12 Concluding Remarks and Discussion

In this chapter, we have presented various theories and techniques for the synthesis of linear arrays with discrete elements. Consideration of sidelobe level, beamwidth, particular pattern shape, and directivity was the main concern. For generality, all the synthesis facets included were given with the idealized isotrophic elements. Actual radiators in free space can easily be incorporated, with minor modifications, when the principle of pattern multiplication applies, as it usually does. Mutual interactions among elements can also be considered if necessary. For example, once the currents are determined by one of the synthesis methods studied, we can then adjust or design the voltage excitation and the actual feeding system, with the knowledge of mutual impedances between elements, to yield such required currents.

Only the pattern inside the visible range of observation received attention in this chapter. The pattern outside the visible range, which is related to the reactive power of the array,[53-55] was totally ignored. Synthesis of continuous sources with a modified requirement on the sidelobe level has been treated by Taylor.[53] An equivalence between arrays and apertures can be easily established.[56]

When an approximation is involved in the course of synthesis, the minimax criterion was mainly used when discussing the methods of interpolation, Haar's theorem, and Bernstein polynomials; synthesis with the other popular criterion of mean-square error can be found elsewhere.[26,54,57] Since space is limited, synthesis by many other valuable methods, notably those by the Poisson sum formula,[58] the Anger function,[59] grating plateau,[60] mechanical quadratures,[61] orthogonal and eigenvalue methods,[62,63] amplitude scanning,[64] potential theory,[65] variational method,[56,66] mode theory,[67] and that yielding a maximally flat sector beam,[68] has been reluctantly omitted here.

All the synthesis approaches discussed in this chapter were, in principle, valid only for a relatively narrow band centered at the designed frequency. The synthesis of arrays good for a much wider band of frequencies has not had much success in the past.[59,69]

PROBLEMS

2.1 Synthesize a six-element, equally spaced ($d=\lambda/2$), broadside array with all its sidelobes equalized at -20 dB, and tabulate the important radiation characteristics.

2.2 What sidelobe level of the array in Problem 2.1 will yield the maximum directivity?

2.3 Obtain an optimum (narrowest beamwidth), five-element, equally spaced, broadside array when $d=0.4\lambda$ and all its sidelobes are at a level of 25 dB below the main beam.

2.4 Synthesize an optimum, six-element, equally spaced, endfire array with $d=\lambda/4$ and the sidelobe level at -20 dB. What will the directivity and first-null beamwidth be if the ordinary endfire condition is imposed?

2.5 Synthesize a few five-element, equally spaced, braodside arays whose patterns will approximate $J_0^2(v)$ in $0\leqslant v\leqslant 5$. What are the maximum deviations between your results and the desired pattern? If a maximum deviation of the order of 0.01 is permitted, what will the minimum number of elements be in your synthesis?

2.6 If the same pattern as that in Problem 2.5 is synthesized by the Bernstein polynomial of order 6, obtain the final required amplitude excitations and the maximum error involved.

2.7 Synthesize a five-element, nonuniformly but symmetrically spaced, broadside array whose overall length is 2λ by using Haar's theorem if $f(u)=J_1(u)/u$ is the specified field pattern. What is the best position of the inner pair of elements in order to minimize the error between the synthesized and desired patterns?

2.8 If the pattern of a five-element uniform array with $d=\lambda/2$ is used as the starting point, apply the perturbation technique discussed in Section 2.8 to obtain an equally spaced array whose sidelobes are equalized at $-20\,\text{dB}$.

2.9 Consider a five-element, nonuniformly but symmetrically spaced, braodside array whose field pattern may be expressed as

$$E(u)=I_0+2I_1\cos b_1u+2I_2\cos b_2u,$$

where $b_i=2d_i/\lambda$, and $u=\pi\cos\theta$. Determine the optimum set of I_i to maximize the array directivity with respect to $b_1=1.4$ and $b_2=2.2$. If b_1 is allowed to change while b_2 is fixed at 2.2, what value of b_1 and the corresponding I_i will yield the absolute maximum directivity?

2.10 What will your answeres be if $E(u)$ in Problem 2.9 is replaced by a difference pattern $\Delta(u)=I_1\cos \mathrm{b}_1u_s\cos b_1u+I_2\cos b_2u_s\cos b_2u$ with $u_s=16°$? How does u_m change with b_1 when $b_2=2.2$?

REFERENCES

1. **Stone, J. S.** *Binomial array*, United States Patents No. 1,643,323 and No. 1,715,433.

2. **Dolph, C. L.** A current distribution for broadside arrays which optimizes the relationship between beam width and side-lobe level, *Proc. IRE*, Vol. 34, No. 6, pp. 335–348, June, 1946.

3. **Brown, Lawrence B. and Glenn A. Scharp.** Tschebyscheff antenna distribution, beamwidth, and gain tables, *Report No. 4629*, Naval Ordnance Laboratory, Corona, Calif., February, 1958.

4. **Magnus, Wilhelm and Fritz Oberhettinger.** *Formulas and Theorems for the Functions of Mathematical Physics*, Chelsea Publishing Co., New York, 1949, p. 78.

5. **Riblet, H. J.** Discussion of Dolph's paper, *Proc. IRE*, Vol. 35, No. 5, pp. 489–492, May, 1947.

6. **Ma, M. T.** A new mathematical approach for linear array analysis and synthesis, Ph.D dissertation, Syracuse University, Syracuse, N.Y., 1961.

7. **Tai, C. T.** The optimum directivity of uniformly spaced broadside arrays of dipoles, *IEEE Trans. Antennas and Propagation*, Vol. AP-12, No. 4, pp. 447–454, July, 1964.

8. **Pritchard, R. L.** Optimum directivity patterns for linear point arrays, *J. Acoust. Soc. Am.*, Vol. 25, No. 5, pp. 879–891, September, 1953.

9. **Pokrovskii, V. L.** On optimum linear antennas, *Radio Engineering and Electronics* (Russian), Vol. 1, No. 5, pp. 593–600, 1956.

10. **Pokrovskii, V. L.** Optimum linear antennas radiating at a given angle to the axis, *Radio Engineering and Electronics* (Russian), Vol. 2, No. 5, pp. 559–565, 1957.

11. **Rhodes, D. R.** The optimum linear array for a single main beam, *Proc. IRE*, Vol. 41, No. 6, pp. 793–794, June e, 1953.

12. **Uzkov, A. I.** An approach to the problem of optimum directive antenna design, *Compt. Rend. Dokl. Acad. Sci. URSS*, Vol. 3, p. 35, 1946.

13. **DuHamel, R. H.** Optimum patterns for endfire arrays, *Proc. IRE*, Vol. 41, No. 5, pp. 652–659, May, 1953.

14. **Pritchard, R. L.** Discussion of DuHamel's paper, *IRE Trans. Antennas and Propagation*, Vol. AP-3, No. 1, pp. 40–43, January, 1955.

15. **Ma, M. T.** Directivity of uniformly spaced optimum endfire arrays with equal sidelobes, *Radio Sci.*, Vol. 69D, No. 9, pp. 1249–1255, September, 1965.

16. **Ma, M. T., and D. C. Hyovalti.** A table of radiation characteristics for uniformly spaced optimum endfire arrays with equal sidelobes, *NBS Monograph No. 95*, December, 1965. Available from U. S. Government Printing Office, Washington, D. C. 20402.

17. **Jordan, E. C.** *Electromagnetic Waves and Radiating Systems*, Prentice-Hall, Inc., New York, pp. 433–440, 1950.

18. **Achieser, N. L.** *Theory of Approximation*, translated from Russian by C. Hyman, Frederick Ungar Publishing Co., New York, 1956.

19. **Jackson, D.** *The Theory of Approximation*, American Mathematical Society Colloquium Publication, New York, 1930, Vol. 11.

20. **Scarborough, J. B.** *Numerical Mathematical Analysis*, The Johns Hopkins Press, Baltimore, Md., 6th Edition, 1966.

21. **Todd, J.** *On Numerical Approximation*, edited by R. E. Langer, The University of Wisconsin Press, Madison, 1959, p. 424.

22. **Buck, R. C.** Survey of recent Russian literature on approximation, in *On Numerical Approximation*, edited by R. E. Langer, The University of Wisconsin Press, Madison, 1959.

23. **Krylov, V. J.** Convergence of algebraic interpolation with respect to roots of Chebyshev polynomials for absolutely continuous functions and functions of bounded variation, *Dokl. Acad. Sci. URSS*, Vol. 107, pp. 362–365, 1956.

24. **Lanczos, C.** *Applied Analysis*, Prentice-Hall, Inc., Englewood Cliffs, N. J., pp. 229–239, 1956.

25. **Ma, M. T.** Application of Bernstein Polynomials and interpolation theory to linear array synthesis, *IEEE Trans. Antennas and Propagation*, Vol. AP-12, No. 6, pp. 668–677, November, 1964.

26. **Hansen, R. C.** Aperture theory, Chap. 1 in *Microwave Scanning Antennas*, Vol. 1, edited by R. C. Hansen, Academic Press, New York, 1964.

27. **Silver, S.** Aperture illumination and antenna patterns, Chap. 6 in *Microwave Antenna Theory and Design*, edited by S. Silver, McGraw-Hill Book Co., New York, 1949.

28. **Lorentz, G.** *Bernstein Polynomials*, Toronto University Press, Toronto, Canada, 1953.

29. **Ma, M. T.** An application of the inverse Z-transform theory to the synthesis of linear antenna arrays, *IEEE Trans. Antennas and Propagation,* Vol. AP-12, No. 6, p. 798, November, 1964.

30. **LePage, W. R.** *Complex Variables and the Laplace Transform for Engineers*, McGraw-Hill Book Co., New York, Chap. 5, 1961.

31. **Jury, E. I.** *Sampled-Data Control Systems*, John Wiley and Sons, Inc., New York, 1958.

32. **Tang, D. T.** The Tchebysheff approximation of a prescribed impulse response with RC network realization, *IRE International Convention Record*, part 4, pp. 214–220, March, 1961.

33. **Ma, M. T.** Another method of synthesizing nonuniformly spaced antenna arrays, *IEEE Trans. Antennas and Propagation,* Vol. AP-13, No. 5, pp. 833–834, September, 1965.

34. **Ma, M. T. and L. C. Walters.** Optimum nonuniformly spaced antenna arrays, *Final Report, AFCRL-65-830*, Contract No. PRO-S65-534, November 30, 1965.

35. **Ma, M. T.** Note on nonuniformly spaced arrays, *IEEE Trans. Antennas and Propagation*, Vol. AP-11, No. 4, pp. 508–509, July, 1963.

36. **Harrington, R. F.** Sidelobe reduction by nonuniform element spacing, *IEEE Trans. Antennas and Propagation*, Vol. AP-9, No. 2, pp. 187–192, March, 1961.

37. **Cheng, D. K. and F. I. Tseng.** Gain optimization for arbitrary antenna arrays, *IEEE Trans. Antennas and Propagation*, Vol. AP-13, No. 6, pp. 973–974, November, 1965.

38. **Gantmacher, F. R.** *The Theory of Matrices*, Vol. 1, Chap. 10, Chelsea Publishing Co., New York (Translated by K. A. Hirsch), 1959.

39. **Hildebrand, F. B.** *Methods of Applied Mathematics*, Prentice-Hall, Inc., Englewood Cliffs, N.J., 1952, p. 46.

40. **Pang, C. C.** Analysis and synthesis of monopulse phased arrays, Ph.D thesis, University of Colorado, Boulder, 1967.

41. **Butler, J. K. and H. Unz.** Optimization of beam efficiency and synthesis of nonuniformly spaced arrays, *Proc. IEEE*, Vol. 54, No. 12, pp. 2007–2008, December, 1966.

42. **Butler, J. K. and H. Unz.** Beam efficiency and gain optimization of antenna arrays with nonuniform spacings, *Radio Sci.*, Vol. 2 (new series), No. 7, pp. 711–720, July, 1967.

43. **Guillemin, E. A.** *Synthesis of Passive Networks*, John Wiley and Sons, Inc., New York, 1957.

44. **Tuttle, D. F.** *Network Synthesis*, John Wiley and Sons, Inc., New York, 1958, Vol. 1.

45. **Brulé, J. D.** Improving the approximation to a prescribed time response, *IRE Trans. Circuit Theory*, Vol. CT-6, No. 4, pp. 355–361, December, 1959.

46. **Kantz, W. H.** Transient synthesis in the time domain, *IRE Trans. Circuit Theory*, Vol. CT-1, No. 3, pp. 29–39, September, 1954.

47. **Darlington, S.** Network synthesis using Chebyshev polynomial series, *BSTJ*, Vol. 31, No. 7, pp. 613–665, July, 1952.

48. **Linvill, J. G.** The approximation with rational function of prescribed magnitude and phase characteristics, *Proc. IRE*, Vol. 40, No. 6, pp. 711–721, June, 1952.

49. **Walsh, J. L.** On interpolation and approximation by rational functions with preassigned poles, *Trans. Am. Math. Soc.*, Vol. 34, pp. 22–74, 1932.

50. **Walsh, J. L.** On approximation to an analytic function by rational functions of best approximations, *Math. Zeit.*, Vol. 38, pp. 163–176, 1934.

51. **Ma, M. T.** Analogies between theories of antenna arrays and passive networks, *IEEE International Convention Record*, part 5, pp. 150–154, March, 1965.

52. **Ma, M. T. and L. C. Walters.** Mathematical aspects of the theory of antenna arrays: viewed from circuit concepts, Tenth Midwest Symposium on Circuit Theory, Paper No. V-5, May 18–19, 1967.

53. **Taylor, T. T.** Design of line source antennas for narrow beamwidth and low sidelobes, *IRE Trans. Antennas and Propagation,* Vol. AP-3, No. 1, pp. 16–28, January, 1955.

54. **Rhodes, D. R.** The optimum line source for the best mean-square approximation to a given radiation pattern, *IEEE Trans. Antennas and Propagation,* Vol. AP-11, No. 4, pp. 440–446, July, 1963.

55. **Woodward, P. M. and J. D. Lawson.** The theoretical precision with which an arbitrary radiation pattern may be obtained from a source of finite size, *Proc. IEE* (London), Vol. 95, part IIIA, pp. 363–369, 1948.

56. **Schell, A. C. and A. Ishimaru.** Antenna pattern synthesis, Chap. 7 in *Antenna Theory*, Part I, edited by R. E. Collin and F. J. Zucker, McGraw-Hill Book Co., New York, 1969.

57. **Wolff, E. A.** Linear antenna array synthesis, Ph.D. Dissertation, University of Maryland, College Park, 1961.

58. **Ishimaru, A.** Theory of unequally spaced arrays, *IRE Trans. Antennas and Propagation*, Vol. AP-10, No. 6, pp. 691–702, November, 1962.

59. **Ishimaru, A. and Y. S. Chen.** Thinning and broadbanding antenna arrays by unequal spacings, *IEEE Trans. Antennas and Propagation,* Vol. AP-13, No. 1, pp. 34–42, January, 1965.

60. **Chow, Y. L.** On grating plateaux of nonuniformly spaced arrays, *IEEE Trans. Antennas and Propagation*, Vol. AP-13, No. 2, pp. 208–215, March, 1965.

61. **Bruce, J. D. and H. Unz.** Mechanical quadratures to synthesize nonuniformly spaced antenna arrays, *Proc. IRE*, Vol. 50, No. 10, p. 2128, October, 1962.

62. **Unz, H.** Nonuniformly spaced arrays: the orthogonal method, *Proc. IEEE*, Vol. 54, No. 1, pp. 53–54, January, 1966.

63. **Unz, H.** Nonuniformly spaced arrays: the eigenvalues method, *Proc. IEEE*, Vol. 54, No. 4, pp. 676–678, April, 1966.

64. **Sletten, C. J., P. Blacksmith, and G. Forbes.** New method of antenna array synthesis applied to generation of double-step patterns, *IRE Trans. Antennas and Propagation*, Vol. AP-5, No. 4, pp. 369–373, October, 1957.

65. **Taylor, T. T. and J. R. Whinnery.** Applications of potential theory to the design of linear arrays, *J. Appl. Phys.*, Vol. 22, No. 1, pp. 19–29, January, 1951.

66. **Morison, J. E. and A. C. Schell.** A technique for power pattern synthesis, *PTGAP International Symposium Digest*, pp. 160–163, July, 1963.

67. **Bates, R. H. T.** Mode theory approach to arrays, *IEEE Trans. Antennas and Propagation*, Vol. AP-13, No. 2, p. 321, March, 1965.

68. **Ksienski, A.** Maximally flat and quasi-smooth sector beams, *IRE Trans. Antennas and Propagation*, Vol. AP-8, No. 5, pp. 476–484, September, 1960.

69. **Bruce, J. D. and H. Unz.** Broadband nonuniformly spaced arrays, *Proc. IRE*, Vol. 50, No. 2, pp. 228–229, February, 1962.

ADDITIONAL REFERENCES

Barbiere, D. A method for calculating the current distribution of Tschebyscheff arrays, *Proc. IRE*, Vol. 40, No. 1, pp. 78–82, January, 1952.

Bracewell, R. N. Tolerance theory of large antennas, *IRE Trans. Antennas and Propagation*, Vol. AP-9, No. 1, pp. 49–58, January, 1961.

Bricout, P. A. Pattern synthesis using weighted functions, *IRE Trans. Antennas and Propagation*, Vol. AP-8, No. 4, pp. 441–444, July, 1960.

Brown, F. W. Note on nonuniformly spaced arrays, *IRE Trans. Antennas and Propagation*, Vol. AP-10, No. 5, pp. 639–640, September, 1962.

Brown, J. L., Jr. A simplified derivation of the Fourier coefficients for Chebyshev patterns, *Proc. IEE* (London), Vol. 105C, pp. 167–168, March, 1958.

Brown, J. L., Jr. On the determination of excitation coefficients for a Tchebycheff pattern, *IRE Trans. Antennas and Propagation*, Vol. AP-10, No. 2, pp. 215–216, March, 1962.

Bloch, A., R. G. Medhurst, and S. D. Pool. A new approach to the design of superdirective aerial arrays, *Proc. IEE* (London), Vol. 100, part 3, pp. 303–314, September, 1953.

Bloch, A., R. G. Medhurst, S. D. Pool, and W. E. Kock. Superdirectivity, *Proc. IRE*, Vol. 48, No. 6, p. 1164, June, 1960.

Cheng, D. K. and B. J. Strait. An unusually simple method for sidelobe reduction, *IEEE Trans. Antennas and Propagation*, Vol. AP-11, No. 3, pp. 375–376, May, 1963.

Cheng, D. K. and F. I. Tseng. Signal-to-noise ratio maximization for receiving arrays, *IEEE Trans. Antennas and Propagation*, Vol. AP-14, No. 6, p. 792, November, 1966.

Chow, Y. L. On the error involved in Poisson's sum formulation of nonuniformly spaced antenna arrays, *IEEE Trans. Antennas and Propagation*, Vol. AP-14, No. 1, p. 101, January, 1966.

Clarke, J. Steering of zeros in the directional pattern of a linear array, *IEEE Trans. Antennas and Propagation*, Vol. AP-16, No. 2, p. 267, March, 1968.

Collin, R. E. and S. Rothschild. Evaluation of antenna *Q*, *IEEE Trans. Antennas and Propagation*, Vol. AP-12, No. 1, pp. 23–27, January, 1964.

Drane, C. J. Derivation of excitation coefficients for Chebyshev arrays, *Proc. IEE* (London), Vol. 110, No. 10, pp. 1755–1758, October, 1963.

Drane, C. J. Dolph-Chebyshev excitation coefficient approximation, *IEEE Trans. Antennas and Propagation*, Vol. AP-12, No. 6, pp. 781–782, November, 1964.

Elliott, R. S. An approximation to Chebyshev distributions, *IEEE Trans. Antennas and Propagation*, Vol. AP-11, No. 6, p. 707, November, 1963.

Galejs, J. Minimization of sidelobes in space tapered linear arrays, *IEEE Trans. Antennas and Propagation*, Vol. AP-12, No. 4, p. 497, July, 1964.

Gilbert, E. N. and S. P. Morgan. Optimum design of directive antenna arrays subject to random variations, *BSTJ*, Vol. 34, No. 5, pp. 637–663, May, 1955.

Goward, F. K. An improvement in endfire arrays, *J. IEE* (London), Vol. 94, part 3, pp. 415–418, November, 1947.

Hannan, P. W. Maximum gain in monopulse difference mode, *IRE Trans. Antennas and Propagation*, Vol. AP-9, No. 3, pp. 314–315, May, 1961.

Hansen, R. C. Gain limitations of large antennas, *IRE Trans. Antennas and Propagation*, Vol. AP-8, No. 5, pp. 490–495, September, 1960.

Hansen, R. C. Gain limitations of large antennas, *IEEE Trans. Antennas and Propagation*, Vol. AP-13, No. 6, pp. 997–998, November, 1965.

Howard, J. E. Statistical patterns of a general array, *IEEE Trans. Antennas and Propagation*, Vol. AP-15, No. 1, pp. 60–65, January, 1967.

Jaeckle, W. G. Antenna synthesis by weighted Fourier coefficients, *IEEE Trans. Antennas and Propagation*, Vol. AP-12, No. 3, pp. 369–370, May, 1964.

Jagermann, D. Cosine sum approximation and synthesis of array antennas, *BSTJ*, Vol. 44, No. 10, pp. 1761–1777, October, 1965.

Kritikos, H. N. Optimal signal-to-noise ratio for linear arrays by the Schwartz inequality, *J. Franklin Inst.*, Vol. 276, No. 10, pp. 295–304, October, 1963.

Ksienski, A. Equivalence between continuous and discrete radiating systems, *Can. J. Phys.*, Vol. 39, No. 2, pp. 335–349, February, 1961.

Lo, Y, T. A spacing weighted antenna array, *IRE International Convention Record*, part 1, pp. 191–195, March, 1962.

Lo, Y. T., S. W. Lee, and Q. H. Lee. Optimization of directivity and signal-to-noise ratio of an arbitrary antenna array, *Proc. IEEE*, Vol. 54, No. 8, pp. 1033–1045, August, 1966.

Ma, M. T. and D. K. Cheng. A critical study of linear arrays with equal sidelobes, *IRE International Convention Record*, part 1, pp. 110–122, March, 1961.

Minnett, H. C. and B. MacA. Thomas. A method of synthesizing radiation patterns with axial symmetry, *IEEE Trans. Antennas and Propagation*, Vol. AP-14, No. 5, pp. 654–656, September, 1966.

Perini, J. Sidelobe reduction by beam shifting, *IEEE Trans. Antennas and Propagation*, Vol. AP-12, No. 6, pp. 791–792, November, 1964.

Powers, E. J. Utilization of the lambda functions in the analysis and synthesis of monopulse antenna difference patterns, *IEEE Trans. Antennas and Propagation*, Vol. AP-15, No. 6, pp. 771–777, November, 1967.

Reuss, M. L. Some design considerations concerning linear arrays having Dolph-Tchebysheff amplitude distribution, US Naval Res. Lab. Report No. 5240, February, 1959.

Rondinelli, L. A. Effects of random errors on the performance of antenna arrays of many elements, *IRE National Convention Record*, part 1, pp. 174–187, March, 1959.

Salzer, H. E. Note on the Fourier coefficients for Chebyshev patterns, *Proc. IEE* (London), Vol. 103C, pp. 286–288, February, 1956.

Schuman, H. K. and B. J. Strait. On the design of unequally spaced arrays with nearly equal sidelobes, *IEEE Trans. Antennas and Propagatio*, Vol. AP-16, No. 4, pp. 493–494, July, 1968.

Sharpe, C. B. and R. B. Crane. Optimization of linear arrays for broadband signals, *IEEE Trans. Antennas and Propagation*, Vol. AP-14, No. 4, pp. 422–427, July, 1966.

Shubert, H. A., J. K. Butler, and H. Unz. Comments on optimization of beam efficiency and synthesis of nonuniformly spaced arrays, *Proc. IEEE*, Vol. 55, No. 7, pp. 1205–1206, July, 1967.

Simon, J. C. Application of periodic functions approximation to antenna pattern synthesis and circuit theory, *IRE Trans. Antennas and Propagation*, Vol. AP-4, No. 3, pp. 429–440, July, 1956.

Skolnik, M. I., G. Neinhauser, and J. W. Sherman. Dynamic programming applied to unequally spaced arrays, *IEEE Trans. Antennas and Propagation*, Vol. AP-12, No. 1, pp. 35–43, January, 1964.

Solymar, L. Maximum gain of a line source antenna if the distribution function is a finite Fourier series, *IRE Trans. Antennas and Propagation*, Vol. AP-6, No. 3, p. 215, July, 1958.

Stearns, C. O. Computed performance of moderate size super-gain antennas, *IEEE Trans. Antennas and Propagation,* Vol. AP-14, No. 2, pp. 241–242, March, 1966.

Stegen, R. J. Gain of Tchebysheff arrays, *IRE Trans. Antennas and Propagation*, Vol. AP-8, No. 6, pp. 629–631, November, 1960.

Stegen, R. J. Excitation coefficients and beamwidth of Tchebyscheff arrays, *Proc. IRE*, Vol. 41, No. 11, pp. 1671–1674, November, 1953.

Strait, B. J. Antenna arrays with partially tapered amplitudes, *IEEE Trans. Antennas and Propagation*, Vol. AP-15, No. 5, pp. 611–617, September, 1967.

Tang, C. H. Approximate method of designing nonuniformly spaced arrays, *IEEE Trans. Antennas and Propagation*, Vol. AP-13, No. 1, p. 177, January, 1965.

Tang, C. H. Design method for nonuniformly spaced arrays, *IEEE Trans. Antennas and Propagation*, Vol. AP-13, No. 4, p. 642, July, 1965.

Tang, C. H. On the optimum performance of nonuniformly spaced arrays, *IEEE Trans. Antennas and Propagation*, Vol. AP-14, No. 5, p. 651, September, 1966.

Taylor, T. T. A discussion of the maximum directivity of an antenna, *Proc. IRE*, Vol. 36, No. 9, p. 1135, September, 1948.

Toraldo di Francia, G. Directivity, super-gain, and information, *IRE Trans. Antennas and Propagation*, Vol. AP-4, No. 3, pp. 473–478, July, 1956.

Tseng, F. I. and D. K. Cheng. Gain optimization for arbitrary antenna arrays subject to random fluctuations, *IEEE Trans. Antennas and Propagation*, Vol. AP-15, No. 3, pp. 356–366, May, 1967.

Unz, H. Linear arrays with arbitrarily distributed elements, *IRE Trans. Antennas and Propagation*, Vol. AP-8, No. 2, pp. 222–223, March, 1960.

Unz, H. Nonuniform arrays with spacings larger than one wavelength, *IRE Trans. Antennas and Propagation*, Vol. AP-10, No. 5, pp. 647–648, September, 1962.

Van der Maas, G. J. A simplified calculation for Dolph-Tchebycheff arrays, *J. Appl. Phys.*, Vol. 25, No. 1, pp. 121–124, January, 1954.

Van der Maas, G. J. and H. Gruenberg. Note on a simplified calculation for Dolph-Tchebycheff arrays, *J. Appl. Phys.*, Vol. 27, No. 8, pp. 962–963, August, 1956.

Willey, R. E. Space tapering of linear and planar arrays, *IRE Trans. Antennas and Propagation*, Vol. AP-10, No. 4, pp. 369–377, July, 1962.

Wilmotte, R. M. Note on practical limitations in the directivity of antennas, *Proc. IRE*, Vol. 36, No. 7, p. 878, July, 1948.

Wolff, I. Determination of the radiating system which will produce a specified directional characteristics, *Proc. IRE*, Vol. 25, No. 5, pp. 630–643, May, 1937.

Yaru, N. A note on supergain antenna arrays, *Proc. IRE*, Vol. 39, No. 9, pp. 1081 – 1085, September, 1951.

Special issue on "Theory of Antenna Arrays," *Radio Sci.*, Vol. 3, No. 5, May, 1968.

CHAPTER 3

ANALYSIS AND SYNTHESIS OF TWO-DIMENSIONAL ARRAYS

In the first two chapters, we analyzed and synthesized linear arrays with discrete elements from different viewpoints. Now we are ready to do the same for two-dimensional arrays. *Rectangular arrays* with constant or variable spacings will be studied first. This is followed by the *ring* (*circular*) and *elliptical arrays*. The same techniques employed previously for the synthesis of linear arrays will also be applied in this chapter.

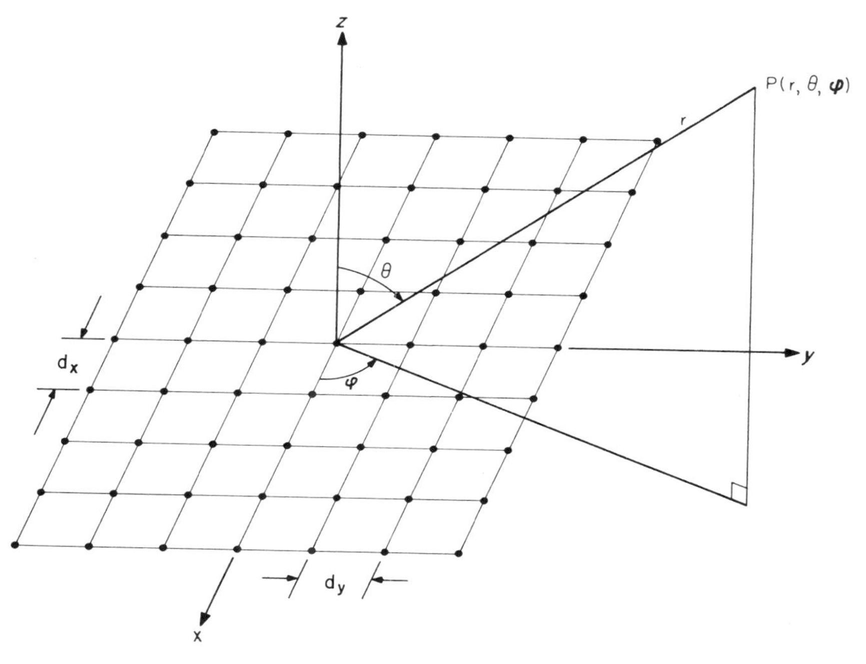

Fig. 3.1 A rectangular array.

3.1 Rectangular Arrays

Consider the rectangular array of discrete elements situated in the xy plane as shown in Fig. 3.1, where the central element is chosen as the coordinate origin. If we have $2N_x+1$ elements in each column parallel to the x axis with an equal spacing d_x and $2N_y+1$ similarly oriented elements in each row parallel to the y axis with a common spacing d_y, the entire array will have $n_x \times n_y$ elements, where $n_x = 2N_x+1$ and $n_y = 2N_y+1$. The array will become a square array when $N_x = N_y$ and $d_x = d_y$.

The field (at a distant point P in free space) contributed by the column coincident with the x axis is then, according to (1.1),

$$E_0 = f(\theta,\varphi) \sum_{m=-N_x}^{N_x} I_{mo} \exp[j(mkd_x \sin\theta \cos\varphi + \alpha_x)], \tag{3.1}$$

where $f(\theta,\varphi)$ is the element pattern function, I_{mo} is the amplitude excitation of the mth element (counting from the origin) in the column $y=0$, α_x is the associated phase excitation, and $k = 2\pi/\lambda$.

Note that the array factor in (3.1) is a function of both θ and φ. Only the pattern in the xz plane ($\varphi = 0$) has an expression comparable to that for linear arrays considered before. In general, the pattern $|E_0|$ consists of a conical main beam and sidelobes.

When amplitude excitations for elements in other columns are proportional to those for corresponding elements on the x axis,

$$I_{mn} = I_{mo} I_{on}, \tag{3.2}$$

we can sum up the total contribution from the array as

$$\begin{aligned} E(\theta,\varphi) &= \sum_{n=-N_y}^{N_y} E_n \\ &= f(\theta,\varphi) \sum_{m=-N_x}^{N_x} \sum_{n=-N_y}^{N_y} I_{mn} \exp[j(mkd_x \sin\theta\cos\varphi + \alpha_x)] \\ &\quad \times \exp[j(nkd_y \sin\theta \sin\varphi + \alpha_y)] \\ &= f(\theta,\varphi) S_x S_y, \end{aligned} \tag{3.3}$$

where

$$S_x = \sum_{m=-N_x}^{N_x} I_{mo} \exp\left[j(mkd_x \sin\theta \cos\varphi + \alpha_x) \right]$$

$$S_y = \sum_{n=-N_y}^{N_y} I_{on} \exp\left[j(nkd_y \sin\theta \sin\varphi + \alpha_y) \right]. \tag{3.4}$$

It is clear that I_{mn} and $\alpha_x + \alpha_y$ may, respectively, be considered the total amplitude and phase excitations of the (m,n)th element in the array.

It is also clear from (3.3) that, under the stated conditions, the pattern of the rectangular array is, besides the element pattern, the product of array factors of two linear arrays, one along the x axis and the other along the y axis. As such, the requirement on the spacings d_x and d_y should remain the same as that for linear arrays (discussed in Chapter 1) in order to avoid the grating lobes.

Although α_x and α_y can be arbitrarily adjusted, in principle, so that the position of the main beam of S_x is not the same as that for S_y in (3.4), or the two conical beams do not "intersect," the practical application does demand that the two beams intersect. This indeed is the principal idea behind the utilization of the rectangular or planar arrays. Assuming that the main beams do point at the same position (θ_0, φ_0) and that the elements are progressively phased, we can determine the required α_x and α_y as follows:

$$\alpha_x = -mkd_x \sin\theta_0 \cos\varphi_0, \tag{3.5}$$

$$\alpha_y = -nkd_y \sin\theta_0 \sin\varphi_0. \tag{3.6}$$

Since $\sin\theta = \sin(\pi - \theta)$, we see from (3.4) that both S_x and S_y are generally *bidirectional* (except of course the endfire array where $\theta_0 = \pi/2$) in any vertical plane given by $\varphi = \text{constant}$. This represents two *pencil beams*, one each above and below the array plane. The one below the array plane can usually be eliminated by a proper choice of directive element pattern function $f(\theta, \varphi)$ or be reflected by the use of a ground plane.

When I_{mo}, I_{on}, N_x, N_y, d_x, d_y, θ_0, and φ_0 are all specified, the characteristics such as the beamwidth, sidelobe levels, and positions can be analyzed in the same manner as that for linear arrays. The results can be presented as a function of θ and φ, or more conveniently as a function of u and v, where

$$u = \sin\theta\cos\varphi, \qquad v = \sin\theta\sin\varphi. \tag{3.7}$$

The directivity of the rectangular array can also be defined in similar fashion:

$$D = \frac{4\pi|E(\theta_0,\varphi_0)|^2}{\int_0^{2\pi}\int_0^{\pi}|E(\theta,\varphi)|^2\sin\theta\, d\theta\, d\varphi}, \tag{3.8}$$

which can readily be calculated once the parameters are known. Therefore, as far as the analysis of a rectangular array is concerned, the task is not much different from that for linear arrays.

For the work of synthesis, the techniques employed before may also be useful here. Once a desired pattern $F(\theta,\varphi)$ is specified, it can be converted into $g(u, v)$ through (3.7). If the desired pattern happens to be separable into two components such as

$$g(u,v) = g_1(u)g_2(v), \tag{3.9}$$

the synthesis methods described in Sections 2.4 through 2.8 can be directly applied to devise $L_1(u)$ and $L_2(v)$, approximating, respectively, $g_1(u)$ and $g_2(v)$. The product, $L(u, v) = L_1(u)L_2(v)$, is then considered as the final synthesized pattern with the error to be estimated by

$$\epsilon(u,v) = g(u,v) - L(u,v), \tag{3.10}$$

or

$$\overline{\epsilon^2} = \int_0^{2\pi}\int_0^{\pi}|g(u,v) - L(u,v)|^2 J\left(\frac{u,v}{\theta,\varphi}\right) d\theta\, d\varphi, \tag{3.11}$$

where

$$J\left(\frac{u,v}{\theta,\varphi}\right) = \begin{vmatrix} \dfrac{\partial u}{\partial\theta} & \dfrac{\partial u}{\partial\varphi} \\ \dfrac{\partial v}{\partial\theta} & \dfrac{\partial v}{\partial\varphi} \end{vmatrix} = \sin\theta\cos\theta.$$

When the desired pattern $g(u,v)$ cannot be separated into two parts such as that in (3.9), the typical method of synthesis is then through a *double Fourier series expansion*:[1,2]

$$g(u,v) \sim \sum_{m=-\infty}^{\infty} \sum_{n=-\infty}^{\infty} a_{mn} \exp[j(mu+nv)\pi]. \tag{3.12}$$

The success of this approach, of course, is limited to the case when $d_x = d_y = \lambda/2$ as we noted in the beginning of Section 2.4, and also depends on whether the double integral,

$$\int\int g(u,v) \exp[-j(mu+nv) \;\; du\,dv,] \tag{3.13}$$

for calculating the Fourier coefficients a_{mn} can be carried out easily or not.

If the elements are not equally but symmetrically spaced, the formulation given above still holds if minor modifications are made in (3.4). First, md_x and nd_y should be respectively replaced by d_m and d_n, with d_m denoting the distance in wavelengths between the mth pair of elements along the x axis and the coordinate origin and d_n denoting the counterpart along the y axis. Second, the summation indices m and n should run, respectively, from $-M$ to M and $-N$ to N, with $(2d_M)\times(2d_N)$ as the entire array size.

Arrays with even number of elements on the rows and columns can be similarly formulated.

3.2 Analysis of Ring Arrays

When an angular symmetry is desired in a two-dimensional operation, ring or circular arrays can be considered to satisfy the requirement. *Ring arrays* have been used in radio direction finding, radar, sonar, and many other system applications.[3,4] Early significant contributions on this subject were made by DuHamel,[5] who synthesized a single-ring array, with or without a concentric cylindrical reflector, to produce a Dolph-Chebyshev type of pattern. Extending the theory developed primarily for nonuniformly spaced linear arrays,[6] Tighe[7] was successful in synthesizing concentric ring arrays yielding a Taylor's type of pattern.[8] The directivity of such an array was not discussed. Tillman[9] and his associates treated the same subject by using a "symmetric component" technique, and obtained the results in terms of sequential currents. Real currents required to excite the elements are then determined from combinations of sequential currents. Their final results demand different current magnitudes on each ring and undergo phase changes from ring to ring. Using a still different approach of Fourier-Bessel expansion as derived originally by LePage, Roys, and Seely,[10] Stearns and Stewart[11] investigated ring arrays primarily for obtaining azimuthal patterns with low sidelobes. Royer[12] and Chu[13] showed that the ring arrays can be used to achieve a nearly omnidirectional pattern. Antennas with circular apertures were studied by Taylor.[8] Extensive numerical results, based on Taylor's formulation, were obtained by Hansen.[14] The problem of corner reflector treated by Wait[15] is also related to the present subject.

In this section we are analyzing not only the horizontal (azimuthal) pattern which has been important in many applications, but also the vertical pattern (beam maximum pointing toward the array normal) in view of its latest application for modifying the ionosphere.[16] When the array is placed above a ground system, the main beam in the elevational pattern can be designed to point at about 10°–20° from the ground for long-distance communication or direction finding. This elevational pattern in the presence of a ground plane will be considered in Chapters 4–6.

Consider a single-ring array of radius a as shown in Fig. 3.2. The ring is in the xy plane. There are N isotropic elements on the circumference of the ring. The free-space far-field pattern function of such an array may be obtained by summing the contribution from each element at some distant point:

$$E(\theta,\varphi)=\sum_{n=1}^{N} I_n \exp\left[jka\sin\theta\cos(\varphi-\varphi_n)+j\alpha_n\right], \tag{3.14}$$

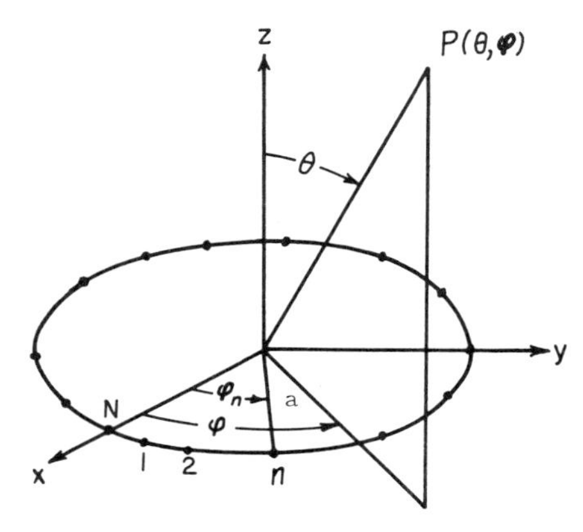

Fig. 3.2 A ring array.

where I_n is the current excitation of the nth element located at $\varphi=\varphi_n$ and α_n is the associated phase excitation (relative to the array center at which an element may or may not be present). For the conventional *cophasal excitation*,

$$\alpha_n = -ka\sin\theta_0\cos(\varphi_0-\varphi_n), \tag{3.15}$$

where (θ_0, φ_0) is the designed position of the beam maximum. By defining a new set of variables ρ and ξ such that

$$\rho = a\Big[(\sin\theta\cos\varphi-\sin\theta_0\cos\varphi_0)^2 + (\sin\theta\sin\varphi-\sin\theta_0\sin\varphi_0)^2\Big]^{1/2},$$

$$\cos\xi = \frac{\sin\theta\cos\varphi - \sin\theta_0\cos\varphi_0}{\Big[(\sin\theta\cos\varphi-\sin\theta_0\cos\varphi_0)^2 + (\sin\theta\sin\varphi-\sin\theta_0\sin\varphi_0)^2\Big]^{1/2}}, \tag{3.16}$$

we can rewrite (3.14) after using (3.15) in a more compact form:

$$E = \sum_{n=1}^{N} I_n \exp[jk\ \cos(\xi-\varphi_n)]. \tag{3.17}$$

Although (3.17) can be used to calculate the pattern as a function of θ and φ when a, N, I_n, φ_n, θ_0, and φ_0 are given, it can be very time consuming even when N is only moderately large.

If the elements are uniformly excited and equally spaced along the circumference in order to be practical and achieve angular symmetry, $I_n=I$, $\varphi_n=2\pi n/N$. Equation (3.17) can then be expanded as [2,17]

$$E = NI\sum_{m=-\infty}^{\infty} \exp[jmN(\pi/2-\xi)]J_{mN}(k\rho), \tag{3.18}$$

where mN means the product of the running index m and the total number of elements N, the term with the zeroth-order Bessel function $J_0(k\rho)$ is called the principal term, and the rest are residuals.

For the *horizontal* (azimuthal) cophasal *pattern* which lies in the array plane, $\theta=\theta_0=\pi/2$. If the beam maximum is designed to point, say, in the x direction, $\varphi_0=0$. Then, we have

$$\alpha_n=\alpha_{nh}=-ka\cos\frac{2\pi n}{N}, \tag{3.19}$$

$$\rho=\rho_h=2a\sin\frac{\varphi}{2}, \qquad 0\leqslant\varphi\leqslant 2\pi,$$

$$\cos\xi=\cos\xi_h=-\sin\frac{\varphi}{2}, \tag{3.20}$$

or

$$\xi_h=\frac{\pi+\varphi}{2}, \tag{3.21}$$

$$E=E_h=I_t\sum_{m=-\infty}^{\infty}\exp\left(\frac{-jmN\varphi}{2}\right)J_{mN}\left(2ka\sin\frac{\varphi}{2}\right), \tag{3.22}$$

where $I_t=NI$ is the total current on the ring.

When considering the *vertical pattern* with the beam maximum pointing toward the z direction ($\theta_0=0$), we have

$$\alpha_n=\alpha_{nv}=0 \qquad \text{(in phase)}, \tag{3.23}$$

$$\rho=\rho_v=a\sin\theta, \tag{3.24}$$

$$\cos\xi=\cos\xi_v=\cos\varphi, \qquad \text{or} \qquad \xi_v=\varphi, \tag{3.25}$$

$$E=E_v=I_t\sum_{m=-\infty}^{\infty}\exp\left[jmN\left(\frac{\pi}{2}-\varphi\right)\right]J_{mN}(ka\sin\theta). \tag{3.26}$$

Here again the situation concerning the range of variation similar to linear arrays arises. Mathematically, J_{mN} is a well-defined function in the entire range of its argument. Physically, the visible ranges for E_h and E_v are, respectively, restricted to $0\leqslant\rho_h\leqslant 2a$ and $0\leqslant\rho_v\leqslant a$. Furthermore, since the value of a Bessel function of large order is very small in its visible range, both E_h in (3.22) and E_v in (3.26), for a large N, reduce approximately to the same mathematical form:

$$E\cong I_tJ_0(k\rho), \tag{3.27}$$

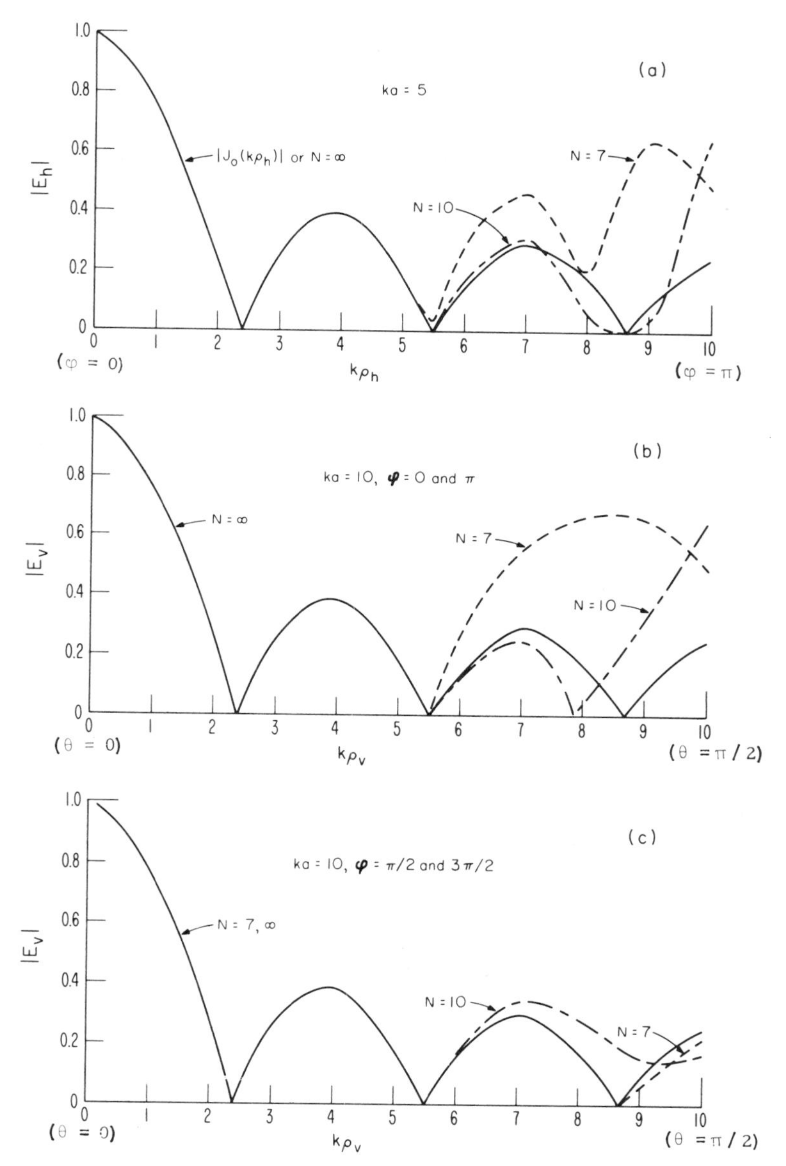

Fig. 3.3 Radiation patterns for a uniform-cophasal ring array: (*a*) horizontal, $\theta=\theta_0=\pi/2, \varphi_0=0$; (*b*) vertical, $\theta_0=0$, $\varphi=0$ and π; (*c*) vertical, $\theta_0=0$, $\varphi=\pi/2$ and $3\pi/2$.

with the understanding that $\rho=\rho_h$ when $E=E_h$, and that $\rho=\rho_v$ when $E=E_v$. The expression (3.27) is exact when N approaches infinity, which represents a continuous distribution on the ring circumference. The accuracy of the approximation for a finite N depends, of course, on N and ka. For a given ka, a reasonable N can always be determined from a table of Bessel functions for a good approximation. Figure 3.3 displays some results in this respect. Actually, the approximation between (3.27) and the exact formulation can be measured by the residual terms:

$$\begin{aligned} R_h &= E_h - I_t J_0(k\rho_h) \\ &= \left[\exp\left(\frac{-jN\varphi}{2}\right) + (-1)^N \exp\left(\frac{jN\varphi}{2}\right) \right] J_N(k\rho_h) \\ &\quad + 2\cos N\varphi J_{2N}(k\rho_h) + \cdots \end{aligned} \tag{3.28}$$

or

$$\begin{aligned} R_v &= E_v - I_t J_0(k\rho_v) \\ &= \left\{ \exp\left[jN\left(\frac{\pi}{2} - \varphi\right)\right] + (-1)^N \exp\left[-jN\left(\frac{\pi}{2} - \varphi\right)\right] \right\} J_N(k\rho_v) \\ &\quad + 2\cos(N\pi - 2N\varphi) J_{2N}(k\rho_v) + \cdots, \end{aligned} \tag{3.29}$$

where the relation $J_{-N}(x)=(-1)^N J_N(x)$ has been used.

In general, the accuracy mentioned above can be improved steadily by making N larger, as can be seen from Figs. 3.3(a) and (b) for E_h and E_v when $\varphi=0$ and π. There is, however, an exception for E_v when $\varphi=\pi/2$ and $3\pi/2$, as is clear from Fig. 3.3(c) where the approximation with $N=7$ is better than that with $N=10$. In fact, in the case when $\varphi=\pi/2$ and $3\pi/2$ (yz plane), an odd N always gives a better approximation for E_v than the even number $N+1$. This can be explained by (3.29). For example, the dominant term in R_v for $N=7$ is $2J_{14}(k\rho_v)$, while that for $N=8$ is $2J_8(k\rho_v)$ [which should have more influence than $2J_{14}(k\rho_v)$ in the range of interest]. In any case, the worst discrepancy between the exact and approximate patterns occurs in the sidelobe region. As far as the main-lobe region is concerned, (3.27) indeed is a simple and good approximation.

When N is large enough to make (3.27) valid so that patterns in both planes can be described as a part of $J_0(k\rho)$, there is no possible presence of a grating lobe with an increasing ka, such as that for a linear array with an increasing kd. This is true because the sidelobe levels of $J_0(k\rho)$ decrease monotonically as ka increases. On the other hand, because the first two sidelobes at $k\rho=3.8$ and 7.0 are relatively high with respective levels of -7.9 and -10.5 dB only, it is clear that the use of a single-ring array is rather limited. If, however, another element is put at the center of the ring with an excitation I_0, the patterns may approximately be expressed as

$$E_{h,v} \cong I_0 + I_t J_0(k\rho_{h,v}). \tag{3.30}$$

Choosing I_0 and I_t with the same sign and adjusting the magnitude of I_0 relative to that of I_t, we are able to reduce the level of the first sidelobe at the expense of the second for this single-ring-plus-one-at-the-center array. If this arrangement is still not enough for some particular applications, we can then use concentric rings where we will have an additional access to control the radii and amplitudes for various rings to produce a favorable result. Details of this from the synthesis viewpoint will be given in Section 3.6.

The directivity of a single-ring array with isotropic elements can be expressed as

$$D = \frac{4\pi |E_{\max}|^2}{\int_0^{2\pi}\int_0^{\pi} |E(\theta,\varphi)|^2 \sin\theta \, d\theta \, d\varphi}, \tag{3.31}$$

which is exact with (3.14) and approximate with (3.27). Equation (3.31) can readily be calculated when I_n, a, and N are known. A simple exact expression for the denominator of (3.31) can be found. Since the typical term in $|E(\theta,\varphi)|^2$ is

$$\begin{aligned} I_m I_n \exp[j(\alpha_m-\alpha_n)] \exp\{jka\sin\theta[\cos(\varphi-\varphi_m)-\cos(\varphi-\varphi_n)]\} \\ = I_m I_n \exp[j(\alpha_m-\alpha_n)] \exp[jk\rho_{mn}\sin\theta\cos(\varphi-\varphi_{mn})], \\ m,n=1,2,\ldots,N, \end{aligned} \tag{3.32}$$

where

$$\begin{aligned} \rho_{mn} &= 2a\sin\frac{\varphi_m-\varphi_n}{2}, \qquad m \neq n \\ &= 0, \qquad m = n, \end{aligned} \tag{3.33}$$

$$\varphi_{mn} = \tan^{-1}\left[\frac{\sin\varphi_m - \sin\varphi_n}{\cos\varphi_m - \cos\varphi_n}\right], \qquad m \neq n, \tag{3.34}$$

the denominator of (3.31) becomes

$$\sum_{m=1}^{N}\sum_{n=1}^{N} I_m I_n \exp[j(\alpha_m-\alpha_n)]\int_0^{2\pi}\int_0^{\pi}\exp[jk\rho_{mn}\sin\theta\cos(\varphi-\varphi_{mn})]\sin\theta\, d\theta\, d\varphi$$

$$=2\pi\sum_m\sum_n I_m I_n \exp[j(\alpha_m-\alpha_n)]\int_0^{\pi} J_0(k\rho_{mn}\sin\theta)\sin\theta\, d\theta$$

$$=4\pi\sum_m\sum_n I_m I_n \exp[j(\alpha_m-\alpha_n)]\int_0^{\pi/2} J_0(k\rho_{mn}\sin\theta)\sin\theta\, d\theta$$

$$=4\pi W, \tag{3.35}$$

where

$$W=\sum_{m=1}^{N}\sum_{n=1}^{N} I_m I_n \exp[j(\alpha_m-\alpha_n)]\left(\frac{\sin k\rho_{mn}}{k\rho_{mn}}\right). \tag{3.36}$$

In obtaining (3.36), we have applied the following relation:[18]

$$\int_0^{\pi/2} J_0(x\sin\theta)\sin\theta\, d\theta=(\pi/2)^{1/2}\frac{J_{1/2}(x)}{(x)^{1/2}}=\frac{\sin x}{x}. \tag{3.37}$$

Substituting (3.36) into (3.31), we have

$$D=\frac{|E(\theta_0,\varphi_0)|^2}{W}. \tag{3.38}$$

Note that while W is a function of φ_m and φ_n through ρ_{mn} defined in (3.33), it is independent of φ_{mn} in (3.34).

Two special cases can be noted. When the radius of the ring approaches zero ($a\to 0$), we have

$$\rho_{mn}\to 0, \qquad \alpha_m\to 0, \qquad \alpha_n\to 0,$$

$$W\to\sum_{m=1}^{N}\sum_{n=1}^{N} I_m I_n=\left(\sum_{n=1}^{N} I_n\right)^2=[E(\theta_0,\varphi_0)]^2, \qquad D\to 1.$$

On the other hand, when a$\rightarrow\infty$,

$$\frac{\sin k\rho_{mn}}{k\rho_{mn}} \rightarrow 0 \qquad \text{for} \qquad m \neq n,$$

$$\frac{\sin k\rho_{mm}}{k\rho_{mm}} = 1,$$

$$W \rightarrow \sum_{n=1}^{N} I_n^2,$$

$$D \rightarrow \frac{\left(\sum_{n=1}^{N} I_n\right)^2}{\sum_{n=1}^{N} I_n^2}.$$

A numerical example for the cophasal ring array with $N=6$, $\varphi_n = \pi n/3$, and $I_n = 1$ is presented in Fig. 3.4 as a function of the radius a. Since $\theta_0 = 0$ is used in Fig. 3.4(a), the result there may be considered as the directivity of the vertical pattern. The two curves in Fig. 3.4(b) are the corresponding results for the horizontal pattern ($\theta_0 = \pi/2$), where one is for the case when the beam maximum points toward $\varphi_0 = 0$ (along an element) and the other is for $\varphi_0 = 30°$ (between two elements). It can be seen that the directivities for all cases approach unity and N, respectively, as the radius approaches zero and infinity. It is also clear that the directivity for $\theta_0 = 0$ reaches its maximum when a is approximately $\frac{7}{8}\lambda$. The maximum directivity for $\theta_0 = \pi/2$ and $\varphi_0 = 0$ occurs approximately when $a = 0.5\lambda$ or 1.75λ, and that for $\theta_0 = \pi/2$ and $\varphi_0 = 30°$ occurs in the neighborhood of $a = 0.75\lambda$. Of course, the above observation is true only for this particular example. When any condition varies, the position of the maximum directivity will change accordingly. Since the case of a uniformly excited cophasal ring array is considered in the above example, the *main-beam radiation efficiency* is unity (100%) according to the definition given in (2.118).

It should be noted that the formulation outlined in (3.33) through (3.38) is also good for arbitrary I_n, φ_n, and α_n. Results on this will be presented later when the synthesis of a single-ring array from the directivity viewpoint is considered.

The same formulation can also be modified to include the possibility when there is an element represented by I_0 at the array center. For this case, all we must do is to change the running indices m, n starting from 0

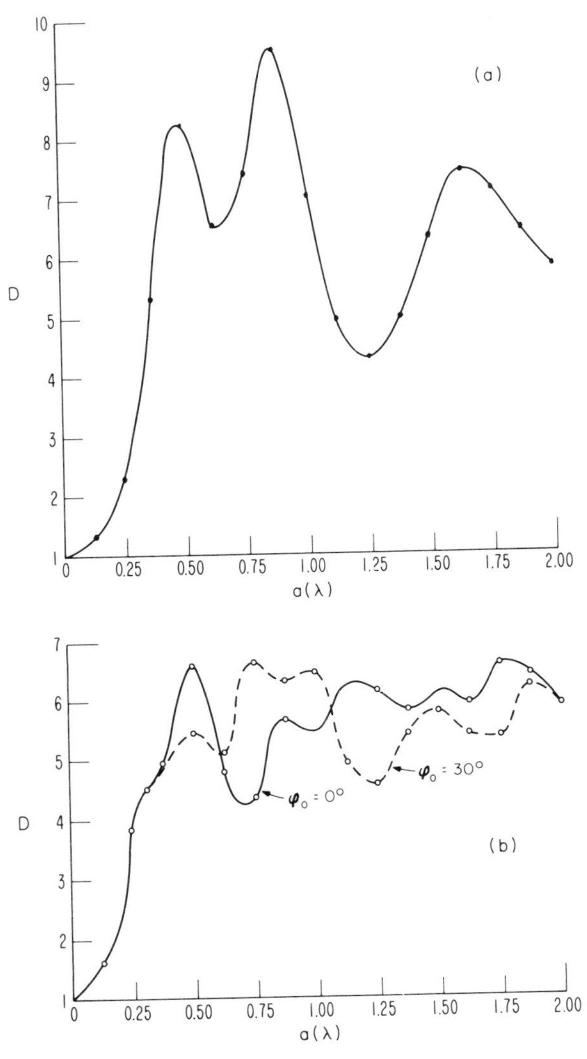

Fig. 3.4 Directivity for a uniform-cophasal ring array with six isotropic elements: (*a*) $\theta_0=0$, (*b*) $\theta_0=\pi/2$.

for the double summation in (3.36) with the following understanding:

$$\alpha_0 = 0, \qquad \rho_{00} = 0, \qquad \rho_{0n} = \rho_{0m} = a. \tag{3.39}$$

When a particular physical element whose pattern can be represented by $f(\theta, \varphi)$ is used in the array, $f^2(\theta, \varphi)$ should be included in the integrand of the denominator of (3.31). The specific result for $f(\theta, \varphi) = \sin^p \theta \cos^{q-1/2} \theta$, $p = 0, 1, \ldots$,and $q \geqslant \frac{1}{2}$, has been obtained by Cheng and Tseng:[19]

$$W = \sum_{m=1}^{N} \sum_{n=1}^{N} I_m I_n \exp[j(\alpha_m - \alpha_n)] S_{mn}, \tag{3.40}$$

where

$$S_{mn} = \tfrac{1}{2}\Gamma(q) \sum_{i=0}^{p} \frac{(-1)^i p! J_{p+q+i}(k\rho_{mn})}{(i!)^2 (p-i)! (k\rho_{mn}/2)^{p+q+i}}, \qquad m \neq n, \tag{3.41}$$

$$S_{nn} = \frac{\Gamma(p+1)\Gamma(q)}{2\Gamma(p+q+1)}, \tag{3.42}$$

and Γ denotes the gamma function.

Clearly, (3.40) reduces to (3.36) when $p = 0$ and $q = \frac{1}{2}$. For the short vertical dipole considered in Fig. 1.11, $f(\theta, \varphi) = \sin\theta$ corresponding to $p = 1$ and $q = \frac{1}{2}$. We then have

$$S_{mn} = \frac{\sin k\rho_{mn}}{k\rho_{mn}} - \frac{1}{(k\rho_{mn})^2} \left[\frac{\sin k\rho_{mn}}{k\rho_{mn}} - \cos k\rho_{mn} \right], \tag{3.43}$$

$$S_{nn} = \tfrac{2}{3}. \tag{3.44}$$

In this case, we can only consider the horizontal pattern with $\theta_0 = \pi/2$. The results of directivity for the six-element ring array with this short dipole as its elements, when $I_n = 1$, $\alpha_n = -ka\cos(\varphi_0 - \pi n/3)$, are given in Fig. 3.5. Now, the directivity approaches 1.5 and 9.0, respectively, as the radius approaches zero and infinity. Since the results appearing in both Figs. 3.4 and 3.5 are high oscillatory, only those calculated values at $a = i\frac{1}{8}\lambda$, $i = 0, 1, \ldots, 16$ are considered accurate.

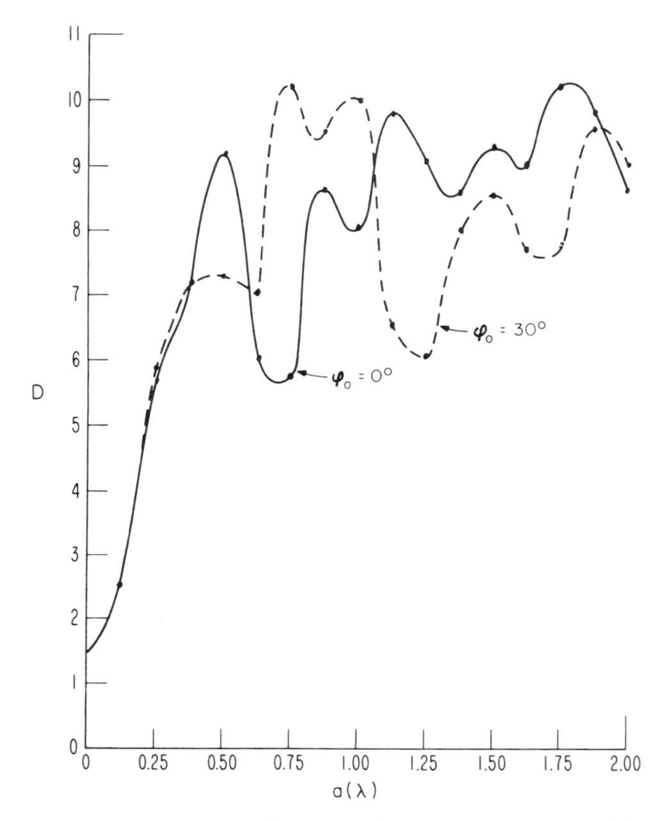

Fig. 3.5 Directivity for a uniform-cophasal ring array with six short dipoles when $\theta_0 = \pi/2$.

3.3 Analysis of Elliptical Arrays

When analyzing the characteristics associated with a ring array, we note from Fig. 3.4 that the directivities in both planes can be very low for a finite size of the ring. For example, the directivity in Fig. 3.4(a) drops to a relative minimum when $a = 1.25\lambda$ or $ka = 2.5\pi$. The reason for this is that the contributions from individual elements add rather destructively for that particular ka to cause either a higher sidelobe level and/or a broader beamwidth in the radiation pattern. In other words, as far as that particular ka and the number of elements are concerned, the arrangement of elements in the form of a ring as discussed in the last section will not yield

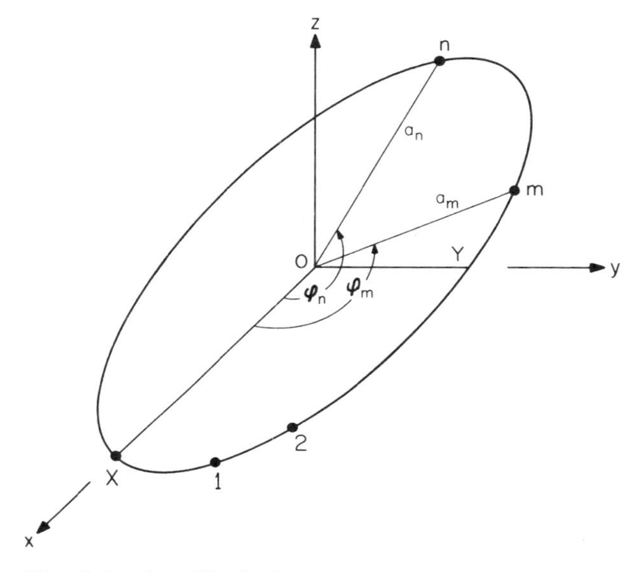

Fig. 3.6 An elliptical array.

a satisfactory directivity. A minor change in element arrangement may improve the result. One of the possible minor changes is to have an *elliptical array*, as shown in Fig. 3.6, where the semimajor and semiminor axes are respectively denoted by X and Y. The distance from the origin to the nth element on the elliptical circumference is now

$$a_n = \frac{Y}{[1-(1-v^2)\cos^2\varphi_n]^{1/2}}, \tag{3.45}$$

where

$$v = \frac{Y}{X} \leqslant 1.$$

When the major and minor axes are interchanged, we have

$$a_n' = \frac{X}{[1-(1-v'^2)\sin^2\varphi_n]^{1/2}}, \tag{3.46}$$

where

$$v' = \frac{X}{Y} \leqslant 1.$$

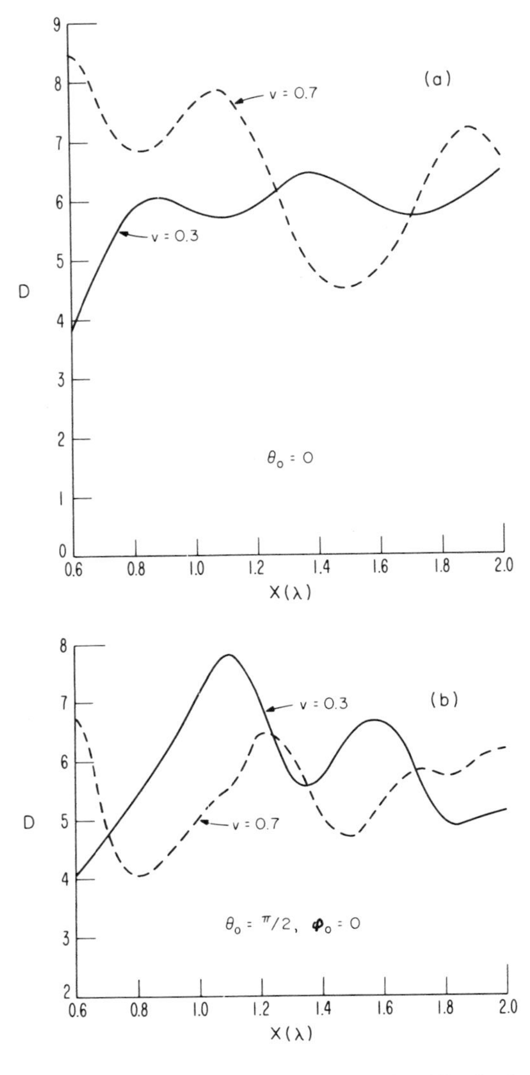

Fig. 3.7 Directivity for a uniform-cophasal elliptical array with six isotropic elements: (*a*) $\theta_0 = 0$, (*b*) $\theta_0 = \pi/2$, $\varphi = 0$.

Formulas (3.14) through (3.17) and (3.35) through (3.44) are still valid for the elliptical array if the radius a of the ring array discussed in the last section is replaced by (3.45) or (3.46), and (3.33) and (3.34) are, respectively, replaced by the following:

$$\rho'_{mn} = [a_m^2 + a_n^2 - 2a_m a_n \cos(\varphi_m - \varphi_n)]^{1/2}, \qquad m \neq n$$

$$= 0, \qquad m = n \qquad (3.47)$$

and

$$\varphi'_{mn} = \tan^{-1}\left[\frac{a_m \sin\varphi_m - a_n \sin\varphi_n}{a_m \cos\varphi_m - a_n \cos\varphi_n}\right]. \qquad (3.48)$$

Numerical results for the cophasal elliptical array with $N=6$, $I_n=1$, and $\varphi_n = n\pi/3$ are presented in Fig. 3.7 for two different values of v. The curves in Fig. 3.7(a) are for $\theta_0=0$, and those in Fig. 3.7(b) are for $\theta_0=\pi/2$ and $\varphi_0=0$. It is interesting to see that, for a given X, the case with a larger value of Y (or v) does not always yield better directivity. By comparing these results with those in Fig. 3.4, we also see that for $\theta_0=0$ the elliptical array with $X=1.25\lambda$, $v=0.3$ is better than the ring array with $a=1.25\lambda$ while the area (πXY) occupied by the former is only 30% of that (πa^2) by the latter. A similar situation is also noted for the case with $\theta_0=\pi/2$ and $\varphi_0=0$.

Since the array considered in the above example is a uniform cophasal one, the *main-beam radiation efficiency* as defined in (2.118) is also unity (100%).

3.4 Equivalence to Linear Arrays

Before we present the synthesis of ring and elliptical arrays, we digress a little here to study the equivalence between the linear array discussed in the first two chapters and the ring array in Section 3.2. To simplify the algebra involved, let us consider the broadside pattern ($\alpha_n=0$) from the exact formula (3.14). If there is also an element at the array center, we can write the broadside pattern as

$$E(\theta.\varphi) = I_0 + \sum_{n=1}^{N} I_n \exp[jka\sin\theta\cos(\varphi - \varphi_n)], \qquad (3.49)$$

which appears more complicated than the broadside pattern for a linear array as it is a function of both θ and φ. However, when the pattern in a vertical plane, given by $\varphi = \varphi_p = \text{constant}$, is of particular interest, we have

$$E(\theta, \varphi_p) = I_0 + \sum_{n=1}^{N} I_n \exp(jka_n \sin\theta), \tag{3.50}$$

where

$$a_n = a\cos(\varphi_p - \varphi_n). \tag{3.51}$$

Clearly, (3.50) has exactly the same mathematical form as that for a linear array of $N+1$ elements with a_n denoting the distance between nth element and the reference element whose amplitude excitation is I_0. This nth element is located on the right side of the reference element if a_n is positive, or on the left side if a_n is negative. When $a_n = 0$, this element coincides with the reference element to reduce the total number of elements by one. We should not be confused by the apparent discrepancy between the factor $\sin\theta$ in (3.50) and $\cos\theta$ involved before for linear arrays. This is due to the fact that θ is used as the angle measured from the array normal in this chapter while it was denoted as the angle from the array axis in the first two chapters.

With this equivalence established, the experience learned previously for linear arrays can be useful in understanding the characteristics associated with the broadside pattern of a ring array. As an example, when $N = 12$, $\varphi_n = \pi n/6$, $\varphi_p = 0$ and π (xz plane), we have for $\varphi_p = 0$,

$$a_{12} = a, \quad a_1 = a_{11} = 0.866a, \quad a_2 = a_{10} = 0.5a, \quad a_3 = a_9 = 0,$$

$$a_4 = a_8 = -0.5a, \quad a_5 = a_7 = -0.866a, \quad a_6 = -a;$$

for $\varphi_p = \pi$,

$$a_6 = a, \quad a_5 = a_7 = 0.866a, \quad a_4 = a_8 = 0.5a, \quad a_3 = a_9 = 0,$$

$$a_2 = a_{10} = -0.5a, \quad a_1 = a_{11} = -0.866a, \quad a_{12} = -a.$$

If, in addition, $I_1 = I_5$, $I_2 = I_4$, $I_6 = I_{12}$, $I_7 = I_{11}$, $I_8 = I_{10}$ as they usually are, (3.50) takes the following special form:

$$E(\theta, 0) = E(\theta, \pi) = I_0 + 2(I_2 + I_{10})\cos(0.5ka\ \sin\theta)$$
$$+ 2(I_1 + I_{11})\cos(0.866ka\sin\theta) + 2I_6\cos(ka\sin\theta), \tag{3.52}$$

which may be pictured in Fig. 3.8 as an equivalent symmetrically spaced seven-element linear array. Once the amplitude excitations are known, the

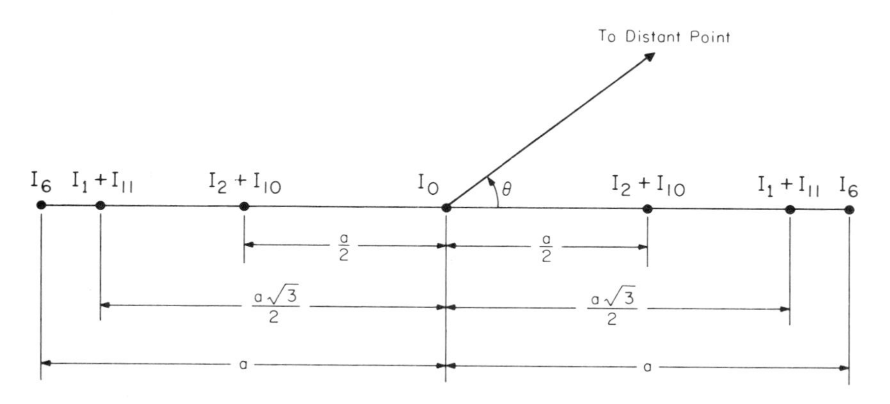

Fig. 3.8 An equivalent symmetrically spaced seven-element linear array.

entire vertical pattern in the xz plane can be analyzed as that of an equivalent linear array. Vertical patterns in other planes, horizontal patterns identified by $\theta_0=\theta=\pi/2$ and $\varphi_0=$constant, and cases with different N and φ_n can be handled in a similar manner. The equivalence between elliptical and linear arrays can also be easily established.

3.5 Synthesis of Ring and Elliptical Arrays—from the Directivity Viewpoint

In Sections 3.2 and 3.3, we formulated the mathematical basis for analyzing characteristics of ring and elliptical arrays and gave a few numerical results of directivity for the uniformly excited cophasal arrays. The directivity obtained for this simple excitation condition is not necessarily always maximized. Now, we are ready to present the synthesis of a single ring or elliptical array with the purpose of maximizing the directivity.[19] The method is essentially the same as that given in Section 2.9, using the quadratic form approach.

In general, the directivity of a single ring or elliptical array can be jointly expressed in the following form:

$$D=\frac{|E(\theta_0,\varphi_0)|^2}{(1/4\pi)\int_0^{2\pi}\int_0^{\pi}|E(\theta,\varphi)|^2\sin\theta\,d\theta\,d\varphi}$$

$$=\frac{[I]^+[A][I]}{[I]^+[B][I]},\tag{3.53}$$

where in the first expression

$$E(\theta,\varphi)=f(\theta,\varphi)\sum_{n=1}^{N} I_n e^{j\alpha_n}\exp\left[jka_n\sin\theta\cos\left(\varphi-\varphi_n\right)\right], \qquad (3.54)$$

with the understanding that $a_n=a$ for the ring array and a_n is given by (3.45) or (3.46) for the elliptical array. The factor $f(\theta,\varphi)$ in (3.54) represents the element pattern with its beam maximum normalized to unity, $f(\theta_0,\varphi_0)=1$. The phase and amplitude excitations are to be determined by maximizing the directivity. For excitations other than uniform-cophasal, as discussed in the previous sections, $I_n \neq I$, $\alpha_n \neq -ka_n\sin\theta_0\cos(\varphi_0-\varphi_n)$.

In the second expression of (3.53), we have

$$[I]=\begin{bmatrix} I_1e^{j\alpha_1} \\ I_2e^{j\alpha_2} \\ \vdots \\ I_Ne^{j\alpha_N} \end{bmatrix}, \qquad (3.55)$$

$[I]^+$ = conjugate transpose (adjoint) of [I],

$$[A]=[e][e]^+=[a_{mn}], \qquad (3.56)$$

$$[e]=\begin{bmatrix} \exp\left[-jka_1\sin\theta_0\cos\left(\varphi_0-\varphi_1\right)\right] \\ \exp\left[-jka_2\sin\theta_0\cos\left(\varphi_0-\varphi_2\right)\right] \\ \vdots \\ \exp\left[-jka_N\sin\theta_0\cos\left(\varphi_0-\varphi_N\right)\right] \end{bmatrix}, \qquad (3.57)$$

$$[B]=[b_{mn}]. \qquad (3.58)$$

The typical element of [A] in (3.56) is

$$a_{mn}=\exp\left[-jk\rho'_{mn}\sin\theta_0\cos\left(\varphi_0-\varphi'_{mn}\right)\right], \qquad m\neq n, \qquad (3.59)$$

$$=1, \qquad m=n, \qquad (3.60)$$

where ρ'_{mn} and φ'_{mn} are given, respectively, in (3.47) and (3.48). The typical element of $[B]$ in (3.58) is

$$b_{mn} = \frac{1}{4\pi} \int_0^{2\pi} \int_0^{\pi} f^2(\theta,\varphi) \exp[-jk\rho'_{mn} \sin\theta \cos(\varphi - \varphi'_{mn})] \sin\theta \, d\theta \, d\varphi, \quad m \neq n, \qquad (3.61)$$

$$b_{mm} = \frac{1}{4\pi} \int_0^{2\pi} \int_0^{\pi} f^2(\theta,\varphi) \sin\theta \, d\theta \, d\varphi. \qquad (3.62)$$

According to the derivation given in Section 3.2, closed forms of (3.61) and (3.62) exist for some simple elements. When $f(\theta,\varphi)=1$ (isotropic element),

$$b_{mn} = \frac{\sin k\rho'_{mn}}{k\rho'_{mn}}, \qquad b_{mm} = 1. \qquad (3.63)$$

For the short dipole where $f(\theta,\varphi)=\sin\theta$, we have

$$\begin{aligned} b_{mn} &= \frac{\sin k\rho'_{mn}}{k\rho'_{mn}} - \frac{1}{(k\rho'_{mn})^2}\left(\frac{\sin k\rho'_{mn}}{k\rho'_{mn}} - \cos k\rho'_{mn}\right), \qquad m \neq n, \\ &= \tfrac{2}{3}, \qquad m = n. \end{aligned} \qquad (3.64)$$

Since (3.53) is in the *quadratic form* where matrices $[A]$ and $[B]$ are both *Hermitian* and $[B]$ is also *positive definite*, the optimization theorem described in Section 2.9 applies. The maximum obtainable directivity should then be given by the largest root (also the only nonzero value) of (2.239), or

$$D_{\max} = [e]^+ [B]^{-1} [e], \qquad (3.65)$$

where $[B]^{-1}$ is the inverse matrix of $[B]$.

The required current excitation matrix should be

$$[I]_{\text{opt}} = [B]^{-1} [e], \qquad (3.66)$$

from which I_n and α_n can be easily identified.

(a) Broadside Optimization. When a *vertical* or *broadside pattern* is desired, $\theta_0 = 0$, Eq. (3.57) becomes

$$[e] = \begin{bmatrix} 1 \\ 1 \\ \vdots \\ 1 \end{bmatrix}, \tag{3.67}$$

and (3.66) always yields $\alpha_n = 0$. This implies that, in order to maximize the directivity for a broadside ring or elliptical array, the elements must be excited in phase (i.e., cophasal). This condition, of course, is consistent with that required for the broadside linear array studied in Chapter 2. The amplitude excitations I_n obtained from (3.66) will depend, however, on (N, a) for the ring array or (N, v, X) for the elliptical array when the values of $\varphi_n, n = 1, 2, \ldots, N$, are specified. Two examples for the optimum broadside elliptical array with $N = 6$ and $\varphi_N = n\pi/3$ are given below:

$N = 6, \quad v = 0.7, \quad X = 2.0\lambda, \quad D_{\max} = 6.7977,$

$I_1 = I_2 = I_4 = I_5 = 1.1205, I_3 = I_6 = 1.1578,$

$I_3/I_1 = 1.0333, \quad \eta = 0.9997;$

$N = 6, \quad v = 0.3, \quad X = 0.6\lambda, \quad D_{\max} = 4.5958,$

$I_1 = I_2 = I_4 = I_5 = 0.5686, \quad I_3 = I_6 = 1.1607,$

$I_3/I_1 = 2.0414, \quad \eta = 0.8828.$

Note that in the above examples the main-beam radiation efficiency is no longer 100%. This is true because the amplitude excitations are no longer uniform. Furthermore, since the value of X is moderate in the first example, the current amplitudes, the maximum directivity obtained, and the final radiation efficiency do not differ much from the uniform-cophasal case presented in Fig. 3.7(a) where $I_n = 1, D = 6.7956$, and $\eta = 1.0$. In the second example, where X is rather small, the maximum directivity obtained represents a good improvement over the uniform-cophasal case with $D = 3.8577$. This improvement is achieved with larger differences in the required current amplitudes and lower radiation efficiency.

Even with this limited calculation, we may conclude that if the size of the elliptical array is moderate for a given number of elements, or more precisely, if the arc distance between two consecutive equally spaced elements on the elliptical circumference is no less than one-half wavelength, the directivity from the uniform-cophasal excitation is indeed very satisfactory as far as the broadside operation is concerned. In other words, under the same condition the improvement in directivity by the optimization procedure just described is only marginal. Only when the distance between two consecutive elements is very small, a larger improvement in directivity is possible by sacrificing the radiation efficiency and current amplitude distribution. This situation is similar to that associated with the

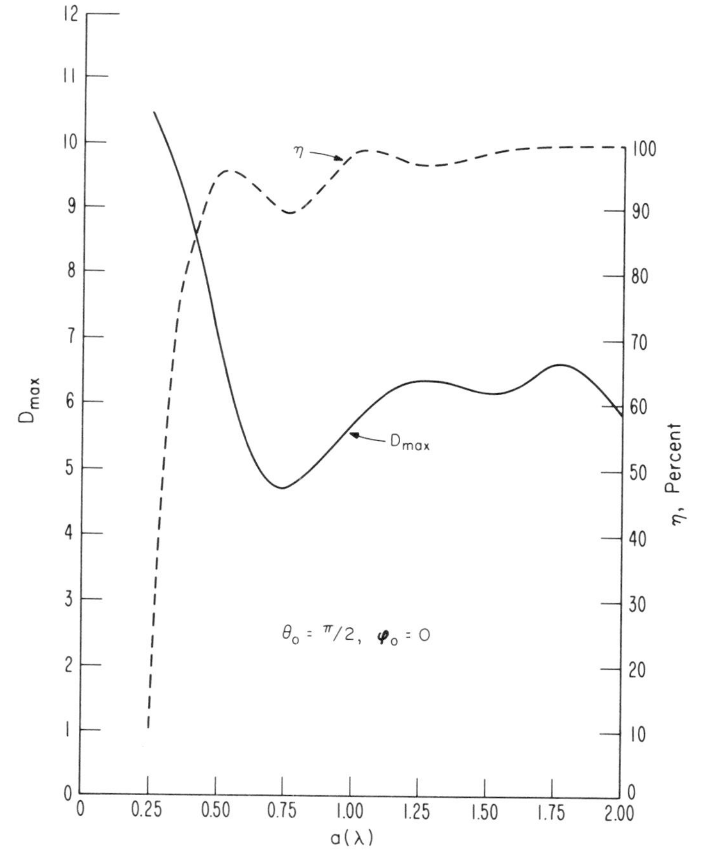

Fig. 3.9 Directivity and radiation efficiency for an optimum ring array with six isotropic elements when $\theta_0 = \pi/2, \varphi_0 = 0$.

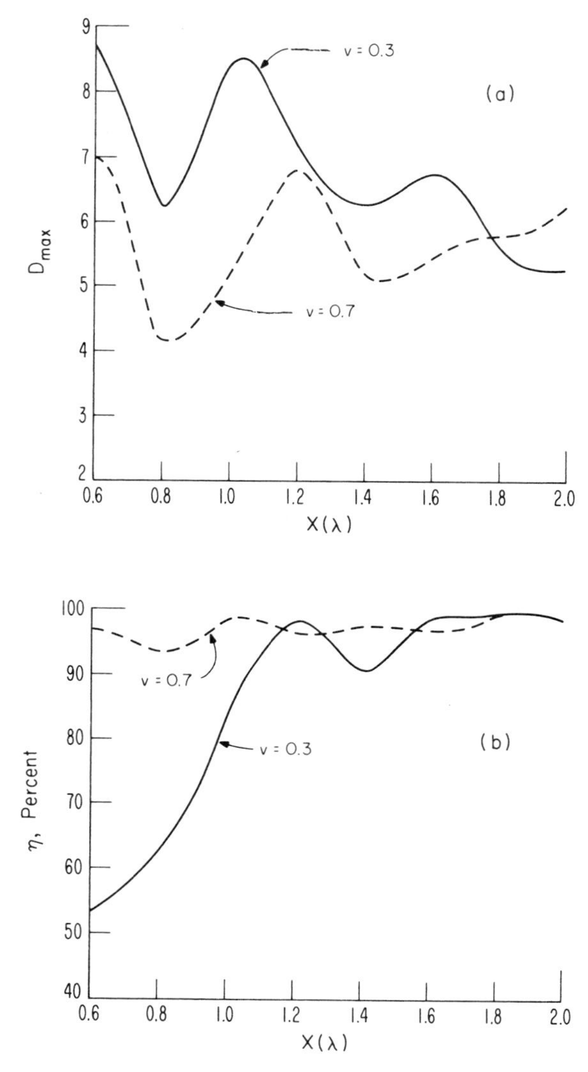

Fig. 3.10 Characteristics for an optimum elliptical array with six isotropic elements when $\theta_0 = \pi/2$ and $\varphi_0 = 0$: (*a*) directivity, (*b*) radiation efficiency.

Table 3.1 Required Excitations for a Ring Array with $N=6$, $f(\theta,\varphi)=1$, $\theta_0=\pi/2$, $\varphi_0=0$, $\varphi_n=\pi n/3$, and $a=0.5\lambda$.

	Optimum	Uniform-Cophasal
$I_1=I_5$	1.2230	1.0
$I_2=I_4$	1.2230	1.0
$I_3=I_6$	1.0972	1.0
$\alpha_1=\alpha_5$	−97.8°	−90°
$\alpha_2=\alpha_4$	97.8°	90°
α_3	−162.4°	−180°
α_6	162.4°	180°

superdirective linear array discussed in Section 2.3 and to that displayed in Table 2.5, where the optimization of the directivity for an equally spaced broadside linear array was presented.

Although the above discussion is specifically directed to the elliptical array, the same conclusion should also hold for the ring array. Since the optimization procedure involved is rather simple, the reader is asked to work a few examples for the ring array with $N=6$ and various values for the ring radius.

(b) Optimization in the Plane of the Array. In this case $\theta_0=\pi/2$. The answer from (3.66) will, in general, not give a uniform-cophasal array. Numerical results for $N=6$ and $\varphi_0=0$ are presented in Figs. 3.9 and 3.10 with the same scale as those in Figs. 3.4(*b*) and 3.7(*b*) for easy comparison.

Table 3.2 Required Excitations for an Elliptical Array with $N=6$, $f(\theta,\varphi)=1$, $\theta_0=\pi/2$, $\varphi_0=0$, $\varphi_n=\pi n/3$, $v=0.3$, and $X=1.0\lambda$.

	Optimum	Uniform-Cophasal
$I_1=I_5$	1.6628	1.0
$I_2=I_4$	1.6628	1.0
$I_3=I_6$	1.2784	1.0
$\alpha_1=\alpha_5$	−73.9°	−61.4°
$\alpha_2=\alpha_4$	73.9°	61.4°
α_3	38.7°	0°
α_6	−38.7°	0°

Since the excitations are no longer uniform and cophasal, the *main-beam radiation efficiency* as defined in (2.118) is also calculated and included in the figures. It is apparent from Fig. 3.9 that the ring array will become superdirective with lower radiation efficiency when its radius is less than about 0.5λ. For this reason, the results when $X < 0.6\lambda$ for the elliptical array are not included in Fig. 3.10. Note that in Fig. 3.9 the required current excitations when $a = 2.0\lambda$ happen to be identical with those for the uniform-cophasal excitation. Therefore, the directivity remains the same as that for the uniform-cophasal array and the *main-beam radiation efficiency* is 100% for $a = 2.0\lambda$.

Specific results of I_n and α_n are given in Table 3.1 for the ring array with $a = 0.5\lambda$. For this case, $D_{max} = 6.9378$, $\eta = 95.63\%$, and the maximum-to-minimum amplitude ratio $= I_1/I_3 = 1.1146$. Similar results for the elliptical array with $v = 0.3$ and $X = 1.0\lambda$ are shown in Table 3.2, where $D_{max} = 8.4864$, $\eta = 83.82\%$, and $I_1/I_3 = 1.3006$.

The normalized radiation patterns in the array plane corresponding to Tables 3.1 and 3.2 are plotted respectively in Figs. 3.11 and 3.12 where uniform-cophasal cases are also given for the purpose of comparison. It is clear from these figures that the increase in directivity for the optimum excitation is mainly achieved by making the main lobe a littler narrower than that with a uniform-cophasal excitation. The sidelobe levels are, however, quite high, as we have expected for the reason discussed in Section 3.2. In fact, while directivity values higher than those given in Tables 3.1 and 3.2 may be realized by a larger array with more elements, the sidelobe levels of the pattern always remain high. Concentric ring or

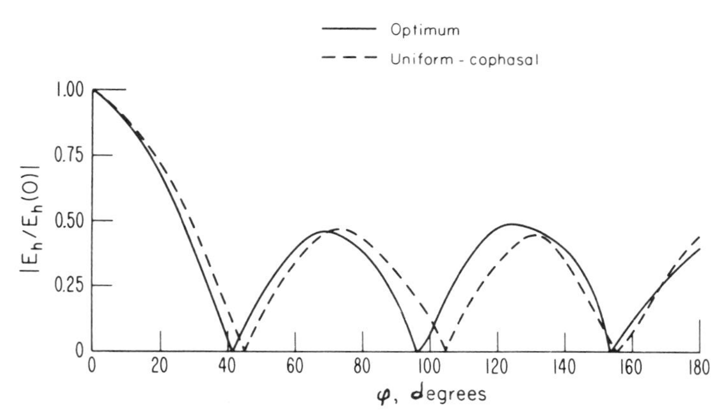

Fig. 3.11 Normalized radiation patterns for a ring array with six isotropic elements when $\theta_0 = \pi/2, \varphi_0 = 0$, and $a = 0.5\lambda$.

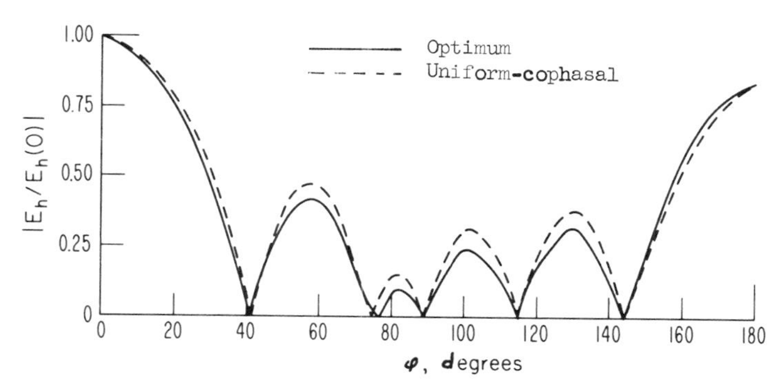

Fig. 3.12 Normalized radiation patterns for an elliptical array with six isotropic elements when $\theta_0 = \pi/2, \varphi_0 = 0, v = 0.3$, and $X = 1.0\lambda$.

elliptical arrays may be syntheiszed to accomplish a substantial reduction in the sidelobe level. The topic on concentric ring arrays is presented in the following section.

3.6 Synthesis of Concentric Ring Arrays—From the Pattern Viewpoint

In Section 3.2 we showed that when the total number of elements equally spaced on the circumference of a cophasal ring array with a given radius a is large and they are equally excited, the pattern from such a ring may be approximated as

$$E \cong I_t J_0(k\rho), \tag{3.68}$$

where I_t is the total amplitude excitation, $\rho = \rho_v \equiv a \sin\theta$ $(0 \leqslant \theta \leqslant \pi)$ for the vertical pattern, and $\rho = \rho_h \equiv 2a \sin(\varphi/2)$ $(0 \leqslant \varphi \leqslant 2\pi)$ for the horizontal pattern.

If there are M *concentric rings* in the array in addition to one element at the array center, and each ring contains a "large number" of elements, the total contribution to the pattern will approximately be[20]

$$E_M \cong \sum_{n=0}^{M} I_{tn} J_0(k\rho_n) = \sum_{n=0}^{M} I_{tn} J_0\left(\frac{kb_n u}{\pi}\right), \tag{3.69}$$

where

$$b_n = \begin{Bmatrix} a_n \\ 2a_n \end{Bmatrix} \quad \text{and} \quad u = \begin{Bmatrix} \pi \sin\theta \\ \pi \sin(\varphi/2) \end{Bmatrix} \text{ for the } \begin{Bmatrix} \text{vertical} \\ \text{horizontal} \end{Bmatrix} \text{ pattern,}$$

and

$$a_0=0<a_1<a_2<\cdots<a_M.$$

If the maximum value of E_M occurs at $u=0$, as it usually does, the directivities in both planes can be respectively expressed as

$$D_v \cong \frac{2E_M^2(0)}{\int_0^{\pi} E_M^2 \sin\theta \, d\theta} \tag{3.70}$$

and

$$D_h \cong \frac{2\pi E_M^2(0)}{\int_0^{2\pi} E_M^2 \, d\varphi}, \tag{3.71}$$

which may be calculated numerically when I_{tn} and b_n are known. It is clear that the directivity expressions given above are only approximate. They become exact when N_n, the number of elements on each ring, is infinitely large. For a finite N_n, the exact results can also be obtained if we use the following:

$$D=\frac{4\pi |E_M|^2{}_{\max}}{\int_0^{2\pi}\int_0^{\pi} |E_M|^2 \sin\theta \, d\theta \, d\varphi}, \tag{3.72}$$

where

$$\begin{aligned} E_M = I_0 &+ \sum_{n=1}^{N_1} I_{1n} \exp\left[\,jka_1 \sin\theta \cos(\varphi-\varphi_{1n}) + j\alpha_{1n}\right] \\ &+ \sum_{n=1}^{N_2} I_{2n} \exp\left[\,jka_2 \sin\theta \cos(\varphi-\varphi_{2n}) + j\alpha_{2n}\right] \\ &+ \cdots \\ &+ \sum_{n=1}^{N_M} I_{Mn} \exp\left[\,jka_M \cos(\varphi-\varphi_{Mn}) + j\alpha_{Mn}\right]. \end{aligned} \tag{3.73}$$

Once again, $|E_M|^2$ will have terms of the same form as (3.32) with ρ_{mn} given by either type of (3.33) or that of (3.47) depending on the individual product involved. Therfore, an exact expression for the directivity similar to that in (3.38) can be derived. The details are omitted since our purpose here is a study of the pattern only.

An examination of (3.69) reveals that the formulation presented in Section 2.7 with *Haar's theorem* should also apply here. This means that, if the conditions outlined previously are satisfied, the set of the required current I_{tn} can be uniquely determined so that the synthesized pattern E_M will be the best approximation to a specified function $f(u)$ in the *minimax* sense. Thus, the maximum magnitude of the deviation function,

$$e(u)=f(u)-E_M(u), \tag{3.74}$$

will be minimized in $0 \leqslant u \leqslant \pi$. The conditions on the basis functions $J_0(kb_n u/\pi)$ and the specified function $f(u)$ required by *Haar's theorem* are

a. $J_0(kb_n u/\pi)$, $n=0, 1, \ldots, M$ are bounded and continuous in $(0, \pi)$;

b. $J_0(kb_n u/\pi)$ are linearly independent real functions;

c. $f(u)$ is also bounded and continuous in $(0, \pi)$;

d. No $E_M = \sum_{n=0}^{M} I_{tn} J_0(kb_n u/\pi)$ should have more than M distinct real zeros in $(0, \pi)$, where the I_{tn}'s are not all zero.

Since we are again using $f(u)=e^{-Au^2}$ (with A as a positive number) as the specified pattern, it is clear that the first three conditions have been satisfied. The fourth condition can also be met by an appropriate choice of b_M (the radius of the largest ring). Let us consider a simple case where $M=1$ to illustrate this point. The array for this case has one ring and one element at the array center, and the pattern function is given by

$$E_1 = I_{t0} + I_{t1} J_0\left(\frac{kb_1 u}{\pi}\right). \tag{3.75}$$

If we choose $kb_1=4$, the part of $J_0(kb_1 u/\pi)$ in the visible region ($0 \leqslant u \leqslant \pi$, or $0 \leqslant kb_1 u/\pi \leqslant 4$) is shown in Fig. 3.13. It is obvious that E_1 cannot have more than one real zero in $0 \leqslant u \leqslant \pi$ no matter what I_{t0} and I_{t1} may be. For this simple case all the conditions required by Haar's theorem are satisfied. We can then apply the "extrema equalization" process, that was successfully used in Section 2.7 when dealing with the synthesis of nonuniformly spaced linear arrays, such that the deviation function in (3.74) attains its maximum magnitude at three ($M+2$) consecutive points in $0 \leqslant u \leqslant \pi$ with the sign of $e(u)$ at these points alternately plus and minus. Under this situation the maximum magnitude of the deviation function, $\max|e(u)|$, is considered minimized. The values of I_{t0} and I_{t1} obtained by solving simultaneous equations as demonstrated in Section 2.7 are also unique with respect to $kb_1=4$.

For arrays of two concentric rings plus one element at the array center, $M=2$. At the beginning we cannot readily see whether the fourth condition required by Haar's theorem can be satisfied by certain choices of b_2. To start with, we generally choose a rather small b_2, say $kb_2=7$, to make sure

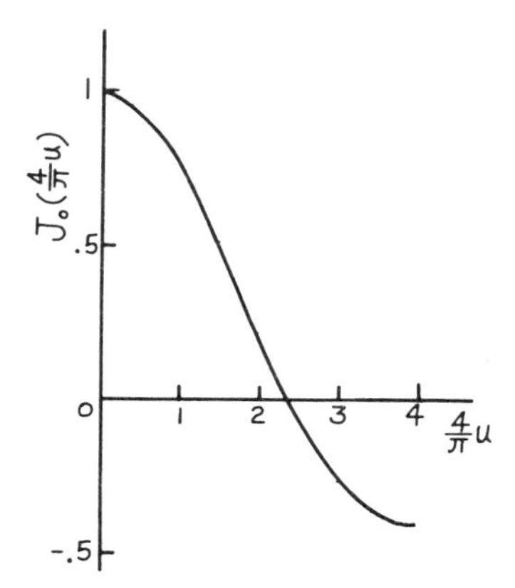

Fig. 3.13 A sketch of $J_0(4u/\pi)$ in $0 \leqslant u \leqslant \pi$.

that $J_0(kb_2u/\pi)$ alone does not have more than two zeros in $0 \leqslant u \leqslant \pi$ and that the fourth condition is "likely" to be satisfied. With another choice of b_1 ($b_1 < b_2$), we can apply the same procedure to determine the required currents I_{t0}, I_{t1}, and I_{t2} such that $\max|f(u) - E_2(u)|$ will be minimized. The solution thus obtained is optimum in Haar's sense with respect to those particular values of b_1 and $b_2 (kb_2 = 7)$. Once the currents are determined, other array characteristics such as the directivity, beamwidth, and sidelobes can also be calculated. The entire procedure can be repeated by varying b_1 and keeping $kb_2 = 7$ to obtain a set of solutions. From these we then select one with the best performance (the narrowest beamwidth, largest directivity, or lowest sidelobe) as the final answer to this synthesis problem. The answer is considered the best one as far as this particular $kb_2 = 7$ (or equivalently the array size) is concerned. The same procedure can be continued by increasing b_2 gradually until the whole process breaks down (i.e., the iteration ceases to converge). It then appears that the maximum array size has been found to give the best array performance under the *minimax* criterion.

For arrays of more concentric rings, the same concept applies although the total number of combinations of b_n becomes more involved.

Thus far we have solved the approximation problem by considering the principal terms only, which corresponds to $N_n \to \infty$. The actual number of elements to be used on each ring can be determined by examining the residual terms in (3.28) and (3.29). With a set of a_n obtained by using Haar's theorem, we can always choose a corresponding set of finite N_n to make the residual terms negligible.

Applying the theory outlined above and the procedure of computation described in Section 2.7, we present in Table 3.3 two numerical examples.

Table 3.3 Characteristics for Arrays of Two and Three Rings.

$f(u)$	M	kb_1	kb_2	kb_3	I_{t0}	I_{t1}	I_{t2}	I_{t3}	$(\mathrm{BW})_v$	S.L.
$e^{-1.5u^2}$	2	1.5	7.5		-0.7532	1.0912	0.5848		21.32	-21.55
e^{-3u^2}	3	4.5	8.1	12.0	0.0477	0.2426	0.2708	0.3752	14.58	-23.34

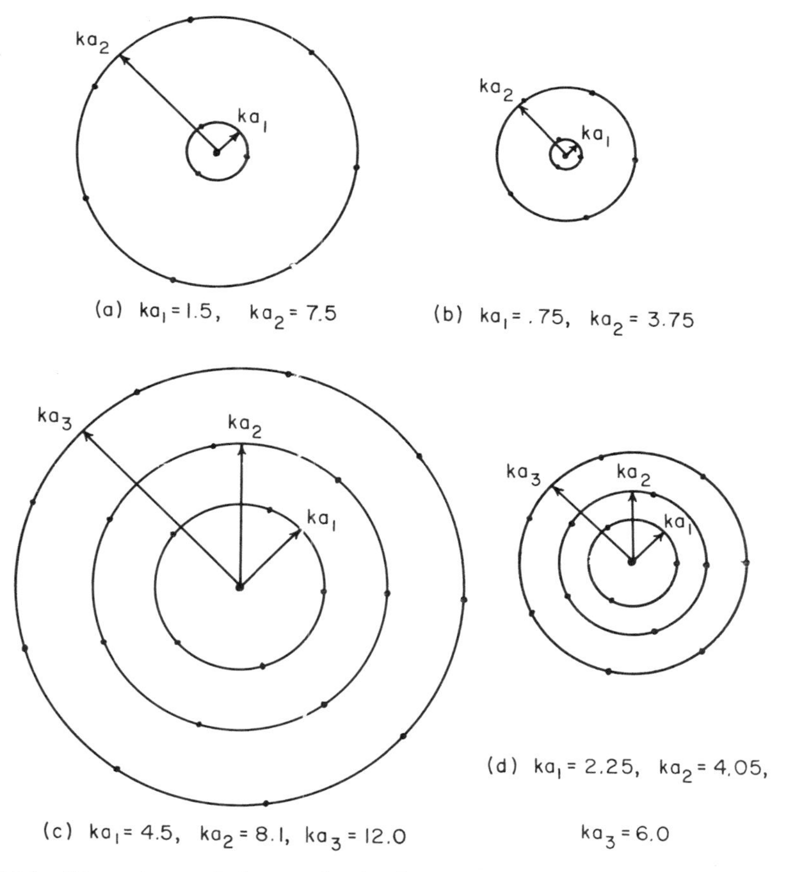

Fig. 3.14 Plan views of the synthesized concentric-ring arrays: (a) $M=2$, vertical, (b) $M=2$, horizontal, (c) $M=3$, vertical, (d) $M=3$, horizontal.

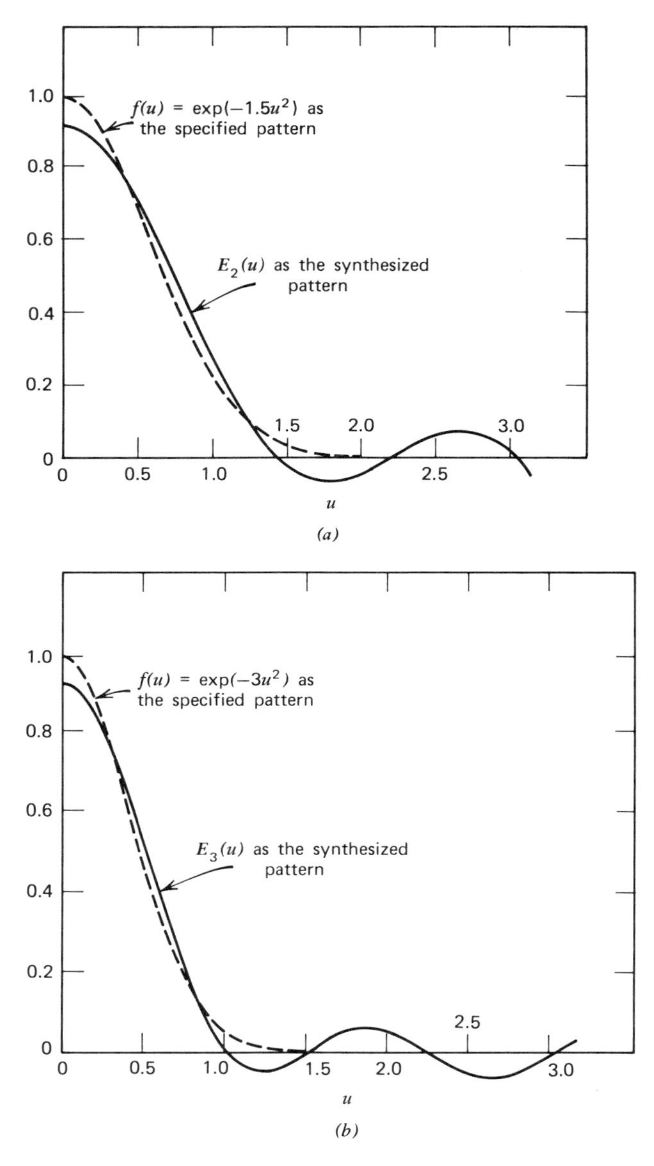

Fig. 3.15 Radiation patterns of the synthesized concentric-ring arrays: (*a*) as a function of $u, M=2$, (*b*) as a function of $u, M=3$, (*c*) as a function of $\theta, M=2$, (*d*) as a function of $\theta, M=3$, (*e*) as a function of $\varphi, M=2$, (*f*) as a function of $\varphi, M=3$.

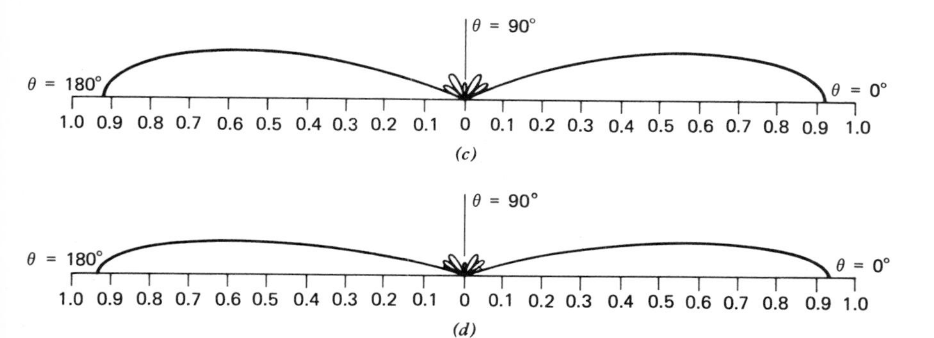
θ = 90°
θ = 180°
θ = 0°
1.0 0.9 0.8 0.7 0.6 0.5 0.4 0.3 0.2 0.1 0 0.1 0.2 0.3 0.4 0.5 0.6 0.7 0.8 0.9 1.0
θ = 90°
θ = 180°
θ = 0°
1.0 0.9 0.8 0.7 0.6 0.5 0.4 0.3 0.2 0.1 0 0.1 0.2 0.3 0.4 0.5 0.6 0.7 0.8 0.9 1.0

(c)

(d)

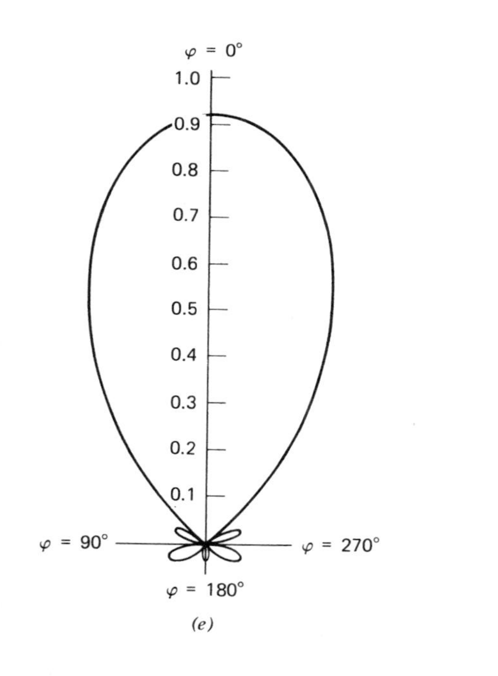
φ = 0°
1.0
0.9
0.8
0.7
0.6
0.5
0.4
0.3
0.2
0.1
φ = 90°
φ = 270°
φ = 180°

(e)

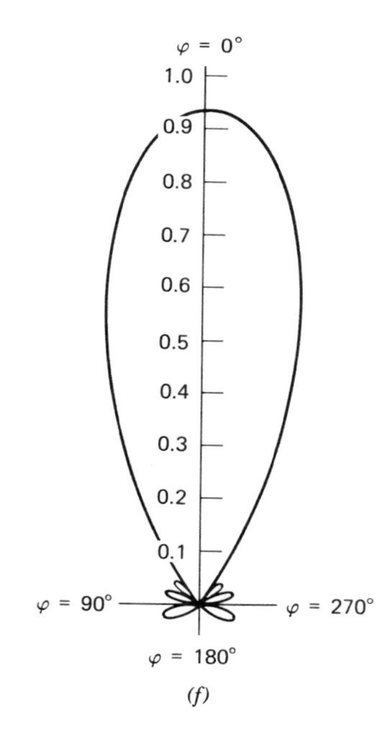
φ = 0°
1.0
0.9
0.8
0.7
0.6
0.5
0.4
0.3
0.2
0.1
φ = 90°
φ = 270°
φ = 180°

(f)

The pattern synthesized according to the current amplitudes given in the table also approximates the specified $f(u)$ in the *minimax* sense. The associated half-power beamwidth for the vertical pattern, $(\mathrm{BW})_v$, is given in degrees, and the approximate sidelobe level in dB (S.L.) is calculated according to $-20\log[E_M(0)/e_{\max}]$, where $e_{\max}=[f(u)-E_M(u)]_{\max}$. Note from the relations following (3.69) that, for a chosen b_n, the radius of the nth ring required for the horizontal pattern is only one-half of the corresponding radius for the vertical pattern. Furthermore, since $\varphi/2$ and θ are equivalent in the mathematical sense, the half-power point, u_h, determined by solving $E_M(u_h)=0.707E_M(0)$ should yield that $\varphi_h=2\theta_h$. This implies that $(\mathrm{BW})_h=2(\mathrm{BW})_v$ for the same set of b_n.

As mentioned above, a study of the sizes of concentric rings and the behavior of J_{mN} should help determine the minimum number of elements required on each ring to justify the neglecting of all the residual terms for a certan accuracy. The actual number of elements for each example in Table 3.3 for an accuracy of the order of 10^{-3} is shown in Fig. 3.14. The detailed synthesized patterns are given in Fig. 3.15.

The synthesis technique of applying Haar's theorem as discussed here for isotropic elements can also be extended to arrays with directive elements, following the same idea cited near the end of Section 2.7. The realization of making this extension possible was essentially presented in (2.209), and is not repeated here.

Although the synthesis of concentric elliptical arrays may be formulated in principle by a parallel approach, the mathematics involved is unfortunately much more complicated because a simple and approximate presentation for the field function, similar to that in (3.68), is not available. Obviously, this complication arises from the fact that the variable ρ in (3.16) has to be replaced by

$$\rho_n=a_n\left[(\sin\theta\cos\varphi-\sin\theta_0\cos\varphi_0)^2+(\sin\theta\sin\varphi-\sin\theta_0\sin\varphi_0)^2\right]^{1/2}$$

for each ellipse, where a_n is given by (3.45), so that a mathematical expansion similar to (3.18) is much involved, if not impossible.

3.7 Concluding Remarks

In this chapter we have presented methods of analyzing and synthesizing two-dimensional arays with detailed attention paid to ring and elliptical arrays. The criteria used for the synthesis of such arrays were in parallel with those for dealing with linear arrays discussed in Chapter 2. Once again we have restricted ourselves to arrays of discrete and simple elements. The mutual impedances between elements have also been ignored.

The antenna of circular aperture type has been discussed elsewhere.[8,14,21-25] This subject can mostly be treated by double Fourier transforms. The impedance problem and the associated matching techniques, especially for large or infinite scanned arrays of waveguides and slots, have thoroughly been studied in the past.[26-38]

Topics on arrays with elements occupying only a part of the circular circumference or other geometry have also received limited attention.[39,40] When only a portion of the elements in the array are excited, the array is said to have parasitic excitation. This subject has been treated by Tillman[9] and his associates.[41] Quasi-ring arrays where the elements are not all identically oriented have also been considered.[42-44] The elements in this special kind of array are usually arranged tangentially or radially (with respect to the ring circumference) to achieve certain characteristics.

The elements of a ring or elliptical array can also be mounted near or on a concentric reflecting structure to increase effectiveness of the system or to facilitate possible airborne applications.[5,45-47] Methods of analysis and synthesis different from those presented in this chapter are also available. These include using a probabilistic approach for analyzing large circular arrays,[48] employing sampling theory to synthesize a prescribed pattern in the plane of the ring for a specified precision,[49] and sidelobe reduction by an iterative technique.[50]

Finally, it should be noted that the problem of spherical or three-dimensional arrays can be formulated as an extension to the materials presented. Limited results in this category have already been published.[5,48,51-54]

PROBLEMS

3.1 Calculate the directivity for a uniform-cophasal ring array with eight elements equally spaced on the circumference and $a=1.0\lambda$: (a)$\theta_0=0$ when the element is isotropic, (b) $\theta_0=\pi/2$ and $\varphi_0=0$ when the element is isotropic, (c)$\theta_0=\pi/2$ and $\varphi_0=0$ when the element is a vertical short dipole.

3.2 Plot the radiation patterns corresponding to cases in Problems 3.1. How do they differ from the ideal $J_0(x)$?

3.3 Calculate the directivity for a uniform-cophasal elliptical array with eight equally spaced isotropic elements, $X=2.0\lambda$ and $v=0.5$: (a)$\theta_0=0$, (b) $\theta_0=\pi/2$ and $\varphi_0=0$.

3.4 Plot the vertical pattern (in the xz plane) of the elliptical array described in Problem 3.3, and find an equivalent linear array.

3.5 Determine the optimum directivity and radiation efficiency for a ring array with eight equally spaced isotropic elements as a function of the array radius when $\theta_0 = \pi/2$ and $\varphi_0 = 0$. What are the required amplitude and phase excitations?

3.6 Plot the radiation pattern from the array in Problem 3.5 with $a = 1.0\lambda$.

3.7 Synthesize an array of three concentric rings to staisfy the following specifications:

(a) the sidelobes in the radiation pattern are approximately at an equal level,

(b) the first-null beamwidth is approximately $2u_1 = 1.8$ rad (where u_1 is the position of the first null), and

(c) $kb_3 = 10$.

The array should also have an element at the array center. Make an arbitrary choice of b_1 and b_2. What are the required total currents on each ring and the final sidelobe level in dB? Based on your method of synthesis, what are the required minimum number of elements for each ring?

3.8 Calculate the directivity for the two concentric-ring array outlined in Table 3.3 when $\theta_0 = 0$.

REFERENCES

1. **Collin, R. E.** Pattern synthesis with nonseparable aperture fields, *IEEE Trans. Antennas and Propagation,* Vol. AP-12, No. 4, pp. 502–503, July, 1964.

2. **Schell, A. C. and A. Ishimaru.** Antenna pattern synthesis, Chap. 7, in *Antenna Theory*, Part I, edited by R. E. Collin and F. J. Zucker, McGraw-Hill Book Co., New York, 1969.

3. **Page, H.** Ring aerial systems, *Wireless Engineers*, Vol. 25, No. 301, pp. 308–314, October, 1948.

4. **Wild, J. P.** Circular aerial arrays for radio astronomy, *Proc. Royal Soc.* (London) Ser. A, Vol. 262, No. 1308, pp. 84–99, June, 1961.

5. **DuHamel, R. H.** Pattern synthesis for antenna arrays on circular, elliptical, and spherical surfaces, *Technical Report No. 16,* EE Research Lab., University of Illinois, Urbana, 1952.

6. **Ishimaru, A.** Theory of unequally spaced arrays, *IRE Trans. Antennas and Propagation,* Vol. AP-10, No. 6, pp. 691–701, November, 1962.

7. **Tighe, R. F.** Nonuniform two dimensional scanning arrays, *IEEE Wescon Record*, Paper No. 104, 1963.

8. **Taylor, T. T.** Design of circular apertures for narrow beam width and low sidelobes, *IRE Trans. Antennas and Propagation,* Vol. AP-8, No. 1, pp. 17–22, January, 1960.

9. **Tillman, J. D., Jr.** *The Theory and Design of Circular Antenna Arrays*, Jniversity of Tennessee Eng. Exp. Station, Knoxville, 1966.

0. **LePage, W. R., C. S. Roys, and S. Seely.** Radiation from circular urrent sheets, *Proc. IRE,* Vol. 38, No. 9, pp. 1069–1072, September, 1950.

1. **Stearns, C. O. and A. C. Stewart** An investigation of concentric ring ntennas with low sidelobes, *IEEE Trans. Antennas and Propagation*, Vol. ΛP-13, No. 6, pp. 856–863, November, 1965.

2. **Royer, G. M.** Directive gain and impedance of a ring array of antenas, *IEEE Trans. Antennas and Propagation*, Vol. AP-14, No. 5, pp. 66–573, September, 1966.

3. **Chu, T. S.** On the use of uniform circular arrays to obtain omnidirecional patterns, *IRE Trans. Antennas and Propagation,* Vol. AP-7, No. 4, p. 36, October, 1959.

4. **Hansen, R. C.** Tables of Taylor distributions for circular aperture ntennas, *IRE Trans. Antennas and Propagation,* Vol. AP-8, No. 1, pp. 3–26, January, 1960.

5. **Wait, J. R.** Theory of an antenna with an infinite corner reflector, *Can. J. Phys.*, Vol. 32, No. 5, pp. 365–371, May, 1954.

6. **Utlaut, W. F.** Some radio and optical observations of ionospheric nodification by very high power HF ground-based transmission, *GAP nternational Symposium Digest*, pp. 208–212, September, 1970.

7. **Watson, G. N.** *Theory of Bessel Functions*, University of Cambridge ress, Cambridge, England, 1958.

8. **Magnus, Wilhelm and Fritz Oberhettinger.** *Formulas and Theorems for he Functions of Mathematical Physics*, Chap. 3, Chelsea Publishing Comany, New York, 1949.

19. **Cheng, D. K. and F. I. Tseng.** Maximisation of directive gain for ircular and elliptical arrays, *Proc. IEE* (London), Vol. 114, No. 5, pp. 89–594, May, 1967.

20. **Ma, M. T. and L. C. Walters.** Synthesis of concentric ring antenna rrays yielding approximately equal sidelobes, *Radio Sci.*, Vol. 3, No. 5, pp. 465–470, May, 1968.

21. **Fante, R. L.** Optimum distribution over a circular aperture for best nean-square approximation to a given radiation pattern, *IEEE Trans. on Antennas and Propagation*, Vol. AP-18, No. 2, pp. 177–181, March, 1970.

22. **Hansen, R. C.** Aperture theory, Chap. 1, Vol. 1 in *Microwave Scanning Antennas*, edited by R. C. Hansen, Academic Press, New York, 1964.

23. **Hu, M. K.** Fresnel region field distributions of circular aperture anennas, *IRE Trans. Antennas and Propagation*, Vol AP-8, No. 3, pp. 344–346, May, 1960.

24. **Rhodes, D. R.** On a fundamental principle in the theory of planar apertures, *Proc. IEEE,* Vol. 52, No. 9, pp. 1013–1021, September, 1964.

25. **Ruze, J.** Circular aperture synthesis, *IEEE Trans. Antennas and Propagation,* Vol. AP-12, No. 6, pp. 691–694, November, 1964.

26. **Allen. J. L.** On surface-wave coupling between elements of large arrays, *IEEE Trans. Antennas and Propagation,* Vol. AP-13, No. 4, pp. 638–639, July, 1965.

27. **Amitay, N., J. S. Cook, R. G. Pecina, and C. P. Wu.** On mutual coupling and matching conditions in large planar phased arrays, *IEEE G-AP International Symposium Digest,* 1964.

28. **Diamond, B. L.** A generalized approach to the analysis of infinite planar array antennas, *Proc. IEEE,* Vol. 56, No. 11, pp. 1837–1851 November, 1968.

29. **Edelberg, S. and A. A. Oliner.** Mutual coupling effects in large antenna arrays: Part I—slot arrays, *IRE Trans. Antennas and Propagation,* Vol. AP-8, N. 3, pp. 286–297, May, 1960.

30. **Farrell, G. F. and D. H. Kuhn.** Mutal coupling effects of triangular-grid arrays by modal analysis, *IEEE Trans Antennas and Propagation,* Vol. AP-14, No. 5, pp. 652–654, September, 1966.

31. **Farrell, G. F. and D. H. Kuhn.** Mutal coupling in infinite planar arrays of rectangular waveguide horns, *IEEE Trans. Antennas and Propagation,* Vol. AP-16, No. 4, pp. 405–414, July, 1968.

32. **Galindo, V. and C. P. Wu.** Numerical solutions for an infinite phased array of rectangular waveguides with thick walls, *IEEE Trans. Antennas and Propagation,* Vol. AP-14, No. 2, pp. 149–158, March, 1966.

33. **Hannan, P. W., D. S. Lerner, and G. H. Knittel.** Impedance matching a phased-array over wide scan angles by connecting circuits, *IEEE Trans. Antennas and Propagation,* Vol. AP-13, No. 1, pp. 28–34, January, 1965.

34. **Lechtreck, L. W.** Effects of coupling accumulation in antenna arrays, *IEEE Trans. Antennas and Propagation,* Vol. AP-16, No. 1, pp. 31–37, January, 1968.

35. **Oliner, A. A.** The impedance properties of narrow radiating slots in the broad face of rectangular waveguide: Part I—Theory, *IRE Trans. Antennas and Propagation,* Vol. AP-5, No. 1, pp. 4–11, January, 1957.

36. **Oliner, A. A. and R. G. Malech.** Mutual coupling in infinite scanning arrays, Chap. 3, and Mutual coupling in finite scanning arrays, Chap. 4 Vol. 2 in *Microwave Scanning Antennas,* edited by R. C. Hansen, Academic Press, New York, 1964.

37. **Wu, C. P. and V. Galindo.** Properties of a phased array of rectangular waveguides with thin walls, *IEEE Trans. Antennas and Propagation,* Vol. AP-14, No. 2, pp. 163–173, March, 1966.

38. **Varon, D. and G. I. Zysman.** On the mismatch of electronically steerable phased-array antennas, *Radio Sci.*, Vol. 3, No. 5, pp. 487–489, May, 1968.

39. **Lee, S. W. and Y. T. Lo.** On the pattern function of circular arc arrays, *IEEE Trans. Antennas and Propagation,* Vol. AP-13, No. 4, pp. 649–650, July, 1965.

40. **Chiang, B. and D. H. S. Cheng.** Curvilinear arrays, *Radio Sci.,* Vol. 3, No. 5, pp. 405–409, May, 1968.

41. **Simpson, T. L. and J. D. Tillman.** Parasitic excitation of circular antenna arrays, *IRE Trans. Antennas and Propagation*, Vol. AP-9, No. 3, pp. 263–267, May, 1961.

42. **Hilburn, J. L.** Circular arrays of radial and tangential dipoles for turnstile antennas, *IEEE Trans. Antennas and Propagation,* Vol. AP-17, No. , pp. 658–660, September 1969.

43. **Hilburn, J. L. and C. E. Hickman.** Circular arrays of tangential dipoles, *J. Appl. Phys.* Vol. 39, No. 12, pp. 5953–5959, December, 1968.

44. **Knudsen, H. L.** Radiation from ring quasi-arrays, *IRE Trans. Antennas and Propagation,* Vol. AP-4, No. 3, pp. 452–472, July, 1956.

45. **Knudsen, H. L.** Antennas on circular cylinders, *IRE Trans. Antennas and Propagation,* Vol. AP-7 (supplement), pp. S361–S370, December, 1959.

46. **Ma, M. T. and L. C. Walters.** Theoretical methods for computing characteristics of Wullenweber antennas, *Proc. IEE* (London), Vol. 117, No. 11, pp. 2095–2101, November, 1970.

47. Proceedings of the Conformal-Array Antenna Conference, held at Naval Electronics Laboratory Center, San Diego, Calif., January 13–15, 1971.

48. **Panicali, A. R. and Y. T. Lo.** A probabilistic approach to large circular and spherical arrays, *IEEE Trans. Antennas and Propagation,* Vol. AP-17, No. 4, pp. 514–522, July, 1969.

49. **Redlich, R. W.** Sampling synthesis of ring arrays, *IEEE Trans. Antennas and Propagation,* Vol. AP-18, No. 1, pp. 116–118, January, 1970.

50. **Coleman, H. P.** An iterative technique for reducing sidelobes of circular arrays, *IEEE Trans. Antennas and Propagation,* Vol. AP-18, No. 4, pp. 566–567, July, 1970.

51. **Chan, A. K., A. Ishimaru, and R. A. Sigelmann.** Equally spaced spherical arrays, *Radio Sci.*, Vol. 3, No. 5, pp. 401–404, May, 1968.

52. **Hoffman, M.** Convention for the analysis of spherical arrays, *IEEE Trans. Antennas and Propagation,* Vol. AP-11, No. 4, pp. 390–393, July, 1963.

53. **MacPhie, R. H.** The element density of a spherical antenna array,

IEEE Trans. Antennas and Propagation, Vol. AP-16, No. 1, pp. 125–126 January, 1968.

54. **Sengupta, D. L., T. M. Smith, and R. W. Larson.** Radiation characteristics of a spherical array of circularly polarized elements, *IEEE Trans. Antennas and Propagation,* Vol. AP-16, No. 1, pp. 2–7, January 1968.

ADDITIONAL REFERENCES

Das, R. Concentric ring array, *IEEE Trans. Antennas and Propagation* Vol. AP-14, No. 3, pp. 398–400, May, 1966.

Ksienski, A. A. Synthesis of nonseparable two-dimensional patterns by means of planar arrays, *IRE Trans. Antennas and Propagation,* Vol. AP-8, No. 2, pp. 224–225, March, 1960.

Lo, Y. T. and H. C. Hsuan. An equivalence theory between elliptical and circular arrays, *IEEE Trans. Antennas and Propagation,* Vol. AP-13 No. 2, pp. 247–256, March, 1965.

Lo, Y. T. and S. W. Lee. Affine transformation and its application to antenna arrays, *IEEE Trans. Antennas and Propagation,* Vol. AP-13 No. 6, pp. 890–896, November, 1965.

Lo. Y. T., S. W. Lee, and Q. H. Lee. Optimization of directivity and signal-to-noise ratio of an arbitrary antenna array, *Proc. IEEE,* Vol. 54, No. 8, pp. 1033–1045, August, 1966.

Mack, R. B. A study of circular arrays, Cruft Lab., *Technical Report* Harvard University, May 1, 1963.

Munger, A. D. and J. H. Provencher. Beam width and current distribution for circular array antennas, Tables, *NELC Report 1522,* October 1967.

Sureau, J. C. and A. Hessel. Element pattern for circular arrays of axial slits on large conducting cylinders, *IEEE Trans Antennas and Propagation,* Vol. AP-17, No. 6, pp. 799–803, November, 1969.

Taylor, T. T. A synthesis method for circular and cylindrical antennas composed of discrete elements, *IRE Trans. Antennas and Propagation* No. 3, pp. 251–261, August, 1952.

Willey, R. E. Space tapering of linear and planar arrays, *IRE Trans. Antennas and Propagation,* Vol. AP-10, No. 4, pp. 369–383, July, 1962

Special issue on "Electronic Scanning," *IEEE Proc.,* Vol. 56, No. 11 November, 1968.

Special issue on "Theory of Antenna Arrays," *Radio Sci.,* Vol. 3, No. 5 May, 1968.

Phased-Array Antenna Symposium Digest, Polytechnic Institute of Brooklyn, 1970.

CHAPTER 4

ARRAYS OF STANDING-WAVE ANTENNAS ABOVE LOSSY GROUND

In the first three chapters, we presented various theoretical aspects of linear and two-dimensional arrays. The arrays considered therein were constructed of isotropic or simple elements in free space, and the associated impedance problem was not duly treated. Now we are ready to formulate the array problem from an application viewpoint. First, the arrays to be considered are made of practical elements and the elements are placed above a lossy ground. Second, the current distribution on the element is assumed according to an existing and authoritative literature.[1] Third, based on the assumed current distribution the self- and mutual impedances of the element will be calculated by the induced emf method.[2] Finally, the total far-field radiated from the array, the input power accepted by the array, and therefore the power gain will also be computed.[3] The particular elements treated in this chapter are various standing-wave types of antennas.

4.1 Current and Impedance of a Center-Fed Dipole

It has been known that the radiation characteristics of an antenna in the presence of a lossy ground can be influenced substantially by the finite ground conductivity and inhomogeneity. The original work on this subject was begun some sixty years ago by Sommerfeld,[4] and was later discussed and extended by many others.[5–8] The problem was conventionally simplified on the basis of a Hertzian dipole with a specified current moment. This simplification is, of course, justifiable when the application is confined to a very low frequency range. For higher frequencies the antenna can no longer be regarded very short in terms of the operating wavelength. Under this condition, a finite length of antenna should be considered.

Recent solutions to current distribution for a single finite dipole or a pair of monopoles in the presence of a homogeneous and dissipative half-space rely almost exclusively on numerical approaches. Basically, the unknown current on the surface of an antenna is first formulated as a Hallén-type integral equation, and then solved numerically when appropriate boundary conditions are imposed.[9-11] For arrays of many elements, the resulting coupled integral equations, if formulated by the same approach as mentioned above, would be very large in number and complicated, if not impossible, to solve. To simplify the situation, we follow the traditional technique of assuming a realistic and approximate current distribution on the surface of an antenna. For example, for the center-fed dipole, the simplest form ever postulated for the current was a sinusoidal distribution.[12] The analysis based on this simple assumption is straightforward, and the approximation is known to have only negligible effect on the far-field pattern. The input impedance calculated accordingly is also considered satisfactory when the half-length of the dipole is no greater than a quarter wavelength. The accuracy, however, begins to deteriorate when the antenna length is increased.[2] Specifically, when the half-length becomes a half wavelength (or any multiple thereof), the assumed sinusoidal distribution gives a value of zero for the current at the feeding point, resulting in an infinite value for the input impedance. Of course, the actual input current for these antenna lengths will be small but not zero, and the input impedance will be large but not infinite. Furthermore, the radius of the antenna does not enter into the analysis with this simple current distribution. To offer a better approximation than the sinusoidal and to overcome the associated difficulties, King and Wu developed a theory consisting of three terms in the assumed current form.[1] Briefly, two additional trigonometric terms, as a consequence of examining the properties of the kernel in the original integral equation, are superposed on the dominant sinusoidal term. The results for the input admittance (or impedance) thus obtained were found to agree more closely with the measured data.[13] This method remains valid only when the half length of the antenna does not exceed roughly $5\lambda/8$, and when applied to arrays, requires that all the elements in the array be symmetrically located.

As an extension to the three-term theory, Chang and King proposed later a five-term method in order to be able to treat arrays whose elements are unsymmetrically distributed.[14,15]

Since most of the arrays considered in this chapter are symmetric and the element half-length is not greater than the limit of $5\lambda/8$, we decide to use King and Wu's three-term assumption for the current in our analysis.[1] That is, for a single center-fed dipole in free space,

$$I(s) = \frac{jV}{60\psi_{dR}\cos kh}\left[\sin k(h-|s|) + T_U(\cos ks - \cos kh) + T_D\left(\cos\frac{ks}{2} - \cos\frac{kh}{2}\right)\right], \qquad kh \neq \frac{\pi}{2}, \tag{4.1}$$

or

$$I(s)=\frac{-jV}{60\psi_{dR}}\left[\sin k|s|-1+T_U'\cos ks-T_D'\left(\cos\frac{ks}{2}-\cos\frac{\pi}{4}\right)\right],\qquad kh=\frac{\pi}{2},\tag{4.2}$$

where V is the applied voltage, s is the distance along the dipole axis measured from the feeding point, h is the half-length of the dipole, $k=2\pi/\lambda$, and the other symbols are defined below.

The function ψ_{dR} in (4.1) and (4.2) is defined as follows[1]:
for $kh\leqslant\pi/2$,

$$\begin{aligned}\psi_{dR}&=\csc kh\int_{-h}^{h}\sin k(h-|s'|)\left[\frac{\cos kr(0)}{r(0)}-\frac{\cos kr(h)}{r(h)}\right]ds'\\&=2(1+\cos kh)\overline{C}_c(ka,kh)-2\cos kh\overline{C}_c(ka,2kh)\\&\quad-2\cot kh(1+\cos kh)C_s(ka,kh)\\&\quad+(\cot kh\cos kh-\sin kh)C_s(ka,2kh),\end{aligned}\tag{4.3}$$

where

$$r(s)=\sqrt{(s-s')^2+a^2}\qquad\text{with } a \text{ being the radius of the dipole;}$$

for $kh\geqslant\pi/2$,

$$\begin{aligned}\psi_{dR}&=\int_{-h}^{h}\sin k(h-|s'|)\left[\frac{\cos kr(h-\lambda/4)}{r(h-\lambda/4)}-\frac{\cos kr(h)}{r(h)}\right]ds'\\&=\overline{C}_c\left(ka,\frac{\pi}{2}\right)+(1+\cos 2kh)\overline{C}_c\left(ka,kh-\frac{\pi}{2}\right)\\&\quad-\cos 2kh\overline{C}_c\left(ka,2kh-\frac{\pi}{2}\right)+\sin 2kh\left[\overline{C}_c(ka,kh)-\overline{C}_c(ka,2kh)\right]\\&\quad-(1+\cos 2kh)C_s(ka,kh)+\cos 2khC_s(ka,2kh)\\&\quad+\sin 2kh\left[C_s\left(ka,kh-\frac{\pi}{2}\right)-C_s\left(ka,2kh-\frac{\pi}{2}\right)\right].\end{aligned}\tag{4.4}$$

Note that it is perhaps more appropriate to use the notation $\psi_{dR}(ka,kh)$ to denote the expression in (4.3) and (4.4) since it is a function of both the radius and half-length of the dipole. To avoid burdensome printing, we adopt the simpler form. It is clear that, when $kh=\pi/2$, both (4.3) and (4.4) reduce to

$$\psi_{dR}\left(kh=\frac{\pi}{2}\right)=2\overline{C}_c\left(ka,\frac{\pi}{2}\right)-C_s(ka,\pi). \tag{4.5}$$

The generalized sine integrals and cosine integrals appearing in (4.3) through (4.5) and in later expressions are defined as

$$S(b,x)=\int_0^x \frac{\sin\sqrt{y^2+b^2}}{\sqrt{y^2+b^2}}\,dy, \tag{4.6}$$

$$C(b,x)=\int_0^x \frac{1-\cos\sqrt{y^2+b^2}}{\sqrt{y^2+b^2}}\,dy, \tag{4.7}$$

$$\overline{C}(b,x)=\int_0^x \frac{\cos\sqrt{y^2+b^2}}{\sqrt{y^2+b^2}}\,dy, \tag{4.8}$$

$$S_s(b,x)=\int_0^x \frac{\sin y\sin\sqrt{y^2+b^2}}{\sqrt{y^2+b^2}}\,dy, \tag{4.9}$$

$$S_c(b,x)=\int_0^x \frac{\cos y\sin\sqrt{y^2+b^2}}{\sqrt{y^2+b^2}}\,dy, \tag{4.10}$$

$$C_s(b,x)=\int_0^x \frac{\sin y\cos\sqrt{y^2+b^2}}{\sqrt{y^2+b^2}}\,dy, \tag{4.11}$$

$$C_c(b,x)=\int_0^x \frac{(1-\cos y)\cos\sqrt{y^2+b^2}}{\sqrt{y^2+b^2}}\,dy, \tag{4.12}$$

$$\overline{C}_c(b,x) = \int_0^x \frac{\cos y \cos\sqrt{y^2+b^2}}{\sqrt{y^2+b^2}}\, dy. \tag{4.13}$$

These integrals can be found in tabular form[16] or computed directly by a numerical procedure.

The complex coefficients, T_U, T_D, T'_U, and T'_D, in (4.1) and (4.2) can be expressed as

$$T_U = Q^{-1}(\psi_{dD}\psi_V - j\psi_{dI}\psi_D), \tag{4.14}$$

$$T_D = -jQ^{-1}[\psi_{dI}(\psi_{dUR}\cos kh - \psi_U) + \psi_{dUI}\psi_V], \tag{4.15}$$

$$T'_U = \frac{\psi_{dD}\psi_{dR}}{\psi_{dD}\psi_V - j\psi_{dI}\psi_D}, \tag{4.16}$$

$$T'_D = \frac{j\psi_{dI}\psi_{dUR}}{\psi_{dD}\psi_V - j\psi_{dI}\psi_D}, \tag{4.17}$$

where

$$Q = \psi_{dD}(\psi_{dUR}\cos kh - \psi_U) + j\psi_{dUI}\psi_D, \tag{4.18}$$

$$\psi_{dUR} = \frac{1}{1-\cos kh}\int_{-h}^{h} (\cos ks' - \cos kh)\left[\frac{\cos kr(0)}{r(0)} - \frac{\cos kr(h)}{r(h)}\right] ds'$$

$$= \frac{2}{1-\cos kh}\overline{C}_c(ka,kh) - \frac{\cos kh}{1-\cos kh}\overline{C}_c(ka,2kh)$$

$$- \frac{2\cos kh}{1-\cos kh}\overline{C}(ka,kh) + \frac{\cos kh}{1-\cos kh}\overline{C}(ka,2kh)$$

$$- \frac{\sin kh}{1-\cos kh}C_s(ka,2kh), \tag{4.19}$$

$$\psi_{dUI}=\frac{1}{1-\cos(kh/2)}\int_{-h}^{h}(\cos ks'-\cos kh)\left[\frac{\sin kr(0)}{r(0)}-\frac{\sin kr(h)}{r(h)}\right]ds'$$

$$=\frac{1}{1-\cos(kh/2)}[-2S_c(ka,kh)+\cos khS_c(ka,2kh)$$

$$+2\cos khS(ka,kh)-\cos khS(ka,2kh)$$

$$+\sin khS_s(ka,2kh)], \tag{4.20}$$

$$\psi_{dI}=\frac{1}{1-\cos(kh/2)}\int_{-h}^{h}\sin k(h-|s'|)\left[\frac{\sin kr(0)}{r(0)}-\frac{\sin kr(h)}{r(h)}\right]ds'$$

$$=\frac{1}{1-\cos(kh/2)}\{-2\sin khS_c(ka,kh)+2\cos khS_s(ka,kh)$$

$$-\sin 2kh[S_c(ka,kh)-S_c(ka,2kh)]$$

$$+(1+\cos 2kh)S_s(ka,kh)-\cos 2khS_s(ka,2kh)\}, \tag{4.21}$$

$$\psi_{dD}=\frac{1}{1-\cos(kh/2)}\int_{-h}^{h}\left(\cos\frac{ks'}{2}-\cos\frac{kh}{2}\right)\left[\frac{e^{-jkr(0)}}{r(0)}-\frac{e^{-jkr(h)}}{r(h)}\right]ds'$$

$$=\frac{1}{1-\cos(kh/2)}(B_1+jC_1), \tag{4.22}$$

with

$$B_1 = 2\sinh^{-1}\left(\frac{h}{a}\right) - C\left(\frac{ka\sqrt{3}}{2}, \frac{kh}{2}\right) - C\left(\frac{ka\sqrt{3}}{2}, \frac{3kh}{2}\right)$$

$$- \sinh^{-1}\left(\frac{2h}{a}\right)\cos\frac{kh}{2} + \tfrac{1}{2}\cos\frac{kh}{2}\, C\left(\frac{ka\sqrt{3}}{2}, 3kh\right)$$

$$- \tfrac{1}{2}\sin\frac{kh}{2}\, S\left(\frac{ka\sqrt{3}}{2}, 3kh\right) + \cos\frac{kh}{2}\, \bar{C}(ka, 2kh)$$

$$+ \tfrac{1}{2}\cos\frac{kh}{2}\, C\left(\frac{ka\sqrt{3}}{2}, kh\right) + \tfrac{1}{2}\sin\frac{kh}{2}\, S\left(\frac{ka\sqrt{3}}{2}, kh\right)$$

$$- 2\cos\frac{kh}{2}\, \bar{C}(ka, kh), \tag{4.23}$$

$$C_1 = -S\left(\frac{ka\sqrt{3}}{2}, \frac{kh}{2}\right) - S\left(\frac{ka\sqrt{3}}{2}, \frac{3kh}{2}\right) + \tfrac{1}{2}\sin\frac{kh}{2}\, C\left(\frac{ka\sqrt{3}}{2}, 3kh\right)$$

$$+ \tfrac{1}{2}\cos\frac{kh}{2}\, S\left(\frac{ka\sqrt{3}}{2}, 3kh\right) + \tfrac{1}{2}\cos\frac{kh}{2}\, S\left(\frac{ka\sqrt{3}}{2}, kh\right)$$

$$- \tfrac{1}{2}\sin\frac{kh}{2}\, C\left(\frac{ka\sqrt{3}}{2}, kh\right) + 2\cos\frac{kh}{2}\, S(ka, kh)$$

$$- \cos\frac{kh}{2}\, S(ka, 2kh), \tag{4.24}$$

$$\psi_V = \int_{-h}^{h} \sin k(h - |s'|)\, \frac{e^{-jkr(h)}}{r(h)}\, ds' = B_2 + jC_2, \tag{4.25}$$

with

$$B_2 = -\sin 2kh\left[\bar{C}_c(ka, kh) - \bar{C}_c(ka, 2kh)\right]$$

$$+ \cos 2kh\left[C_s(ka, kh) - C_s(ka, 2kh)\right] + C_s(ka, kh), \tag{4.26}$$

$$C_2=\sin 2kh[S_c(ka,kh)-S_c(ka,2kh)]$$

$$-\cos 2kh[S_s(ka,kh)-S_s(ka,2kh)]-S_s(ka,kh), \tag{4.27}$$

$$\psi_U=\int_{-h}^{h}(\cos ks'-\cos kh)\frac{e^{-jkr(h)}}{r(h)}ds'$$

$$=\cos kh\Big[\bar{C}_c(ka,2kh)-\bar{C}(ka,2kh)\Big]+\sin kh C_s(ka,2kh)$$

$$+j\{\cos kh[S(ka,2kh)-S_c(ka,2kh)]-\sin kh S_s(ka,2kh)\}, \tag{4.28}$$

$$\psi_D=\int_{-h}^{h}\left(\cos\frac{ks'}{2}-\cos\frac{kh}{2}\right)\frac{e^{-jkr(h)}}{r(h)}ds'=B_3+jC_3, \tag{4.29}$$

with

$$B_3=\cos\frac{kh}{2}\left[\sinh^{-1}\left(\frac{2h}{a}\right)-\tfrac{1}{2}C\left(\frac{ka\sqrt{3}}{2},3kh\right)\right.$$

$$\left.-\tfrac{1}{2}C\left(\frac{ka\sqrt{3}}{2},kh\right)-\bar{C}(ka,2kh)\right]$$

$$+\tfrac{1}{2}\sin\frac{kh}{2}\left[S\left(\frac{ka\sqrt{3}}{2},3kh\right)-S\left(\frac{ka\sqrt{3}}{2},kh\right)\right], \tag{4.30}$$

$$C_3=\cos\frac{kh}{2}\left[S(ka,2kh)-\tfrac{1}{2}S\left(\frac{ka\sqrt{3}}{2},3kh\right)-\tfrac{1}{2}S\left(\frac{ka\sqrt{3}}{2},kh\right)\right]$$

$$+\tfrac{1}{2}\sin\frac{kh}{2}\left[C\left(\frac{ka\sqrt{3}}{2},kh\right)-C\left(\frac{ka\sqrt{3}}{2},3kh\right)\right]. \tag{4.31}$$

With a set of values for a and h given, all the functions outlined abov can be calculated. Consequently, we have, from (4.1) and (4.2), tl self-impedance of the dipole,

$$Z_s = \frac{-j60\psi_{dR}\cos kh}{\sin kh + T_U(1-\cos kh) + T_D[1-\cos(kh/2)]}, \qquad \text{for} \qquad kh \neq \frac{\pi}{2}, \tag{4.32}$$

or

$$Z_s = \frac{j60\psi_{dR}}{-1 + T'_U - (1-\sqrt{2}/2)T'_D}, \qquad \text{for} \qquad kh = \frac{\pi}{2}. \tag{4.33}$$

The open-circuit mutual impedance between two parallel dipoles of half-lengths h_1 and h_2 spaced a distance d apart, as shown in Fig. 4.1, may be calculated by

$$Z_m = \frac{-1}{I_1(0)I_2(0)} \int_{-h_2}^{h_2} E_{s1}(s) I_2(s)\, ds, \tag{4.34}$$

where $I_1(0)$ and $I_2(0)$ are input currents of No. 1 and No. 2 antennas, respectively. The s component electric field $E_{s1}(s)$, produced at No. 2 antenna by the current on No. 1 antenna, is given by[6]

$$E_{s1} = \frac{-j30}{k}\left(k^2 + \frac{\partial^2}{\partial s^2}\right) \int_{-h_1}^{h_1} I_1(s') K(s,s')\, ds', \tag{4.35}$$

where

$$K(s,s') = \frac{e^{-jkR}}{R}, \tag{4.36}$$

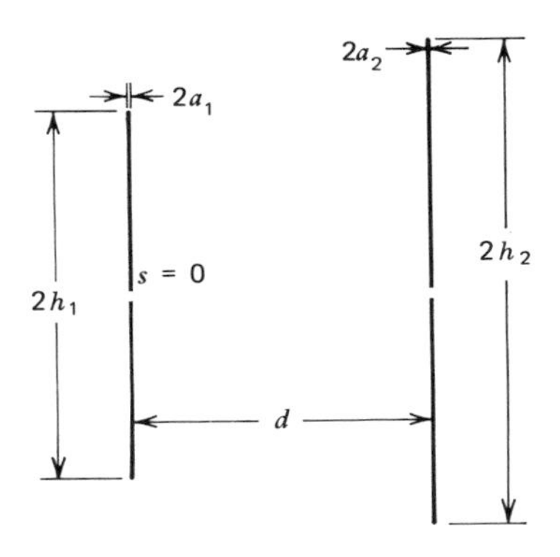

Fig. 4.1 Two parallel dipoles when their feed points are both at $s=0$.

$$R=\left[(s-s')^2+d^2\right]^{1/2}. \tag{4.37}$$

Note that in (4.35) we have suppressed the time-dependence factor $e^{j\omega t}$, and that in (4.36) we have ignored the contribution from the radius of the dipole. The currents $I_2(s)$ in (4.34) and $I_1(s')$ in (4.35) can take the assumed form (4.1) or (4.2) depending on kh. By doing this we have also assumed that the current distribution on one antenna is not affected by the presence of another antenna.

Substituting (4.35) into (4.34) and using (4.1) and (4.2) for $I_1(s')$ and $I_2(s)$, we have:

(a) for $kh_1 \neq \pi/2$ and $kh_2 \neq \pi/2$,

$$Z_m=\frac{-j30\{kI_a+2(\cos kh_1)I_b-[1+T_{U1}\sin kh_1+\tfrac{1}{2}T_{D1}\sin(kh_1/2)]I_c\}}{D_aD_b}, \tag{4.38}$$

where

$$D_a=\sin kh_1+T_{U1}(1-\cos kh_1)+T_{D1}\left(1-\cos\frac{kh_1}{2}\right), \tag{4.39}$$

$$D_b=\sin kh_2+T_{U2}(1-\cos kh_2)+T_{D2}\left(1-\cos\frac{kh_2}{2}\right), \tag{4.40}$$

$$I_a=\int_{-h_2}^{h_2}\left[\sin k(h_2-|s|)+T_{U2}(\cos ks-\cos kh_2)+T_{D2}\left(\cos\frac{ks}{2}-\cos\frac{kh_2}{2}\right)\right]$$
$$\times\int_0^{h_1}\left(T_{U1}\cos kh_1-\tfrac{3}{4}T_{D1}\cos\frac{ks'}{2}+T_{D1}\cos\frac{kh_1}{2}\right)[K(s,s')+K(s,-s')]\,ds'\,ds \tag{4.41}$$

$$I_b=\int_{-h_2}^{h_2}K(s,0)\left[\sin k(h_2-|s|)+T_{U2}(\cos ks-\cos kh_2)\right.$$
$$\left.+T_{D2}\left(\cos\frac{ks}{2}-\cos\frac{kh_2}{2}\right)\right]ds, \tag{4.42}$$

$$I_c = \int_{-h_2}^{h_2} [K(s,h_1) + K(s,-h_1)] \bigg[\sin k(h_2 - |s|) + T_{U2}(\cos ks - \cos kh_2) + T_{D2}\left(\cos \frac{ks}{2} - \cos \frac{kh_2}{2}\right) \bigg] ds; \tag{4.43}$$

(b) for $kh_1 = \pi/2$ and $kh_2 \neq \pi/2$,

$$Z_m = \frac{j30\{kI_d + 2I_b + [T'_{U1} - \frac{1}{2}\sin(\pi/4)T'_{D1}]I_c\}}{D_c D_b}, \tag{4.44}$$

where

$$D_c = -1 + T'_{U1} - T'_{D1}\left(1 - \cos\frac{\pi}{4}\right), \tag{4.45}$$

$$I_d = \int_{-h_2}^{h_2} \left[\sin k(h_2 - |s|) + T_{U2}(\cos ks - \cos kh_2) + T_{D2}\left(\cos \frac{ks}{2} - \cos \frac{kh_2}{2}\right) \right] \times \int_0^{h_1} \left(-1 + \tfrac{3}{4} T'_{D1} \cos \frac{ks'}{2} + \cos \frac{\pi}{4} T'_{D1} \right) [K(s,s') + K(s,-s')] \, ds' \, ds; \tag{4.46}$$

(c) for $kh_1 = kh_2 = \pi/2$,

$$Z_m = \frac{j30\{kI_e + 2I_f + [T'_{U1} - \frac{1}{2}\sin(\pi/4)T'_{D1}]I_g\}}{D_c D_d}, \tag{4.47}$$

where

$$D_d = -1 + T'_{U2} - T'_{D2}\left(1 - \cos\frac{\pi}{4}\right) \tag{4.48}$$

(which will be identical to D_c if $a_2 = a_1$),

$$I_e = \int_{-h_2}^{h_2} \left[\sin k|s| - 1 + T'_{U2} \cos ks - T'_{D2}\left(\cos \frac{ks}{2} - \cos \frac{\pi}{4}\right) \right] \times \int_0^{h_1} \left(-1 + \tfrac{3}{4} T'_{D1} \cos \frac{ks'}{2} + \cos \frac{\pi}{4} T'_{D1} \right) [K(s,s') + K(s,-s')] \, ds' \, ds, \tag{4.49}$$

$$I_f = \int_{-h_2}^{h_2} K(s,0)\left[\sin k|s| - 1 + T'_{U2}\cos ks - T'_{D2}\left(\cos\frac{ks}{2} - \cos\frac{\pi}{4}\right)\right] ds, \tag{4.50}$$

$$I_g = \int_{-h_2}^{h_2} [K(s,h_1) + K(s,-h_1)] \times \left[\sin k|s| - 1 + T'_{U2}\cos ks - T'_{D2}\left(\cos\frac{ks}{2} - \cos\frac{\pi}{4}\right)\right] ds. \tag{4.51}$$

Since T_U, T_D, T'_U, and T'_D are given in (4.14) through (4.17), the integrals I_a through I_g can all be computed numerically once the values for a_1, a_2, h_1, h_2, and d are specified. Therefore, the mutual impedance in (4.38), (4.44), or (4.47) can also be calculated. Obviously, when $T_{U1} = T_{U2} = T_{D1} = T_{D2} = 0$, Eq. (4.38) reduces to the familiar expression for the case of a simple sinusoidal current distribution.[2]

4.2 Fields and Power Gain

When a current distribution such as (4.1) or (4.2) is assumed on a wire antenna situated in a coordinate system, as shown in Fig. 4.2, with the xy plane representing the flat, homogeneous, and lossy earth, we can derive

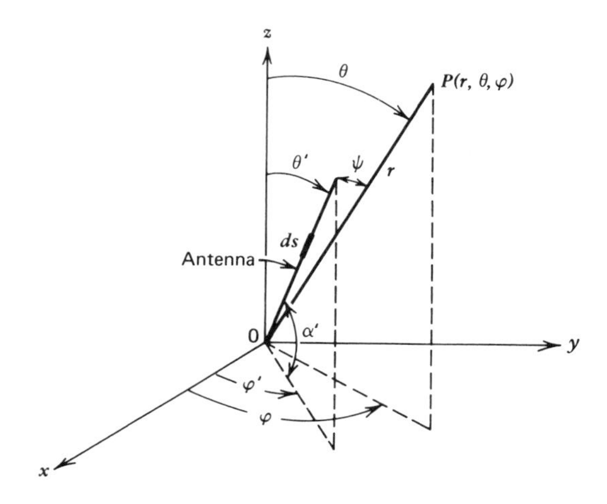

Fig. 4.2 Arbitrarily oriented linear antenna above flat earth.

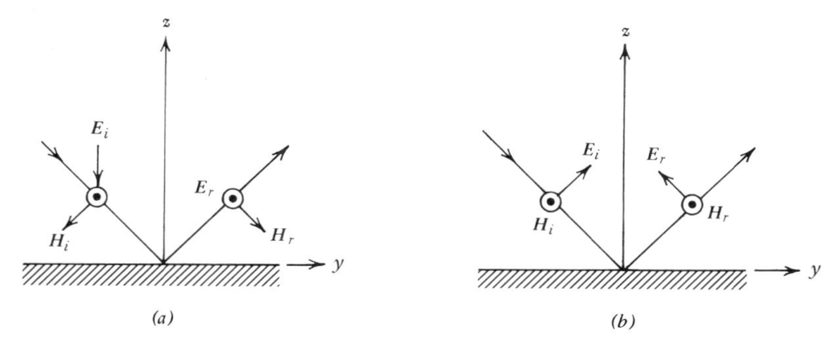

Fig. 4.3 Incident and reflected fields that have (*a*) horizontal polarization, and (*b*) vertical polarization.

the far-field produced at a distant point $P(r,\theta,\varphi)$, including the ground-reflected part. Before doing so, we must consider two separate cases depending on the polarization. The first of these is the case in which the incident (direct) electric vector is parallel to the reflecting surface (xy plane) or perpendicular to the plane of incidence. This case is often referred to as horizontal polarization. In the second case called vertical polarization, the incident electric vector is parallel to the plane of incidence (or the incident magnetic vector is parallel to the reflecting surface). These two cases are shown in Fig. 4.3, where the directions of the incident electric (E_i) and magnetic vectors (H_i) have been, respectively, assumed in the positive x direction. The terms "horizontally and vertically polarized fields" are sometimes confusing. While the electric vectors (E_i and E_r) of the horizontally polarized case are horizontal with respect to the ground [as are clearly shown in Fig. 4.3(a)], the electric vectors in the vertically polarized case are not entirely vertical, as can be seen from Fig. 4.3(b). In the latter case, the electric vectors, in fact, have both the horizontal and vertical components.

With this preliminary explanation, we obtain the total far-field components above the ground as follows:

(a) for horizontal polarization,

$$E_h = \text{total horizontal component} = E_i(r_1) + E_r(r_2)$$

$$= E_i(r_1) + R_h E_i(r_2), \tag{4.52}$$

where r_1 is the direct distance between the antenna feed and the far-field point, r_2 represents the distance between the same points but through the reflecting path, E_i and E_r are, respectively, the direct and reflected electric vectors, and R_h is the complex reflection coefficient for horizontal polarization which is defined as[17]

$$R_h = \frac{\cos\theta - (k'/k)\left\{1 - [(k/k')\sin\theta]^2\right\}^{1/2}}{\cos\theta + (k'/k)\left\{1 - [(k/k')\sin\theta]^2\right\}^{1/2}}, \tag{4.53}$$

with

$$k' \cong k\left[\epsilon_r - j\frac{18\sigma(10)^3}{f_{\mathrm{MHz}}}\right]^{1/2}, \tag{4.54}$$

$$\begin{aligned} \epsilon_r &= \text{relative permittivity (or dielectric constant) of the earth,} \\ \sigma &= \text{conductivity of the earth in mho/m,} \\ f_{\mathrm{MHz}} &= \text{operating frequency in MHz;} \end{aligned} \tag{4.55}$$

(b) for vertical polarization,

$$\begin{aligned} E'_h &= \text{total horizontal component [see Fig. 4.3 (b)]} \\ &= E_i(r_1) - R_v E_i(r_2), \end{aligned} \tag{4.56}$$

$$\begin{aligned} E_v &= \text{total vertical component} \\ &= E_i(r_1) + R_v E_i(r_2), \end{aligned} \tag{4.57}$$

where the reflection coefficient for vertical polarization is given by

$$R_v = \frac{\cos\theta - (k/k')\left\{1 - [(k/k')\sin\theta]^2\right\}^{1/2}}{\cos\theta + (k/k')\left\{1 - [(k/k')\sin\theta]^2\right\}^{1/2}}. \tag{4.58}$$

To emphasize the fact that E_i depends on r_1 while E_r depends on r_2, the variables r_1 and r_2 are specifically introduced as arguments of the field components in (4.52), (4.56), and (4.57). The dependence of the fields on many other parameters is not explicitly expressed in order to simplify the notation.

Applying these field equations to Fig. 4.2 and expressing the result in terms of spherical coordinates, we have[3]

$$E_\theta = j30k\frac{e^{-jkr_1}}{r_1}$$

$$\times\Big\{\cos\alpha'\cos(\varphi-\varphi')\cos\theta\int I(s)\exp(jks\cos\psi)$$

$$\times[1-R_v\exp(-j2kH_s\cos\theta)]\,ds$$

$$-\sin\alpha'\sin\theta\int I(s)\exp(jks\cos\psi)[1+R_v\exp(-j2kH_s\cos\theta)]\,ds\Big\}, \quad (4.59)$$

and

$$E_\varphi = -j30k\frac{e^{-jkr_1}}{r_1}\cos\alpha'\sin(\varphi-\varphi')$$

$$\times\int I(s)\exp(jks\cos\psi)[1+R_h\exp(-j2kH_s\cos\theta)]\,ds, \quad (4.60)$$

where the term $\exp(jks\cos\psi)$ is the phase advance of the current element ds at s from the feeding point; H_s, the height of the current element ds above the ground; α', the angle between the antenna and its projection on the ground; and ψ, the angle between the antenna and the direction of $P(r,\theta,\varphi)$ at which the field components are calculated. The expression of $\cos\psi$ is similar to (1.2),

$$\cos\psi = \cos\theta\cos\theta' + \sin\theta\sin\theta'\cos(\varphi-\varphi'), \quad (4.61)$$

where the prime coordinates refer to the source direction (antenna coordinates).

Note that the integrations in (4.59) and (4.60) should be performed along the antenna length. The integration limits are therefore determined by the actual antenna geometry. In (4.59) and (4.60) we have also applied the conventional far-field approximation for r_2. This means that $r_2 \cong r_1$ is used for the inverse-distance factor, and that $r_2 \cong r_1 + 2H_s\cos\theta$ is used for the phase factor. In addition, we have also neglected the surface wave component because it contributes only near the grazing angle ($\theta=\pi/2$) and attenuates very rapidly with distance.[2,17]

The directive gain can be calculated if we insert

$$|E|^2 = |E_\theta|^2 + |E_\varphi|^2 \quad (4.62)$$

into (1.7) and (1.8). However, because of the complexity of (4.59) and (4.60), it is quite involved, if not impossible, to calculate analytically (1.8) which represents the total power radiated by the subject antenna:

$$W_0 = \int_0^{2\pi} \int_0^{\pi} |E|^2 \sin\theta \, d\theta \, d\varphi. \tag{4.63}$$

In the practical application, we sometimes prefer to calculate the power gain instead.[18,19] This latter quantity, according to the standard set by the Institute of Electrical and Electronics Engineers (IEEE), is defined as[20]

$$G(\theta,\varphi) = \frac{4\pi r_1^2 |E \times H^*|}{W_{in}} = \frac{r_1^2 (|E_\theta|^2 + |E_\varphi|^2)}{30 W_{in}}, \tag{4.64}$$

or

$$10 \log G \text{ in dB}, \tag{4.65}$$

where

$$|E \times H^*| = \frac{1}{120\pi} |E|^2 = \frac{1}{120\pi} (|E_\theta|^2 + |E_\varphi|^2), \tag{4.66}$$

$$W_{in} = |I_{in}|^2 R_{in}, \tag{4.67}$$

$$I_{in} = \text{input current at the feeding point}, \tag{4.68}$$

$$R_{in} = \text{input resistance}. \tag{4.69}$$

The ratio of power gain to directive gain, or W_0 / W_{in}, should represent the radiation efficiency of the antenna.

The input current and resistance required in (4.64) can be determined easily for an antenna in free space with the assumed current distribution (4.1) or (4.2). In fact, under these conditions, R_{in} will be equal to the real part of the self-impedance given in (4.32) or (4.33). On the other hand, the input impedance of an antenna above an imperfect plane earth can also be calculated by viewing the antenna and its imperfect image to form a coupled two-port network[21,22]:

$$V_1 = Z_{11} I_1 + Z_{12} I_2, \tag{4.70}$$

where V_1 in volts and I_1 in amperes are, respectively, the impressed voltage and input current of the antenna, Z_{11} in ohms is the self-impedance [Z_{11} is the same as Z_s in (4.32) or (4.33)], Z_{12} in ohms is the mutual impedance between the antenna and its image, and I_2 in amperes is the image current of I_1 with respect to the imperfect ground. According to the literature[3,21,22]

$$I_2 = CI_1, \tag{4.71}$$

where, for the horizontal antenna ($\alpha' = 0$, see Fig. 4.2),

$$\begin{aligned} C &= R'_h = R_h \text{ of (4.53) calculated at } \theta = 0 \\ &= \frac{k - k'}{k + k'}, \end{aligned} \tag{4.72}$$

and for the vertical antenna ($\alpha' = \pi/2$),

$$\begin{aligned} C &= R'_v = R_v \text{ of (4.58) calculated at } \theta = 0 \\ &= \frac{k' - k}{k' + k}. \end{aligned} \tag{4.73}$$

It can be verified that when the earth is perfectly conducting ($\sigma = \infty$, $k' = \infty$), C approaches respectively -1 and $+1$ for horizontal and vertical antennas.

The input impedance of the antenna above a ground plane is, then,

$$Z_{\text{in}} = \frac{V_1}{I_1} = Z_{11} + Z_{12}\left(\frac{I_2}{I_1}\right) = Z_{11} + CZ_{12}, \tag{4.74}$$

from which we have

$$R_{\text{in}} = R_{11} + \text{Re}(CZ_{12}), \tag{4.75}$$

with R_{11} representing the real part of Z_{11}. The mutual impedance required here can be calculated according to (4.34) for the horizontal antenna, or approximately evaluated by other means to be discussed in later sections for other antenna orientations.

The formulation developed here for a single antenna above a lossy ground can be extended to calculate the impedances and power gain of an array with many dipoles and monopoles as its elements. Now, it should be instructive, based on what we have outlined, to give quantitative results in detail for a simple case. This is presented in the next section.

4.3 Single Horizontal Dipole

The geometry for this problem is depicted in Fig. 4.4. In this case, $\alpha'=0$, $\theta'=\pi/2$, $\varphi'=\pi/2$, $H_s=H$, $s=y$, $\cos\psi=\sin\theta\sin\varphi$, and

$$I(s)=I\left\{\sin k(h-|y|)+T_U(\cos ky-\cos kh) + T_D(\cos \tfrac{1}{2}ky-\cos \tfrac{1}{2}kh)\right\}, \qquad kh\neq\frac{\pi}{2}, \tag{4.76}$$

or

$$I(s)=I'\left\{\sin k|y|-1+T'_U\cos ky-T'_D\left(\cos \tfrac{1}{2}ky-\cos\frac{\pi}{4}\right)\right\}, \qquad kh=\frac{\pi}{2}, \tag{4.77}$$

where T_U, T_D, T'_U, and T'_D are given in (4.14) through (4.17). Note that here we have implicitly assumed that the presence of earth will not affect the distribution form for the current. In reality, this is definitely not true. In view of the complicated derivations experienced in Section 4.2 and the fact that the effect will probably not be too substantial if the antenna is not too close to the ground, we are satisfied with this assumption without overburdening the mathematics involved. For a more rigorous formulation, the reader is referred to other existing works.[9–11]

Substituting (4.76) into (4.59) and (4.60), we obtain

(a) for $kh\neq\pi/2$,

$$\begin{aligned} E_\theta &= j60I\frac{e^{-jkr_1}}{r_1}\sin\varphi\cos\theta\left[1-R_v\exp(-j2kH\cos\theta)\right]A_1,\\ E_\varphi &= j60I\frac{e^{-jkr_1}}{r_1}\cos\varphi\left[1+R_h\exp(-j2kH\cos\theta)\right]A_1 \end{aligned} \tag{4.78}$$

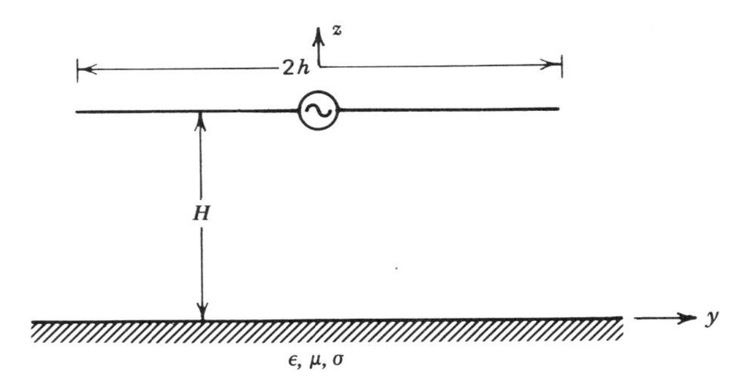

Fig. 4.4 A horizontal dipole above flat earth.

where

$$A_1 = \frac{\cos(kh\cos\psi) - \cos kh}{\sin^2\psi} - \frac{T_U \cos kh \sin(kh\cos\psi)}{\cos\psi}$$

$$+ \frac{T_U}{\sin^2\psi}[\sin kh \cos(kh\cos\psi) - \cos\psi \cos kh \sin(kh\cos\psi)]$$

$$+ \frac{T_D}{\frac{1}{4} - \cos^2\psi}[\tfrac{1}{2}\sin\tfrac{1}{2}kh\cos(kh\cos\psi) - \cos\psi\cos\tfrac{1}{2}kh\sin(kh\cos\psi)]$$

$$- \frac{T_D \cos\frac{1}{2}kh \sin(kh\cos\psi)}{\cos\psi}, \tag{4.79}$$

and R_v and R_h are, respectively, given in (4.58) and (4.53). The input current and resistance in this case are

$$I_{\text{in}} = I[\sin kh + T_U(1 - \cos kh) + T_D(1 - \cos\tfrac{1}{2}kh)], \tag{4.80}$$

$$R_{\text{in}} = \text{Re}(Z_{11} + R'_h Z_{12}) = \text{Re}\left(Z_{11} + \frac{k - k'}{k + k'} Z_{12}\right), \tag{4.81}$$

where

$$Z_{11} = -\frac{j60\psi_{dR}\cos kh}{\sin kh + T_U(1 - \cos kh) + T_D(1 - \cos\frac{1}{2}kh)}, \tag{4.82}$$

$$Z_{12} = Z_m \text{ in (4.38) with } h_1 = h_2 = h \text{ and } d = 2H, \tag{4.83}$$

k' is given in (4.54), and ψ_{dR} can be found in (4.3) or (4.4).

(b) for $kh = \pi/2$,

$$E_\theta = j60I'\frac{e^{-jkr_1}}{r_1}\sin\varphi\cos\theta[1 - R_v\exp(-j2kH\cos\theta)]A'_1,$$

$$E_\varphi = j60I'\frac{e^{-jkr_1}}{r_1}\cos\varphi[1 + R_h\exp(-j2kH\cos\theta)]A'_1, \tag{4.84}$$

$$I_{\text{in}} = I'\left[-1 + T'_U - T'_D\left(1 - \cos\frac{\pi}{4}\right)\right], \tag{4.85}$$

$$Z_{11}=\frac{j60\psi_{dR}}{-1+T'_U-T'_D(1-\cos\pi/4)}, \tag{4.86}$$

$$Z_{12}=Z_m \text{ in (4.47) with } d=2H. \tag{4.87}$$

The factor ψ_{dR} in (4.86) is given in (4.5), and A'_1 in (4.84) is

$$A'_1=\frac{1-\cos\psi\sin[(\pi/2)\cos\psi]}{\sin^2\psi}-\frac{\sin[(\pi/2)\cos\psi]}{\cos\psi}$$
$$+\frac{T'_U\cos[(\pi/2)\cos\psi]}{\sin^2\psi}+\frac{T'_D\cos(\pi/4)\sin[(\pi/2)\cos\psi]}{\cos\psi}$$
$$-\frac{T'_D}{\frac{1}{4}-\cos^2\psi}\left[\tfrac{1}{2}\sin(\pi/4)\cos\left(\frac{\pi}{2}\cos\psi\right)-\cos\psi\cos\left(\frac{\pi}{4}\right)\sin\left(\frac{\pi}{2}\cos\psi\right)\right]. \tag{4.88}$$

Substituting the above formulas into (4.64), we can present some numerical results for power gain by a series of figures. In Fig. 4.5, the power gain of a half-wavelength horizontal dipole ($2h=\lambda/2$) above two different grounds (sea water and poor ground) with $f=10$ MHz, $h/a=4680$, $H=\lambda/4$ is shown as functions of θ and φ. If the ground were perfectly conducting, the maximum power gain for this height ($H=\lambda/4$) should occur at $\theta=0°$. It is also true for the two grounds considered here, except the performance for the poor ground is distorted somewhat in the *xz* plane ($\varphi=0°$) so that the position of the maximum power gain occurs at $\theta\cong 37°$. The difference in the maximum power gain for these two grounds can be as much as approximately 2.5~3.0 dB in favor of sea water because of its higher conductivity.

The results for the same antenna, grounds, and frequency with H changed to $3\lambda/8$ are shown in Fig. 4.6. The major difference is that the position of the maximum gain shifts to $\theta=20°$–$55°$ from $\theta=0°$. If the height of the antenna above ground is increased further to $H=\lambda/2$, the maximum gain point will be shifted further toward the ground ($\theta=90°$) as evidence by most of the curves in Fig. 4.7. This kind of change in characteristics should, of course, be expected from the elementary array theory discussed in Chapter 1, because the situation here corresponds to a two-element array with the element spacing represented by $2H$. Note that there is an exception for the dashed curve in Fig. 4.7(*b*) (corresponding to $\varphi=90°$ and poor ground). The explanation for this is that the imperfect

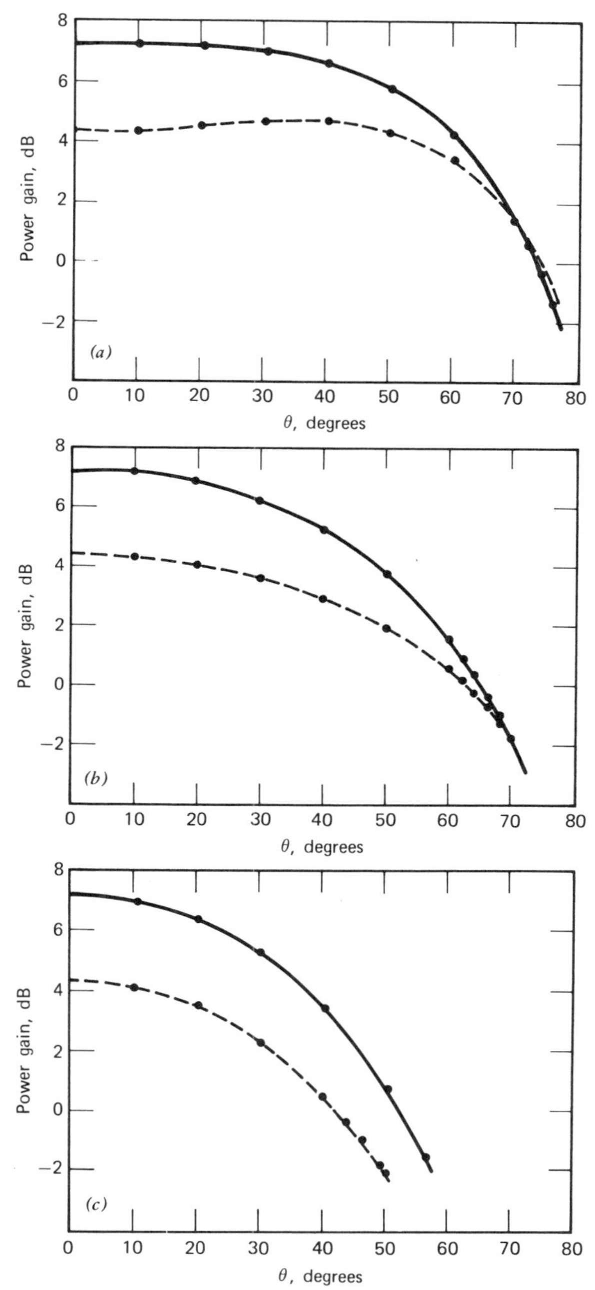

Fig. 4.5 Power gain of a horizontal dipole above a flat lossy ground with $h=\lambda/4$, $H=\lambda/4$, $h/a=4680$, and $f=10$ MHz: (*a*) $\varphi=0°$, (*b*) $\varphi=45°$, (*c*) $\varphi=90°$. Solid curves are for sea water ($\epsilon_r=80$, $\sigma=5$ mho/m); dashed curves are for poor ground ($\epsilon_r=4$, $\sigma=0.001$ mho/m).

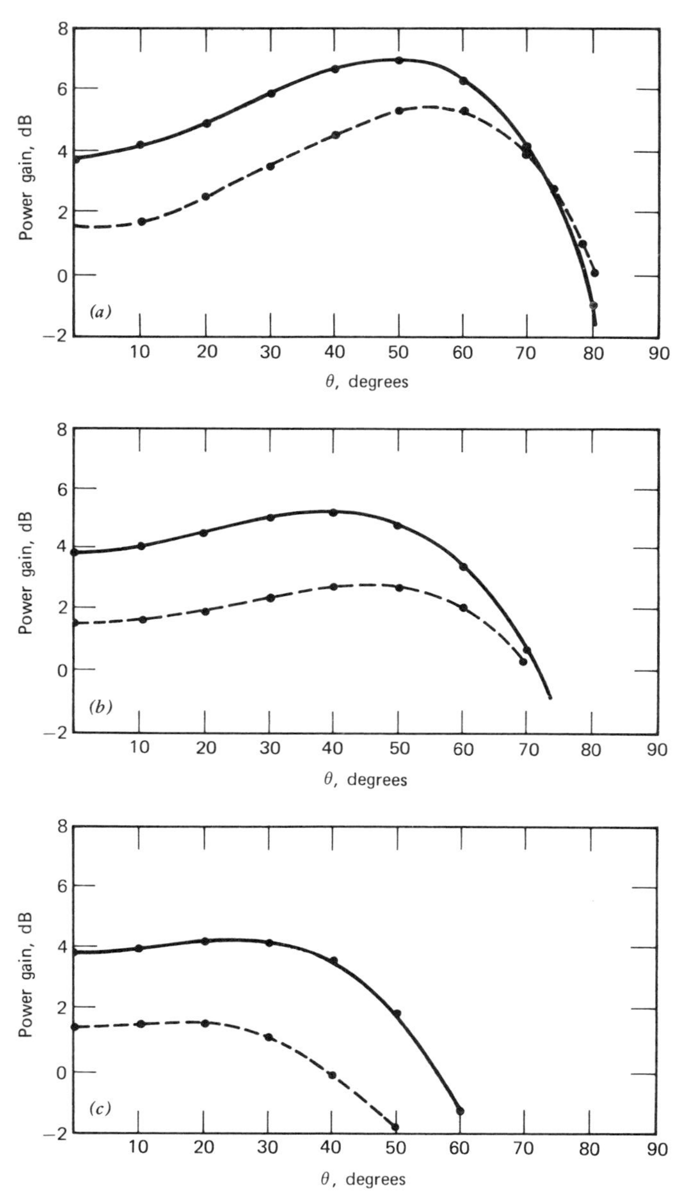

Fig. 4.6 Power gain of a horizontal dipole above a flat lossy ground with $h=\lambda/4$, $H=3\lambda/8$, $h/a=4680$, and $f=10$ MHz: (*a*) $\varphi=0°$, (*b*) $\varphi=45°$, (*c*) $\varphi=90°$. Solid and dashed curves are for sea water and poor ground, respectively.

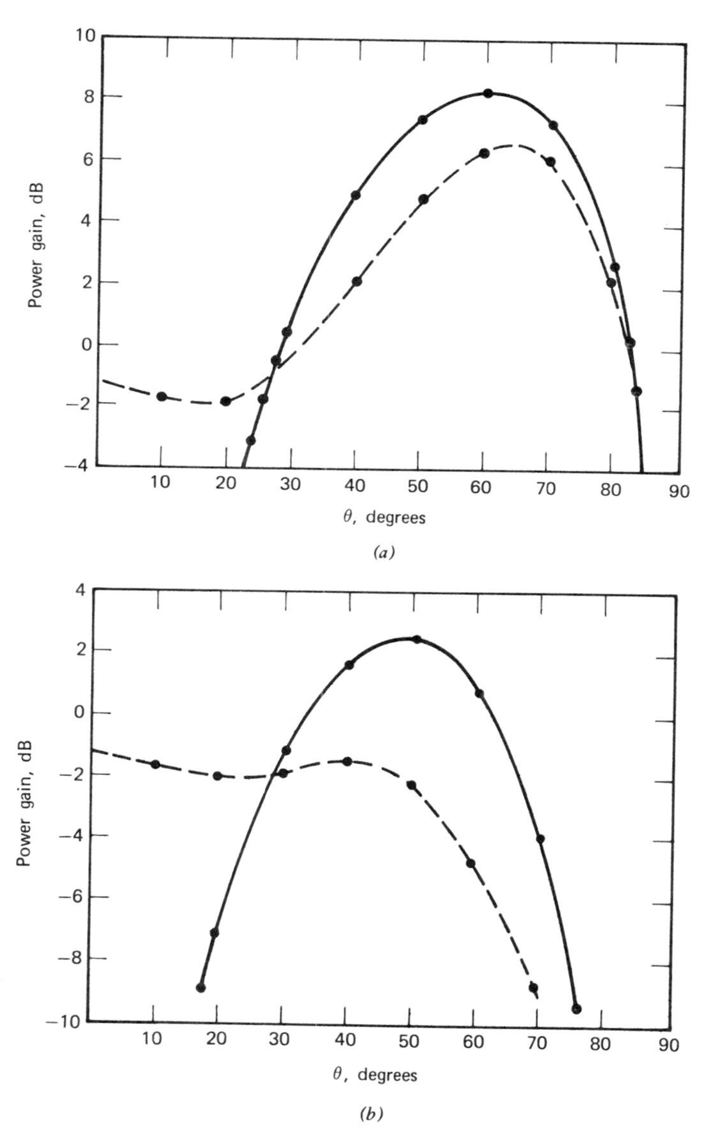

Fig. 4.7 Power gain of a horizontal dipole above a flat lossy ground with $h=\lambda/4$, $H=\lambda/2$, $h/a=4680$, and $f=10$ MHz: (*a*) $\varphi=0°$, (*b*) $\varphi=90°$. Solid and dashed curves are for sea water and poor ground, respectively.

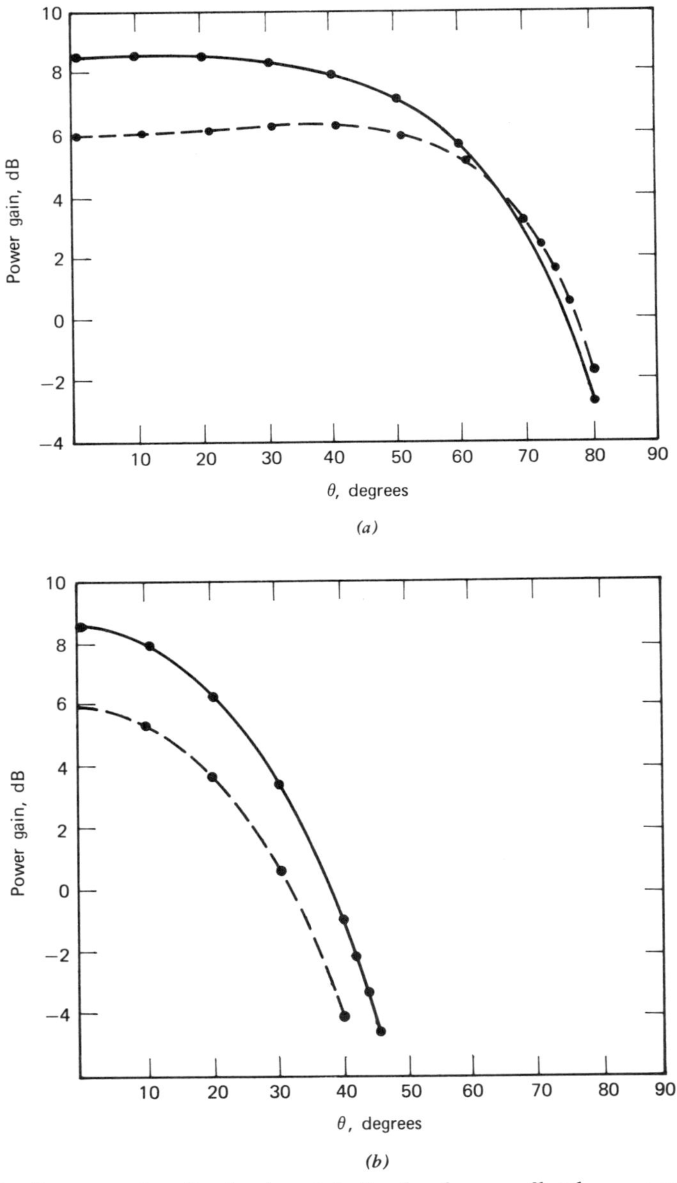

Fig. 4.8 Power gain of a horizontal dipole above a flat lossy ground with $h=\lambda/2$, $H=\lambda/4$, $h/a=9375$, and $f=10$ MHz: (*a*) $\varphi=0°$, (*b*) $\varphi=90°$. Solid and dashed curves are for sea water and poor ground, respectively.

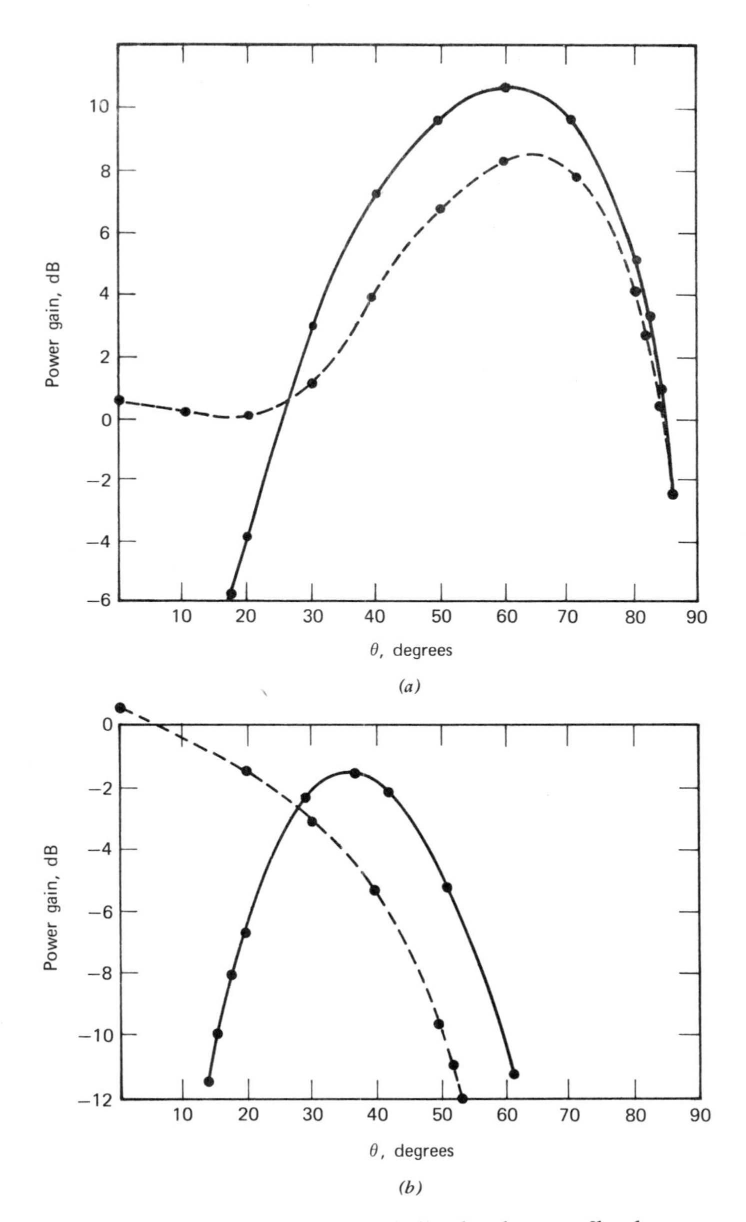

Fig. 4.9 Power gain of a horizontal dipole above a flat lossy ground with $h=\lambda/2$, $H=\lambda/2$, $h/a=9375$, and $f=10$ MHz: (*a*) $\varphi=0°$, (*b*) $\varphi=90°$. Solid and dashed curves are for sea water and poor ground, respectively.

Table 4.1 Change of R_{in} at $f = 10$ MHz when the Antenna Geometry is as Shown in Fig. 4.4.

h/λ	H/λ	h/a	Z_{11} (ohms)	Z_{12} (ohms)	R_{in} (ohms) Sea water	R_{in} (ohms) Poor ground
$\frac{1}{4}$	$\frac{1}{4}$	4680	$76.34 + j41.04$	$-17.47 - j32.82$	94.04	85.63
$\frac{1}{4}$	$\frac{3}{8}$	4680	$76.34 + j41.04$	$-24.72 + j9.91$	100.55	84.25
$\frac{1}{4}$	$\frac{1}{2}$	4680	$76.34 + j41.04$	$6.44 + j19.53$	69.72	72.23
$\frac{1}{2}$	$\frac{1}{4}$	9375	$3631.53 - j2356.47$	$-1218.05 - j1316.13$	4850.59	4188.56
$\frac{1}{2}$	$\frac{3}{8}$	9375	$3631.53 - j2356.47$	$-1099.75 + j898.26$	4701.68	3941.02
$\frac{1}{2}$	$\frac{1}{2}$	9375	$3631.53 - j2356.47$	$692.72 + j962.98$	2935.11	3294.57

earth does not have substantial influence on the antenna performance in the yz plane any more because of the increased height. The final characteristics remain essentially the same as that obtainable in free space.

When the antenna length and the ratio (h/a) are changed, the results for the corresponding heights and grounds do not vary substantially. In general, the power gain in the principal plane ($\varphi = 0°$), when the other parameters are kept unchanged, becomes larger for a longer antenna. Specific results for $h = \lambda/2$, $H = \lambda/4$, $\lambda/2$ are displayed in Figs. 4.8 and 4.9.

Naturally, all the results presented above depend heavily on R_{in} of (4.75) and R_h of (4.53). To give the reader the related quantitative information in this regard, a short table for R_{in} as functions of h, H, and h/a is prepared in Table 4.1. Of course, R_{in} is independent of θ and φ. The variation of R_h with θ at a frequency, say, $f = 10$ MHz, is shown in Fig. 4.10. For completeness and for later application in Section 4.6, corresponding curves for R_v are presented in Fig. 4.11. Note that R_v and R_h are independent of the antenna configurations. As can be seen from (4.53) and (4.58), we have $R_v = R_h = -1$ at $\theta = 90°$, and $R_v = -R_h$ at $\theta = 0°$ for all types of grounds. While both the amplitude and phase of R_h change gradually with θ, the amplitude of R_v reaches a minimum and the phase of R_v undergoes a rapid change at an angle known as the Brewster angle.[2]

The power gain for the same dipole in free space with a simple sinusoidal current distribution (except $h = \lambda/2$, λ, $3\lambda/2, \ldots$) can be obtained as a special case by setting $R_v = R_h = 0$ ($k = k'$) and $T_U = T_D = 0$ in (4.76) through (4.82).

Before proceeding to the next subject, a note about the suitability and therefore the accuracy of using the reflection-coefficient approach as presented here seems appropriate. Definitely, the results for the far-field components are very good because the surface wave component attenuates rapidly at a large distance from the source, as shown by a more rigorous approach using the Sommerfeld integral.[2] The input impedances are also considered reasonably accurate if the antenna is not very close to the ground, as verified by recent works in this field.[23,24]

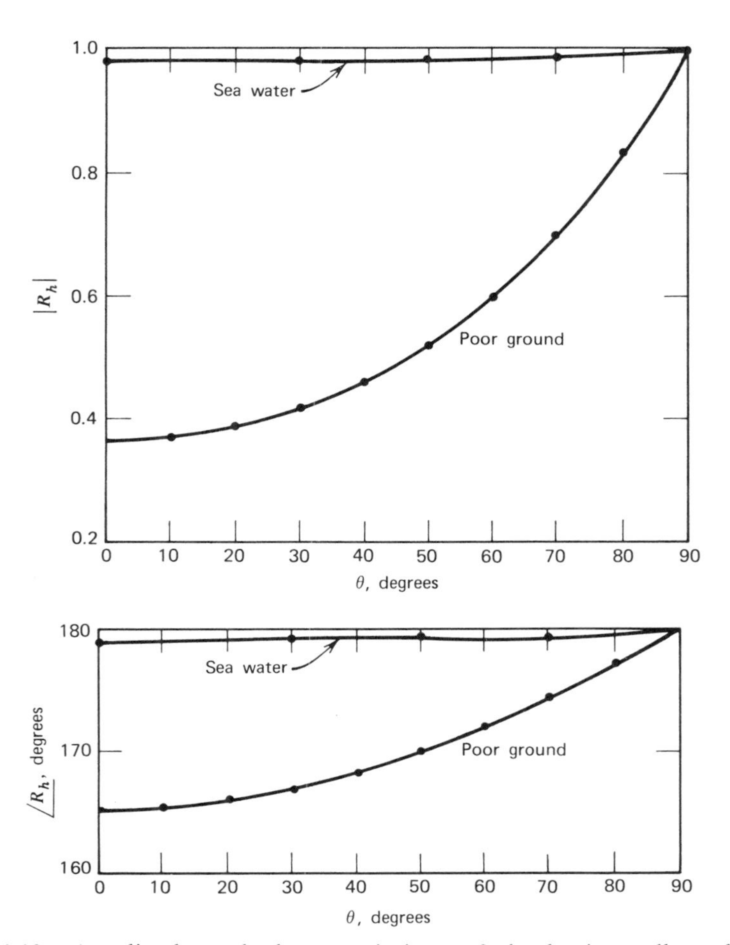

Fig. 4.10 Amplitude and phase variations of the horizontally polarized ground reflection coefficient at $f = 10$ MHz.

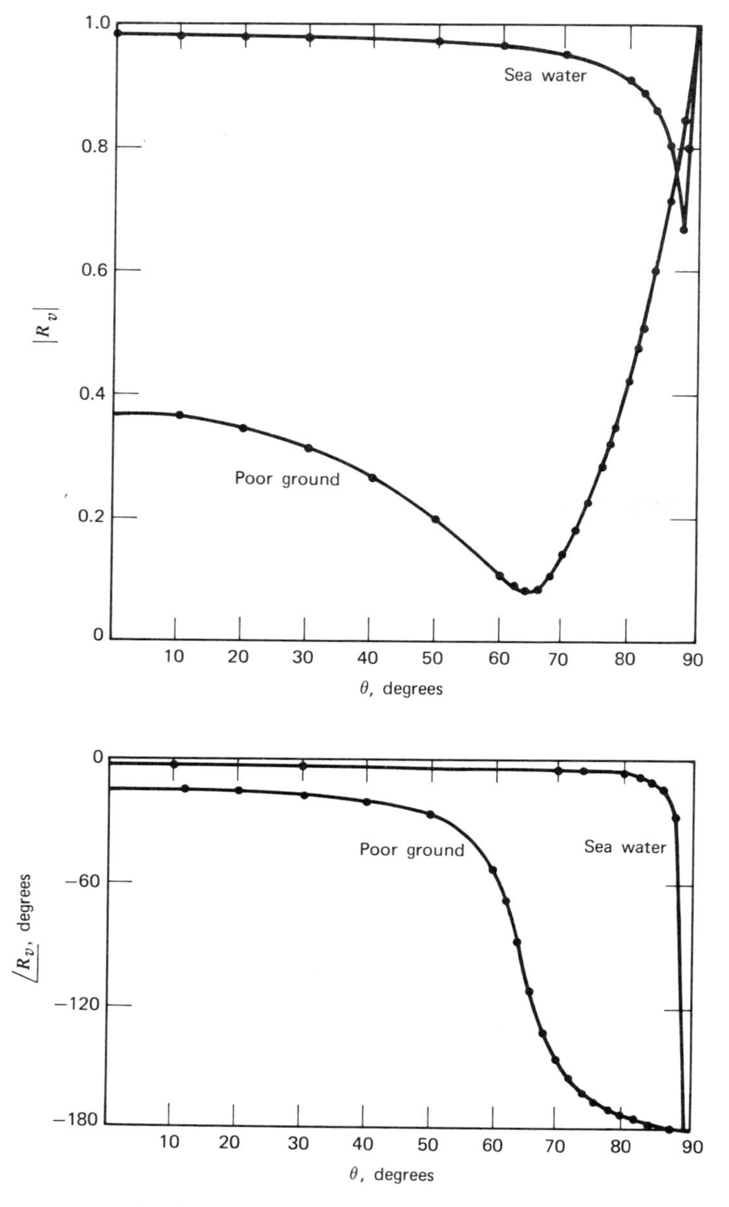

Fig. 4.11 Amplitude and phase variations of the vertically polarized ground reflection coefficient at $f = 10$ MHz.

4.4 Yagi-Uda Antenna

The formulation for the simple case considered in the previous section can readily be extended to Yagi-Uda antenna with the same philosophy in approximation. This antenna is essentially an elevated array of many horizontal dipoles with different lengths and spacings as shown in Fig. 4.12. It consists of a reflector, a driven element, and several directors. The reflector is the longest element, and the directors are shorter than the driven element. Both the reflector and directors are parasitic. Referring to Fig. 4.12, we have

$$\theta' = \pi/2, \qquad \varphi' = 0, \qquad \alpha' = 0, \qquad H_s = H, \qquad s = x, \qquad \cos\psi = \sin\theta\cos\varphi,$$

$$I_i(s) = I_i\left\{\sin k(h_i - |x|) + T_{Ui}(\cos kx - \cos kh_i) + T_{Di}(\cos\tfrac{1}{2}kx - \cos\tfrac{1}{2}kh_i)\right\}, \qquad kh_i \neq \pi/2,$$

$$i = 1, 2 \ldots N. \tag{4.89}$$

The corresponding expression for $kh_i = \pi/2$ is temporarily omitted here. It can be added accordingly, if required.

The field components contributed by the ith element (counting the reflector as the first element) are

$$\begin{aligned} E_{\theta i} &= j60 I_i \frac{e^{-jkr_i}}{r_i}\cos\theta\cos\varphi[1 - R_v\exp(-j2kH\cos\theta)]A_i, \\ E_{\varphi i} &= -j60 I_i \frac{e^{-jkr_i}}{r_i}\sin\varphi[1 + R_h\exp(-j2kH\cos\theta)]A_i, \end{aligned} \tag{4.90}$$

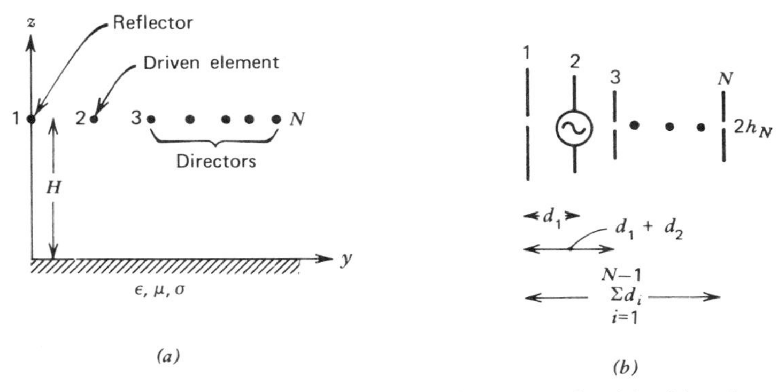

Fig. 4.12 Yagi-Uda antenna above a plane lossy earth: (*a*) side view, (*b*) top view.

where A_i will be identical to A_1 in (4.79) if we replace h, T_U, T_D, and dos $\psi = \sin\theta \sin\varphi$ in (4.79), respectively, by h_i, T_{Ui}, T_{Di}, and $\cos\psi = \sin\theta \cos\varphi$. In the exponential terms in (4.90), the factor r_i representing the distance between the base of the ith element and the far-field point is related to r_1 by

$$r_i \cong r_1 - y_i \sin\theta \sin\varphi, \tag{4.91}$$

where y_i is the coordinate position of the base of the ith element with the understanding that $y_1 = 0$.

The total field components can be obtained by summing up the contributions from all the elements:

$$\begin{aligned} E_\theta &= \sum_{i=1}^{N} E_{\theta i} \cong j60 \frac{e^{-jkr_1}}{r_1} \cos\theta \cos\varphi \left[1 - R_v \exp(-j2kH\cos\theta)\right] S, \\ E_\varphi &= \sum_{i=1}^{N} E_{\varphi i} = -j60 \frac{e^{-jkr_1}}{r_1} \sin\varphi \left[1 + R_h \exp(-j2kH\cos\theta)\right] S, \end{aligned} \tag{4.92}$$

where

$$S = \sum_{i=1}^{N} I_i \exp(jky_i \sin\theta \sin\varphi) A_i. \tag{4.93}$$

The input impedance looking into the base of the driven element ($i = 2$) can be determined by considering the circuit relations:

$$\begin{aligned} V_1 &= \text{voltage applied at the base of the reflector} \\ &= Z_{11} I_{b1} + Z_{12} I_{b2} + \cdots + Z_{1N} I_{bN} \\ &\quad + Z'_{11} I'_{b1} + Z'_{12} I'_{b2} + \cdots + Z'_{1N} I'_{bN} \\ &= (Z_{11} + R'_h Z'_{11}) I_{b1} + (Z_{12} + R'_h Z'_{12}) I_{b2} + \cdots + (Z_{1N} + R'_h Z'_{1N}) I_{bN} \\ &= 0, \end{aligned}$$

$$\begin{aligned} V_2 &= \text{voltage applied at the base of the driven element} \\ &= (Z_{21} + R'_h Z'_{21}) I_{b1} + (Z_{22} + R'_h Z'_{22}) I_{b2} + \cdots + (Z_{2N} + R'_h Z'_{2N}) I_{bN}, \end{aligned}$$

$_N$ = voltage applied at the base of the last director

$$= (Z_{N1} + R'_h Z'_{N1}) I_{b1} + (Z_{N2} + R'_h Z'_{N2}) I_{b2} + \cdots + (Z_{NN} + R'_h Z'_{NN}) I_{bN} \tag{4.94}$$

ith the factor R'_h in (4.72), and the self-impedance of the ith real element, $_{ii}$, given by

$$Z_{ii} = \frac{-j60\psi_{dRi}\cos kh_i}{\sin kh_i + T_{Ui}(1-\cos kh_i) + T_{Di}(1-\cos\frac{1}{2}kh_i)}. \tag{4.95}$$

he open-circuit mutual impedance between the ith and jth real elements, $_{ij} = Z_{ji}$, can be calculated by (4.38) with $h_i = h_1$, $h_j = h_2$, and $d = |y_i - y_j|$; ıe open-circuit mutual impedance between the ith real element and its wn image, Z'_{ii}, can be calculated by (4.38) with $h_1 = h_2 = h_i$ and $d = 2H$; nd finally the open-circuit mutual impedance between the ith real element nd the image of the jth element, $Z'_{ij} = Z'_{ji}$, can also be calculated by (4.38) ith $h_i = h_1$, $h_j = h_2$, and $d = [4H^2 + (y_i - y_j)^2]^{1/2}$.

When V_2, h_i, a_i, y_i, H, ϵ_r, σ, and f_{MHz} are specified, all the impedances sted above will be known quantities. We can then solve for I_{bi} from 4.94), calculate the current maxima,

$$I_i = \frac{I_{bi}}{[\sin kh_i + T_{Ui}(1-\cos kh_i) + T_{Di}(1-\cos\frac{1}{2}kh_i)]}, \qquad i = 1, 2, \ldots, N \tag{4.96}$$

eeded for (4.93), and determine the input resistance according to

$$R_{\text{in}} = \text{Re}\left(\frac{V_2}{I_{b2}}\right). \tag{4.97}$$

Iote that, since the second element alone is excited, the answer for I_{b2} btained in the above process should be the dominant one. Relative mplitudes of the other I_{bi}'s ($i \neq 2$) depend naturally on the array geometry nd ground constants.

Using the equations thus developed, we have the following expression for he power gain for the Yagi-Uda antenna:

$$G = \frac{120|S|^2}{|I_{b2}|^2 R_{\text{in}}} \{\cos^2\theta\cos^2\varphi[1 + |R_v|^2 - 2|R_v|\cos(\psi_v - 2kH\cos\theta)]$$
$$+ \sin^2\varphi[1 + |R_h|^2 + 2|R_h|\cos(\psi_h - 2kH\cos\theta)]\}, \tag{4.98}$$

vhere $R_v = |R_v|e^{j\psi_v}$ and $R_h = |R_h|e^{j\psi_h}$.

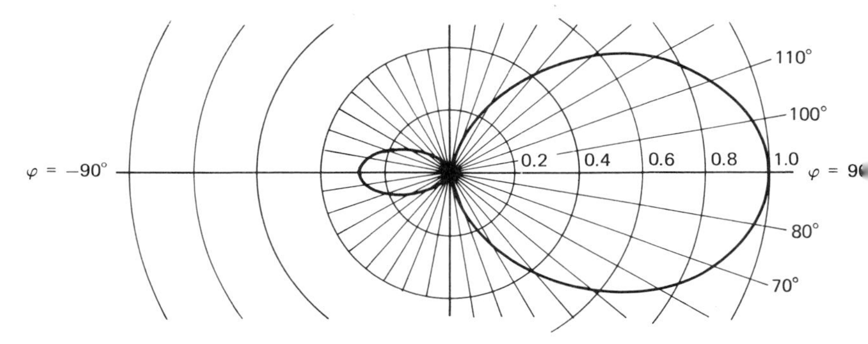

Fig. 4.13 Normalized free-space horizontal pattern of a three-eleme Yagi-Uda antenna with $h_1=7.8$ m, $h_2=7.5$ m, $h_3=7.0$ m, $a_1=$ $=a_3=0.001$ m, $y_2=7.5$ m, $y_3=13.5$ m, $\theta=90°$, and $f=10$ MHz

Up to this point all the derivations in this section are based on th current form (4.89) for $kh_i \neq \pi/2$. Whenever one of the element lengths such that $kh_i=\pi/2$, which often applies to the driven element, we shoul make minor changes following the parallel presentation outlined in Se tions 4.1 and 4.3.

Typical numerical results are now presented in graphical form. First, th normalized free-space horizontal pattern, $|E(\theta=90°,\ \varphi)|$ with $N=3$,

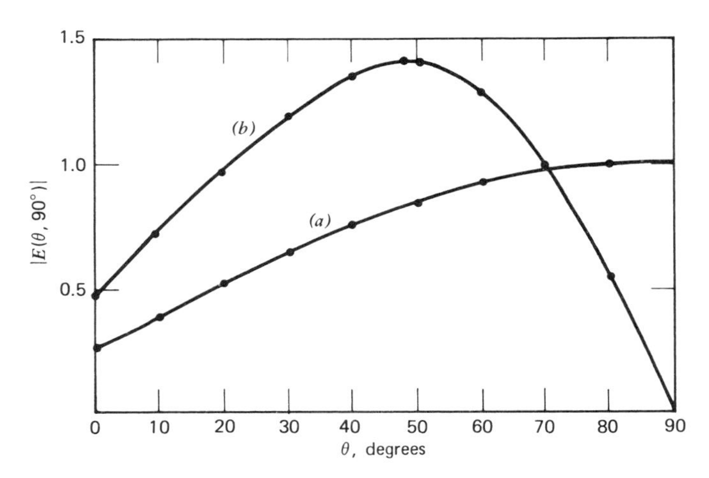

Fig. 4.14 Vertical patterns for the same antenna and frequency as in Fi 4.13: (a) free space, (b) antenna is $\lambda/4$ above the sea wat (normalized to the free-space maximum).

ven in Fig. 4.13 to show that the Yagi-Uda antenna is normally designed s an endfire array (beam maximum at $\varphi=90°$). The associated free-space ertical pattern in the yz plane, $|E(\theta, 90°)|$, is given as curve a in Fig. 4.14 ith the maximum at $\theta=90°$. When the same antenna is placed about $\lambda/4$ bove a plane lossy ground, the beam maximum will be shifted away from ne grazing direction as shown by curve b in Fig. 4.14. Thus, because of the resence of the two parasitic elements, the maximum power gain for a agi-Uda antenna no longer occurs at $\theta=0°$ even when $H=\lambda/4$ as the orresponding single horizontal dipole considered in the previous section.

Results for impedances, base currents, and other related information for ne antenna described in Figs. 4.13 and 4.14 are detailed as follows:

$$Z_{11}=86.1527+j105.5823,$$

$$Z_{22}=79.0692+j38.5374,$$

$$Z_{33}=62.0726-j63.7130,$$

$$Z_{12}=Z_{21}=43.2222-j29.8405,$$

$$Z_{23}=Z_{32}=46.5984-j17.9325,$$

$$Z_{13}=Z_{31}=-3.7859-j33.4458, \qquad \text{ohms}$$

$$Z'_{11}=-14.2381-j33.5649,$$

$$Z'_{22}=-12.5321-j29.9286,$$

$$Z'_{33}=-10.1279-j24.6912,$$

$$Z'_{12}=Z'_{21}=-13.3421-j31.7058,$$

$$Z'_{23}=Z'_{32}=-11.2283-j27.2106,$$

$$Z'_{13}=Z'_{31}=-11.9060-j28.8597,$$

$$T_{U1}=-0.58498+j0.26464, \qquad T_{D1}=-0.03852+j0.05039,$$

$$T'_{U2}=6.43175+j10.11997, \qquad T'_{D2}=1.32795-j0.77386,$$

$$T_{U3}=-0.20264-j0.94152, \qquad T_{D3}=0.10149+j0.09957,$$

$$\frac{I_{b1}}{V_2}=\begin{cases}0.00132+j0.00360 & \text{in free space}\\ 0.00044+j0.00395\ (\text{mhos}) & \text{antenna } \lambda/4 \text{ above sea water}\\ 0.00030+j0.00420 & \text{antenna } \lambda/4 \text{ above poor ground,}\end{cases}$$

$$\frac{I_{b2}}{V_2}=\begin{cases}0.01061-j0.00995 & \text{in free space}\\ 0.01005-j0.00923\ (\text{mhos}) & \text{antenna } \lambda/4 \text{ above sea water}\\ 0.01006-j0.00896 & \text{antenna } \lambda/4 \text{ above poor ground,}\end{cases}$$

$$\frac{I_{b3}}{V_2}=\begin{cases}-0.00912+j0.00211 & \text{in free space}\\ -0.00975+j0.00045\ (\text{mhos}) & \text{antenna } \lambda/4 \text{ above sea water}\\ -0.01021+j0.00019 & \text{antenna } \lambda/4 \text{ above poor ground,}\end{cases}$$

$$R_{\text{in}}=\left\{\begin{array}{ll}50.15 \text{ ohms} & \text{in free space}\\ 53.98 \text{ ohms} & \text{antenna } \lambda/4 \text{ sea water}\\ 55.43 \text{ ohms} & \text{antenna } \lambda/4 \text{ above poor ground.}\end{array}\right\}$$

Instead of presenting detailed variations of power gain with θ fo numerous cases, we will show the maximum power gain occurring at som $\theta_{\max}$ as a function of element lengths and separations, the array height, an the ground constants. In Fig. 4.15(a), three sets of $G_{\max}$ are plotted versu y_3 (position of the director) with two different values of y_2 (position of th driven element). The two curves in set A are for $N=3, f=10$ MHz ($\lambda=3$ m), $h_1=7.8$ m, $h_2=7.5$ m, $h_3=7.0$ m, $a_1=a_2=a_3=0.001$ m, $H=7.5$ m $\varphi=90°$, $\epsilon_r=80$, and $\sigma=5$ mho/m (sea water). In this case, $\theta_{\max}\cong 48°$. If w vary h_3 from 7.0 to 6.75 m and keep the other parameters the same, th results are shown by set B with $\theta_{\max}\cong 44°$. Curves in set C correspond t the case when h_3 is decreased further to 6.5 m, where $\theta_{\max}\cong 42°$. B comparison, we see that the maximum power gain in each case change little with y_2 and y_3 when other parameters are fixed. Also, it is clear that decrease of h_3 from 7.0 to 6.5 m changes the maximum power gain only b a fraction of a decibel. The only significant change there is perhaps th position of the beam maximum. On the other hand, when h_3 alone increased by the same amount to 7.5 m (the same length as the drive element), the performance will deteriorate sharply even with the same se of y_2 and y_3. An example showing this effect is presented in Fig. 4.15(b While the results presented for this case are by no means complete, it fair to conclude that, as far as the three-element Yagi-Uda antenna wit $h_1=7.8$ m, $h_2=7.5$ m, and $a=0.001$ m (which is $\lambda/4$ above the sea wate is concerned, the absolute maximum power gain we can expect from it approximately 10.4 dB, which roughly occurs at $h_3=7.0$ m with $y_2=7.5$ m

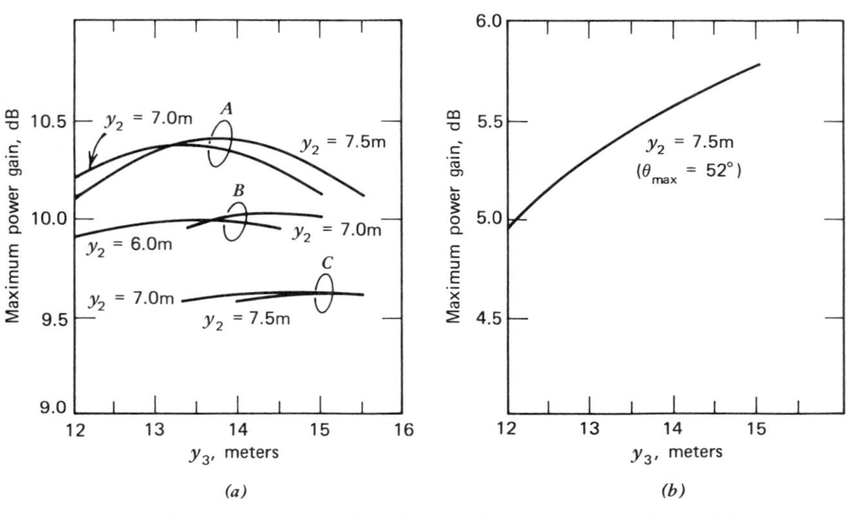

Fig. 4.15 Maximum power gain for a three-element Yagi-Uda antenna with $h_1=7.8$ m, $h_2=7.5$ m, $a_1=a_2=a_3=0.001$ m, $H=7.5$ m, $\varphi=90°$, $f=10$ MHz, $\epsilon_r=80$, and $\sigma=5$ mho/m: (*a*) $h_3=7.0$, 6.75, and 6.5 m, (*b*) $h_3=7.5$ m.

$y_3=13.5$ m, and $\theta_{max}\cong 48°$. In fact, when h_1 alone is varied by a reasonable amount while everything else is kept unchanged, we also find that the maximum power gain is always lower than 10.4 dB (results are too numerous to present here).

Detailed variations of power gain in the *yz* plane ($\varphi=90°$) for the "optimum" case discussed above are shown in Fig. 4.16, where the case with a poor ground is also added for comparison purposes. Of course, sea water and poor ground are two extreme types of ground. The performance of the same antenna with the same height above an "average" ground may be predicted to lie between the two curves in Fig. 4.16.

It is also constructive to compare the above "optimum" case with the similar case in Section 4.3 (simple horizontal dipole, $h=H=\lambda/4$, sea water). In essence, two improvements are made by adding two parasitic elements in the Yagi-Uda antenna. First, the maximum power gain is increased by about 3.2dB. Second, the position of the beam maximum is shifted from $\theta_{max}=0°$ to 48°. This second improvement is considered an important factor when the antenna is used for the purpose of long-distance communication or direction-finding.

Naturally, if we want to have a still larger θ_{max} (beam maximum closer to the ground), the only effective parameter under our control is to

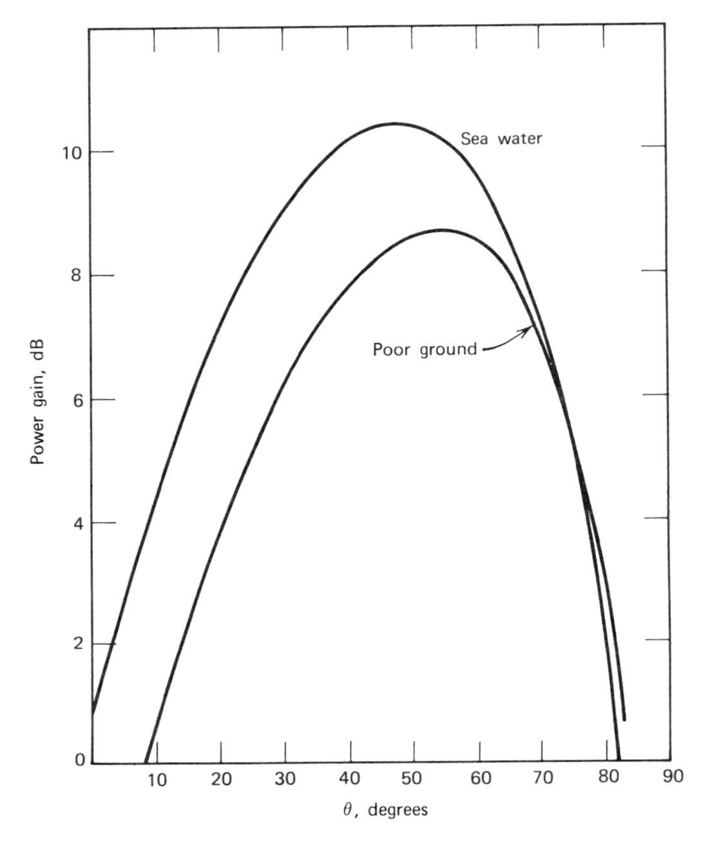

Fig. 4.16 Power gain of the same Yagi-Uda antenna as that described in Fig. 4.13: (*a*) $\lambda/4$ above sea water ($\epsilon_r = 80$, $\sigma = 5$ mho/m), (*b*) $\lambda/4$ above poor ground ($\epsilon_r = 4$, $\sigma = 0.001$ mho/m).

increase the array height H. Results for a comparable case with $H = 15.0$ m ($\lambda/2$) are given in Fig. 4.17. It is seen there that not only $\theta_{\max}$ is moved to 64° (or 26° above the ground), the maximum power gain in this case is also increased somewhat ($G_{\max} = 12.35$ dB). In general, $\theta_{\max}$ always increases with H for a given Yagi-Uda antenna, but it never reaches $\pi/2$ because of the ground loss. The final value for $G_{\max}$, however, depends on the free-space pattern, actual height, frequency, and ground constants.

If more directors are used in the antenna, an increase in power gain in the amount of approximately 1.2 dB per element can be achieved without substantial change in $\theta_{\max}$. The general shape of G versus θ remains almost the same as the basic three-element Yagi-Uda antenna with a comparable

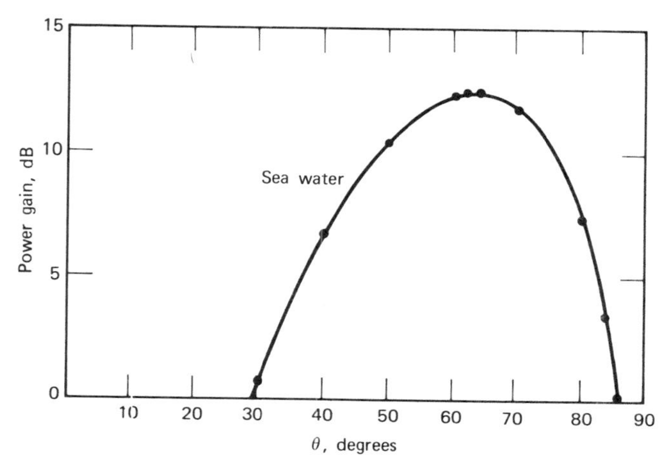

Fig. 4.17 Power gain of a three-element Yagi-Uda antenna with $h_1 = 7.8$ m, $h_2 = 7.5$ m, $h_3 = 7.0$ m, $a = 0.001$ m, $y_2 = 5.5$ m, $y_3 = 12.0$ m, $H = 15.0$ m, $\varphi = 90°$, and $f = 10$ MHz.

geometry and under a similar environment. An example for a four-element antenna confirming the aspects just mentioned is shown in Fig. 4.18. The number of directors we can add to the structure is limited by the cost and weight considerations, although antennas with thirty or more elements have been built.[25]

In principle, more reflectors can also be added to the antenna. However, experience and existing measured results show that practically nothing can be gained by increasing the number of reflectors.

4.5 Curtain Arrays

In this chapter, we have studied the characteristics of a single horizontal dipole above a lossy ground, and then those of a "one-dimensional" array such as the Yagi-Uda antenna. For the case of a single horizontal dipole located in the yz plane ($\varphi = 90°$), we learned that the pattern in a vertical plane ($\varphi =$ constant) is heavily controlled by the height, H, of the antenna above ground. When H is no greater than $\lambda/4$, the position of the beam maximum generally occurs at $\theta_{max} = 0°$. As H is increased, θ_{max} shifts gradually toward the grazing direction. The horizontal pattern along a surface of $\theta =$ constant always has its maximum pointing at $\varphi = 0°$.

For the case of the Yagi-Uda antenna whose axis coincides with the y axis while the elements are parallel to the x axis, we found that the position of beam maximum in a vertical pattern is not only dictated by the antenna

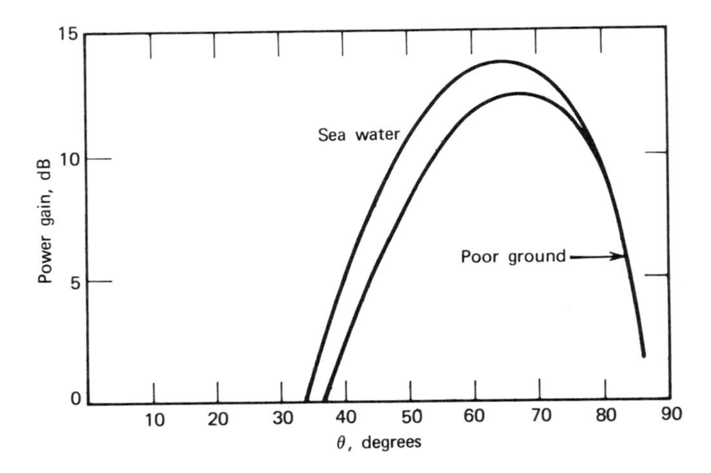

Fig. 4.18 Power gain of a four-element Yagi-Uda antenna with $h_1=7.8$ m, $h_2=7.5$ m, $h_3=h_4=7.0$ m, $a=0.001$ m, $y_2=7.5$ m, $y_3=15.5$ m, $y_4=25.0$ m, $H=15.0$ m, $\varphi=90°$, and $f=10$ MHz.

height but also by the relative positions and lengths of the reflector and directors, although the position of the beam maximum in an azimuthal surface is still fixed (i.e., $\varphi=90°$ for the geometry displayed in Fig. 4.12).

Now we are ready to extend the study to the case of two-dimensional arrays consisting of horizontal dipoles such that positions of both the maxima in the vertical and horizontal patterns can be changed by designing. In particular, we will study the curtain arrays as shown in Fig. 4.19. This type of array has been used extensively by Voice of America.[26]

The entire array is in the yz plane. Generally speaking, there can be N bays arranged, say, along the y axis with each bay containing a stack of M elements along the z axis. In total, there are $M \times N$ identical and parallel horizontal dipoles. There is also a vertical screen at a distance $(x=-x_1)$ from the array to serve as a reflector. Let the height of the ith element on each bay be z_i, $i=1,2,\ldots,M$, and the coordinates of bays be y_n, $n=1,2,\ldots,N$ with $y_1=0$. The field components radiated by the first element in the first bay, including the ground-reflected part, are identical to those given in (4.78) or (4.84) with the understanding that $z_1=H$. The contributions by the first bay as a whole are, therefore,

$$E_{\theta,1}=j60I_1\frac{e^{-jkr_0}}{r_0}(\sin\varphi\cos\theta)A_1B_1,$$
$$E_{\varphi,1}=j60I_1\frac{e^{-jkr_0}}{r_0}(\cos\varphi)A_1B_2, \tag{4.99}$$

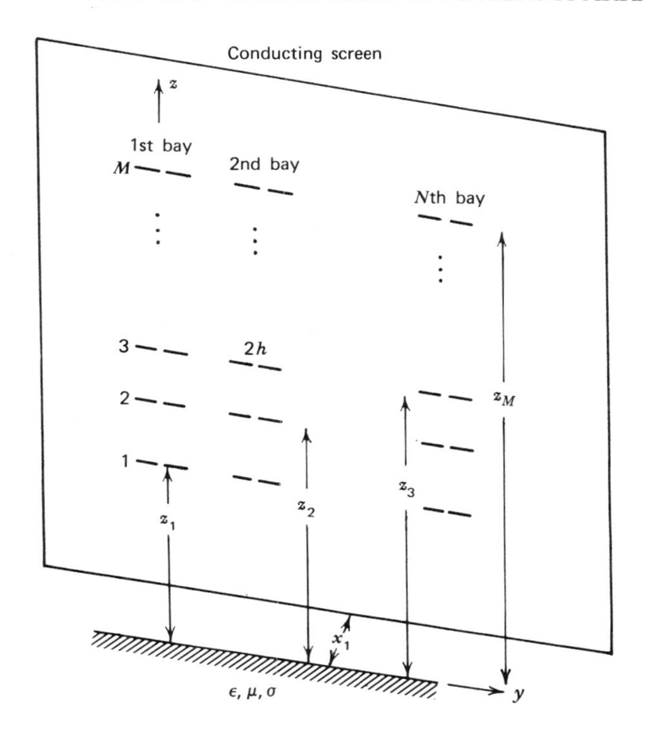

Fig. 4.19 A curtain array above a plane lossy earth.

where

$$B_1 = \sum_{i=1}^{M} C_i \exp\left[jkz_i(\cos\theta - \cos\theta_0)\right]\left[1 - R_v \exp(-j2kz_i\cos\theta)\right], \quad (4.100)$$

$$B_2 = \sum_{i=1}^{M} C_i \exp\left[jkz_i(\cos\theta - \cos\theta_0)\right]\left[1 + R_h \exp(-j2kz_i\cos\theta)\right], \quad (4.101)$$

r_0 is the distance between $(y,z)=(0,0)$ and the far-field point, and A_1 is the same as that in (4.79) if $kh \neq \pi/2$. Of course, A_1 should be replaced by A'_1 in (4.88) if $kh = \pi/2$.

In (4.100) and (4.101), C_i represents the relative amplitude excitation of the ith dipole with $C_1 = 1$, and θ_0 is the desired position of the beam maximum.

The array factor for N bays, according to the basic formulation presented in Chapter 1, is

$$S_y = \sum_{n=1}^{N} e^{jky_n \sin\theta\ (\sin\varphi - \sin\varphi_0)} \tag{4.102}$$

where φ_0 is to determine the position of the beam maximum in an azimuthal surface. Here we have assumed that the amplitude excitations for corresponding elements in each bay are the same.

The array factor resulting from the array of the real elements and its perfect image with respect to the screen at $x = -x_1$ is, for $0° \leqslant \theta \leqslant 90°$ and $-90° \leqslant \theta \leqslant 90°$,

$$S_x = 1 - \exp(-j2kx_1 \cos\psi_x) = 2j\exp(-jkx_1 \cos\psi_x)\sin(kx_1 \cos\psi_x), \tag{4.103}$$

where

$$\cos\psi_x = \sin\theta\cos\varphi. \tag{4.104}$$

The total field components are then

$$\begin{aligned} E_{\theta,t} &= E_{\theta,1} S_x S_y, \\ E_{\varphi,t} &= E_{\varphi,1} S_x S_y. \end{aligned} \tag{4.105}$$

The input resistance of the individual element can be expressed as

$$R_i^n = \mathrm{Re}(Z_i^n), \qquad i = 1, 2, \ldots, M; \qquad n = 1, 2, \ldots, N \tag{4.106}$$

where the subscript i is the position of the element within a bay while the superscript n is the position of the bay. The symbol Z_i^n should be explained as the input impedance at the base of the ith dipole in the nth bay, which can be calculated by considering the following four groups:

$$\left.\begin{aligned} Z_i^n = Z_{i1}^{n1} &+ Z_{i2}^{n1} + \cdots + Z_{ii}^{n1} + \cdots + Z_{iM}^{n1} \\ &+ Z_{i1}^{n2} + Z_{i2}^{n2} + \cdots + Z_{ii}^{n2} + \cdots + Z_{iM}^{n2} \\ &+ \cdots \\ &+ Z_{i1}^{nn} + Z_{i2}^{nn} + \cdots + Z_{ii}^{nn} + \cdots + Z_{iM}^{nn} \\ &+ \cdots \\ &+ Z_{i1}^{nN} + Z_{i2}^{nN} + \cdots + Z_{ii}^{nN} + \cdots + Z_{iM}^{nN} \end{aligned}\right\} \text{first group}$$

$$
\left.
\begin{aligned}
&+R'_h\big(Z^{n1}_{i1p}+Z^{n1}_{i2p}+\cdots+Z^{n1}_{iip}+\cdots+Z^{n1}_{iMp}\\
&+\cdots\\
&+Z^{nn}_{i1p}+Z^{nn}_{i2p}+\cdots+Z^{nn}_{iip}+\cdots+Z^{nn}_{iMp}\\
&+\cdots\\
&+Z^{nN}_{i1p}+Z^{nN}_{i2p}+\cdots+Z^{nN}_{iip}+\cdots+Z^{nN}_{iMp}\big)
\end{aligned}
\right\} \text{second group}
$$

$$
\left.
\begin{aligned}
&-\big(Z^{n1}_{i1d}+Z^{n1}_{i2d}+\cdots+Z^{n1}_{iid}+\cdots+Z^{n1}_{iMd}\\
&+\cdots\\
&+Z^{nn}_{i1d}+Z^{nn}_{i2d}+\cdots+Z^{nn}_{iid}+\cdots+Z^{nn}_{iMd}\\
&+\cdots\\
&+Z^{nN}_{i1d}+Z^{nN}_{i2d}+\cdots+Z^{nN}_{iid}+\cdots+Z^{nN}_{iMd}\big)
\end{aligned}
\right\} \text{third group} \qquad (4.107)
$$

$$
\left.
\begin{aligned}
&-R'_h\big(Z^{n1}_{i1t}+Z^{n1}_{i2t}+\cdots+Z^{n1}_{iit}+\cdots+Z^{n1}_{iMt}\\
&+\cdots\\
&+Z^{nn}_{i1t}+Z^{nn}_{i2t}+\cdots+Z^{nn}_{iit}+\cdots+Z^{nn}_{iMt}\\
&+\cdots\\
&+Z^{nN}_{i1t}+Z^{nN}_{i2t}+\cdots+Z^{nN}_{iit}+\cdots+Z^{nN}_{iMt}\big).
\end{aligned}
\right\} \text{fourth group}
$$

All the impedances in the first group of (4.107) are either open-circuit self- or mutual impedances associated with the real elements. More specifically, Z^{nn}_{ii} is designated as the self-impedance of the ith dipole in the nth bay; and $Z^{nn'}_{ii'}$, when $n \neq n'$, and/or $i \neq i'$, as the mutual impedance between the ith dipole in the nth bay and the i'th dipole in the n'th bay.

The impedances in the second group of (4.107) are all open-circuit mutual impedances between the real elements and their imperfect images with respect to the ground. As an example, $Z^{nn'}_{ii'p}$ should be interpreted as the mutual impedance between the ith dipole in the nth bay and the imperfect image of the i'th dipole in the n'th bay. In a similar fashion, the impedances in the third group of (4.107) are all open-circuit mutual impedances between the real elements and their perfect images with respect to the vertical screen, and those in the fourth group of (4.107) are the ones between the real elements and the imperfect images of perfect images.

All these impedances can readily be calculated in terms of the dipole length, z_i, y_n, and x_1, depending on relative positions of the elements involved. The self-impedance, Z_{ii}^{nn}, can simply be computed according to (4.32) or (4.33); and the mutual impedances, $Z_{ii'}^{nn}$ $(i \neq i')$, $Z_{ii'p}^{nn}$, $Z_{ii'd}^{nn}$, and $Z_{ii't}^{nn}$, by (4.38) or (4.47). All the remaining mutual impedances are those between dipoles arranged in echelon or collinear. Since the geometry for this latter case is no longer symmetric, the mutual impedances should, strictly speaking, be calculated by a more refined approach such as the five-term theory.[14, 15] However, in view of the complication of the system already experienced, we are reluctantly satisfied by using expressions similar to (4.38) and (4.47) as a further approximation. Of course, these expressions can be derived in a completely parallel manner with the relative antenna positions shown in Fig. 4.20. In addition to the horizontal separation of d between the two antennas, there is also a vertical distance l which is measured from the feeding point of one antenna to the nearest end of the other. Note that when $l = -h_2$, the situation reduces to that in Fig. 4.1. Now, we have

$$I_1(z') = I_1 \{ \sin k(h_1 - |z'|) + T_{U1}(\cos kz' - \cos kh_1) + T_{D1}(\cos \tfrac{1}{2}kz' - \cos \tfrac{1}{2}kh_1) \}, \qquad kh_1 \neq \frac{\pi}{2} \tag{4.108}$$

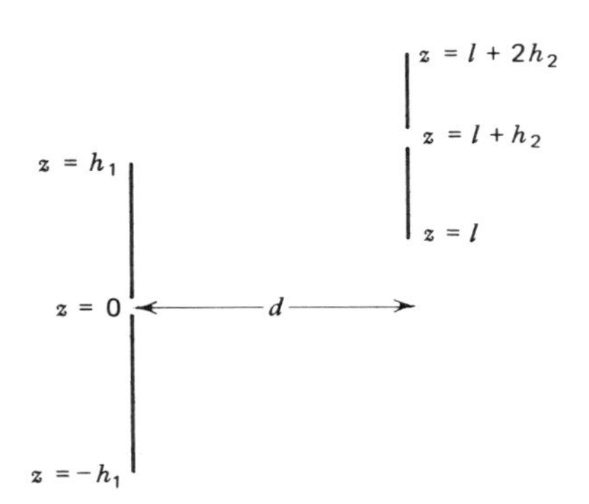

Fig. 4.20 Two parallel dipoles of arbitrary lengths arranged in echelon.

$$E_{z1}(z)=\frac{-j30}{k}\left(k^2+\frac{\partial^2}{\partial z^2}\right)\int_{-h_1}^{h_1} I_1(z')K(z,z')\,dz'$$

$$=j30kI_1\int_0^{h_1}(T_{U1}\cos kh_1-\tfrac{3}{4}T_{D1}\cos\tfrac{1}{2}kz'+T_{D1}\cos\tfrac{1}{2}kh_1)$$

$$\times[K(z,z')+K(z,-z')]\,dz'$$

$$+j60I_1\cos kh_1K(z,0)-j30I_1(1+T_{U1}\sin kh_1+\tfrac{1}{2}T_{D1}\sin\tfrac{1}{2}kh_1)$$

$$\times[K(z,h_1)+K(z,-h_1)], \tag{4.109}$$

where

$$K(z,z')=\frac{e^{-jkR}}{R}, \qquad R=\sqrt{(z-z')^2+d^2}\,. \tag{4.110}$$

In the above, we once again have assumed a very thin dipole ($a_1 \to 0$). The current distribution on the second antenna referred to the feeding point of the first antenna is

$$I_2(z)=I_2'(z)=I_2\{\sin k(2h_2+l-z)+T_{U2}[\cos k(z-l-h_2)-\cos kh_2]$$

$$+T_{D2}[\cos\tfrac{1}{2}k(z-l-h_2)-\cos\tfrac{1}{2}kh_2]\},$$

$$\text{in}\qquad l+h_2\leqslant z\leqslant l+2h_2\qquad\text{for}\qquad kh_2\neq\frac{\pi}{2}, \tag{4.111a}$$

or

$$I_2(z)=I_2''(z)=I_2\{\sin k(z-l)+T_{U2}[\cos k(z-l-h_2)-\cos kh_2]$$

$$+T_{D2}[\cos\tfrac{1}{2}k(z-l-h_2)-\cos\tfrac{1}{2}kh_2]\},$$

$$\text{in}\qquad l\leqslant z\leqslant l+h_2\qquad\text{for}\qquad kh_2\neq\frac{\pi}{2}. \tag{4.111b}$$

Note that $I_2'(l+2h_2)=I_2''(l)=0$ and $I_2'(l+h_2)=I_2''(l+h_2)$.

The open-circuit mutual impedance is then

$$Z_m = -\frac{1}{I_1(0)I_2(l+h_2)} \int_l^{l+2h_2} E_{z1}(z) I_2(z)\, dz$$

$$= -\frac{1}{I_1(0)I_2(l+h_2)} \left[\int_{l+h_2}^{l+2h_2} E_{z1}(z) I'_2(z)\, dz + \int_l^{l+h_2} E_{z1}(z) I''_2(z)\, dz \right], \tag{4.112}$$

which can reduce to an expression similar to (4.38). In fact, (4.112) will be identical to (4.38) when $l = -h_2$. It also simplifies to the expression obtained elsewhere[27] with a simple sinusoidal assumption for the current distribution by setting $T_{UI} = T_{DI} = T_{U2} = T_{D2} = 0$. The numerical result for Z_m in (4.112) will, of course, depend on h_1, h_2, l, and d in terms of the operating wavelength λ.

Expressions corresponding to $h_1 = h_2 = \pi/2$ and $h_1 = \pi/2$, $h_2 \neq \pi/2$ can also be derived accordingly. Note that the special case, $h_1 = h_2$, will apply for curtain arrays considered in this section, although (4.112) is derived for the general case where the lengths of the antennas are not necessarily equal. Equation (4.112) is also good for $d = 0$ and $l > h_1$ when the two antennas are collinear. Note also that we deal only with a passive system; the commonly recognized reciprocal relation for the mutual impedance, $z_{ij} = z_{ji}$, should be true here. However, because numerical procedures are involved for computing (4.112), the degree of computational accuracy carried may cause a small discrepancy between z_{ij} and z_{ji}. In this case, we will use the arithmetic mean of them as the final mutual-impedance term in (4.107). The total power supplied to the entire system, when all the elements are of equal length ($2h$), is

$$\begin{aligned} W_{\text{in}} = |I_1(0)|^2 \{ & R_1^1 + C_2^2 R_2^1 + \cdots + C_M^2 R_M^1 \\ & + R_1^2 + C_2^2 R_2^2 + \cdots + C_M^2 R_M^2 \\ & + \cdots \\ & + R_1^N + C_2^2 R_2^N + \cdots + C_M^2 R_M^N \} \\ = |I_1(0)|^2 & \sum_{n=1}^{N} \sum_{i=1}^{M} C_i^2 R_i^n \qquad [C_1 = 1, \text{ see Eq. (4.101)}]. \end{aligned} \tag{4.113}$$

Here again we should be aware that the superscripts and subscripts associated with the resistances above should be interpreted according to the paragraph following (4.106).

With (4.105) and (4.113) we obtain the final expression for the power gain for the curtain array:

$$G=\frac{480|A_1|^2|S_y|^2\sin^2(kx_1\cos\psi_x)[\sin^2\varphi\cos^2\theta|B_1|^2+\cos^2\varphi|B_2|^2]}{|\sin kh+T_U(1-\cos kh)+T_D(1-\cos\frac{1}{2}kh)|^2\sum_{n=1}^{N}\sum_{i=1}^{M}C_i^2R_i^n}. \tag{4.114}$$

Before we present detailed numerical results, it should be anticipated that the direction of the maximum radiation, θ_{max}, will not occur precisely at θ_0 because of the complicated effects of R_v, R_h, and A_1 (or A'_1) in (4.99).

Typical results for power gain in the vertical plane by a curtain array of two half-wave dipoles in one bay ($h=\lambda/4$, $M=2$, $N=1$) over sea water are shown in Fig. 4.21. The solid curve corresponds to the case with $z_1=\lambda/4$

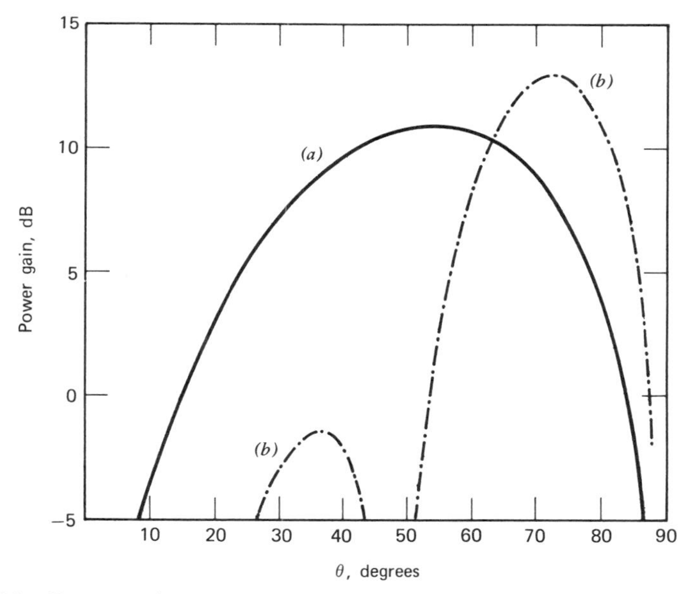

Fig. 4.21 Power gain of a curtain array over sea water with $M=2$, $N=1$, $h=7.5$ m, $a=0.0016$ m, $x_1=7.5$ m, $\theta_0=90°$, $\varphi=\varphi_0=0°$, $C_1=C_2=1$, and $f=10$ MHz: (*a*) $z_1=7.5$ m, $z_2=15.0$ m; (*b*) $z_1=15.0$ m, $z_2=30.0$ m.

and $z_2 = \lambda/2$, and the dashed curve is for $z_1 = \lambda/2$ and $z_2 = \lambda$, both with $\theta_0 = \pi/2$, $\varphi_0 = \varphi = 0$, $C_1 = C_2 = 1$, $x_1 = \lambda/4$, and $f = 10$ MHz. Obviously, both G_{max} and θ_{max} for the dashed curve are higher than the corresponding values for the solid curve. This is because of that the array yielding the results in the dashed curve has a wider space between the dipoles and a higher distance above the ground. In addition, we can also gain insight by comparing the results presented here with those for a single horizontal dipole studied in Section 4.3 where no vertical screen was present. For example, by comparing the dashed curve in Fig. 4.21 with the solid curve in Fig. 4.7(*a*) where a single horizontal dipole of the same length is of the same height above the same ground, we see that the general shape remains essentially unchanged, although the additional element and the vertical screen included here contribute approximately 4 dB to the maximum power gain, make the pattern narrower, and shift θ_{max} toward the grazing direction. On the other hand, the solid curve in Fig. 4.21 differs substantially from the result presented in Fig. 4.5(*a*). The reason that a minimum (in Fig. 4.21) rather than a maximum [in Fig. 4.5(*a*)] radiation occurs at $\theta = 0$ is mainly due to the array factor (4.103). The extra element and the vertical screen in the curtain array also increase the maximum power gain by roughly 4 dB although the position of the maximum power gain has changed.

The associated azimuthal pattern at $\theta = 54°$ corresponding to the geometry represented by the solid curve in Fig. 4.21 is shown in Fig. 4.22. Of course, the parameter φ_0 is not involved in this case since we consider the curtain array with only one bay.

In the process of computation we found that the effect of θ_0 in the practical range, $60° \leqslant \theta_0 \leqslant 90°$, is very negligible as far as the result in the plane $\varphi = 0$ is concerned. Important characteristics such as the value and position of the maximum power gain and the pattern shape remain essentially the same for a given set of array geometry, frequency, and the type of ground. Although lower values of θ_0 $(0° \leqslant \theta_0 \leqslant 60°)$ do have substantial effect on B_1 and B_2 in (4.100) and (4.101), and tend to shift the position of maximum radiation toward the zenith, this change in θ_{max}, B_1, and B_2, however, has little effect on the final field expressions in (4.105) because of the strong damping effect by the array factor (4.103) for smaller values of θ. Therefore, the influence of θ_0 on the overall array performance is not very sensitive. An example showing this fact is illustrated in Fig. 4.23, where θ_{max} occurs near 54° for all values of θ_0 concerned.

The role played by x_1 can be seen from Fig. 4.24. It is apparent that, for the particular example shown in the figure, an increase of x_1 tends to decrease both the values of G_{max} and θ_{max}. Values of x_1 smaller than $\lambda/4$

will cause stronger coupling between the array and the screen, and are often considered impractical.

Although sea water ($\epsilon_r = 80$, $\sigma = 5$ mho/m) is chosen as the ground in all of the examples given in this section, the formulation also applies to other types of ground. For poor ground ($\epsilon_r = 4$, $\sigma = 0.001$ mho/m) with the same array geometry and frequency as the sea water case described, the maximum power gain is about 1.5 dB lower and the position of the maximum power gain is about 2° smaller.

When there are more elements in the bay, we should expect an increase in both $G_{\max}$ and $\theta_{\max}$ (closer to the ground) and a decrease in beamwidth. An example for $M = 4$, $N = 1$, $z_i = i \times 7.5$ m, $h = x_1 = 7.5$ m, $\theta_0 = 90°$, $\varphi_0 = \varphi = 0°$, $C_i = 1$, and $f_{\text{MHz}} = 10$ over sea water is shown in Fig. 4.25. The difference between Fig. 4.25 and the solid curve in Fig. 4.21 represents the improvement by the two extra dipoles.

When two bays are used, an increase in $G_{\max}$ (relative to that with only one bay) is also expected. The exact amount of increase depends primarily on the number of elements in each bay, the dipole length, and the distance between the bays. In addition, an improvement in the azimuthal pattern should also be evident. An example of the vertical pattern for $M = N = 2$, $y_1 = 0$, and $y_2 = 15.6$ m with other parameters the same as for the solid curve in Fig. 4.21 is presented in Fig. 4.26, from which we see that $G_{\max} = 12.41$ dB, and that the beamwidth and $\theta_{\max}$ in the vertical plane are almost unchanged as compared with those associated with the solid curve in Fig. 4.21. The corresponding azimuthal pattern at $\theta = 54°$ with $\varphi_0 = 0°$ is displayed in Fig. 4.27(*a*). Clearly, the beamwidth here is narrower than that in Fig. 4.22. Since we now have two bays in the array, the parameter φ_0 can be used to control the position of the beam maximum in an azimuthal surface. Figure 4.27(*b*) shows the result with $\varphi_0 = 30°$ when all the other parameters are identical to those in Fig. 4.27(*a*). Evidently, the precise position of the beam maximum in Fig. 4.27(*b*) is not quite at $\varphi = 30°$ as it is supposed to be. This is mainly due to the influence of other factors in (4.99) and (4.103).

Before concluding this section, an explanation of the value for y_2 used in Figs. 4.26 and 4.27 is in order. The value of $y_2 = 15.6$ m gives a center-to-center distance of 0.52λ (at $f_{\text{MHz}} = 10$) between the bays. First, we note the fact that the half-length of the dipole is $\lambda/4$. This requires the center-to-center distance between consecutive bays be at least $\lambda/2$. Second, because of the extra factor $\sin\theta$ appearing in the exponent of (4.102), the element factor in (4.99), and the screen factor (4.103), the use of a much larger value for y_2 will not cause appearance of any grating lobe as we learned from the theory in Chapter 1. A bay-to-bay distance of the order of a full wavelength or larger is commonly adopted in practice.[18]

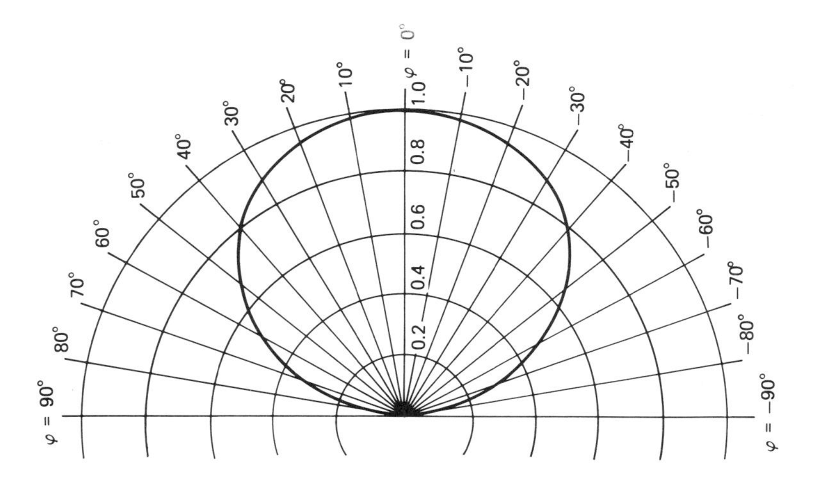

Fig. 4.22 Azimuthal pattern at $\theta = 54°$ of the same curtain array represented by the solid curve in Fig. 4.21.

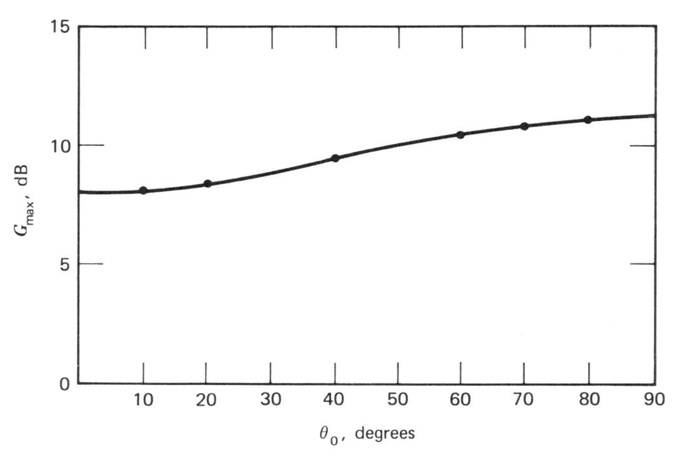

Fig. 4.23 Maximum power gain as a function of θ_0 for the same curtain array represented by the solid curve in Fig. 4.21.

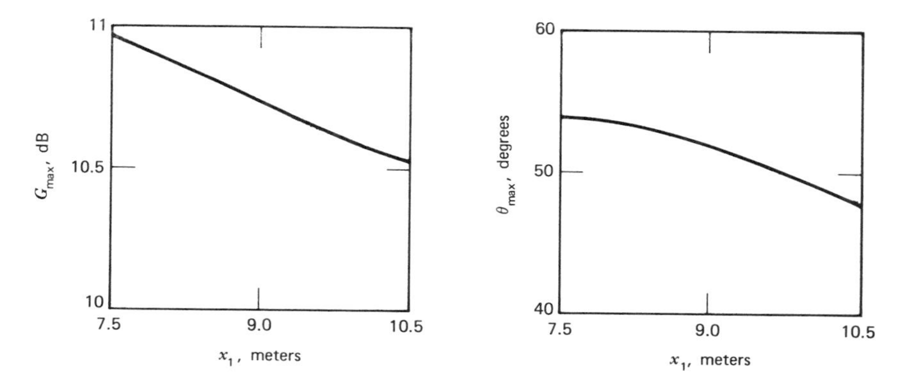

Fig. 4.24 Characteristics variations versus array-screen distance for the same curtain array represented by the solid curve in Fig. 4.21: (*a*) maximum power gain, (*b*) position of the beam maximum.

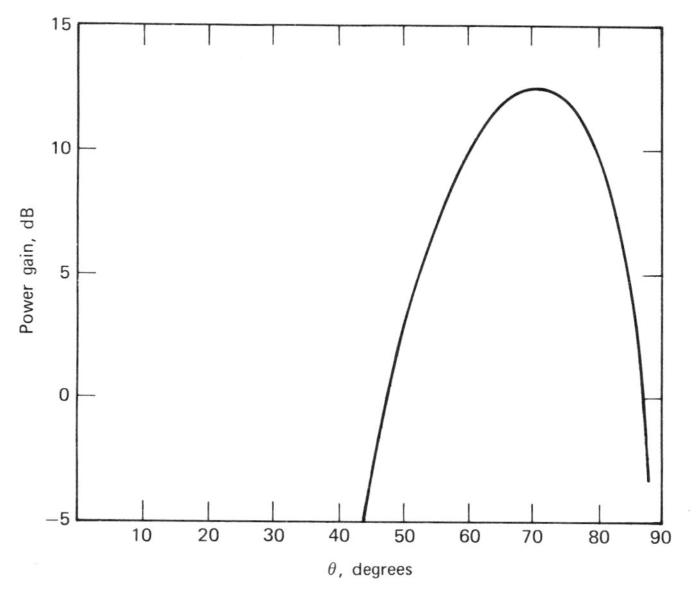

Fig. 4.25 Power gain of a curtain array over sea water with $M=4$, $N=1$, $h=x_1=7.5$ m, $a=0.0016$ m, $z_i=i\times 7.5$ m, $\theta_0=90°$, $\varphi=\varphi_0=0°$, $C_i=1$, and $f=10$ MHz.

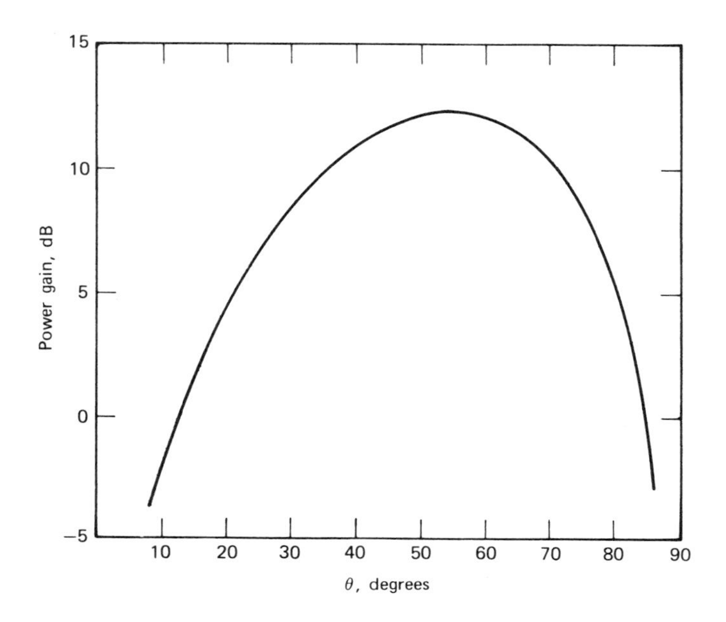

Fig. 4.26 Power gain of a curtain array over sea water with $M=N=2$, $h=x_1=7.5$ m, $a=0.0016$ m, $z_i=i\times 7.5$ m, $y_1=0$, $y_2=15.6$ m, $\theta_0=90°$, $\varphi=\varphi_0=0°$, $C_i=1$, and $f=10$ MHz.

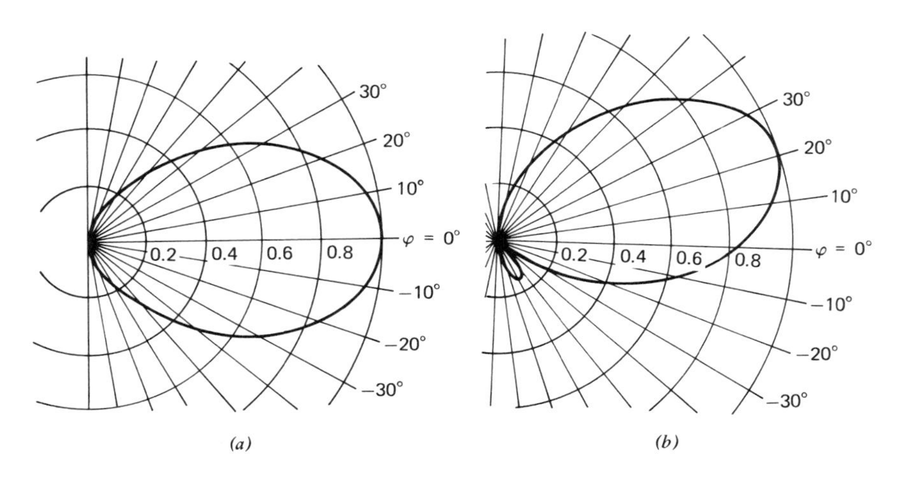

Fig. 4.27 Azimuthal pattern at $\theta=54°$ of the same array as in Fig. 4.26: (*a*) $\varphi_0=0°$, (*b*) $\varphi_0=30°$.

4.6 Vertical Monopole

In the previous sections we have analyzed and discussed characteristics of a single horizontal dipole and arrays of horizontal dipoles, where both the field components, E_θ and E_φ, were involved. In the remaining sections of this chapter, we will be concerned with vertically polarized antennas. As we see later, only one component of the electric field, namely E_θ, is present in these cases.

Let us consider first the commonly known whip antenna, or base-fed vertical monopole. In general, there always is a piece of metal between the ground and the base of the monopole for the purpose of excitation, whether it is a solid disk or a special screen system. The geometry of the problem may be depicted in Fig. 4.28, where a circular screen system consisting of N equally spaced radial wires is specifically indicated. We may consider the situation where a solid circular disk is put on the surface of the earth as the limiting case when $N \to \infty$.

For the moment, let us ignore the presence of the screen. In this case, we have (according to Fig. 4.2)

$$\alpha' = 90°, \qquad \theta' = 0°, \qquad H_s = z, \qquad s = z, \qquad \psi = \theta,$$

$$I(z) = I\{\sin k(h-z) + T_U(\cos kz - \cos kh) + T_D(\cos \tfrac{1}{2}kz - \cos \tfrac{1}{2}kh)\}, \qquad z \geqslant 0, \qquad kh \neq \frac{\pi}{2}, \tag{4.115}$$

or

$$I(z) = I'\left\{\sin kz - 1 + T'_U \cos kz - T'_D\left(\cos \tfrac{1}{2}kz - \cos\frac{\pi}{4}\right)\right\}, \qquad z \geqslant 0, \qquad kh = \frac{\pi}{2}. \tag{4.116}$$

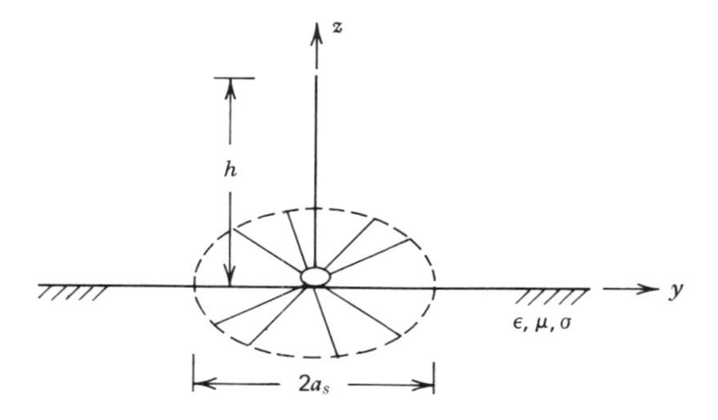

Fig. 4.28 Vertical antenna over a radial conductor ground screen.

The field components can be obtained from (4.59) and (4.60). They are

$$E_{\varphi}^{0}=0,$$

$$\begin{aligned}
E_{\theta}^{0}&=-j30k\frac{e^{-jkr_0}}{r_0}\sin\theta\int_0^h I(z)\exp(jkz\cos\theta)[1+R_v\exp(-j2kz\cos\theta)]\,dz\\
&=-j30I\frac{e^{-jkr_0}}{r_0}\left\{\frac{1}{\sin\theta}[A_2+jB_2+R_v(A_2-jB_2)]\right.\\
&+\frac{T_U}{\sin\theta}[A_3+jB_3+R_v(A_3-jB_3)]\\
&+\frac{T_D\sin\theta}{\frac{1}{4}-\cos^2\theta}[A_4+jB_4+R_v(A_4-jB_4)]\\
&\left.-(T_U\cos kh+T_D\cos\tfrac{1}{2}kh)\tan\theta[A_5+jB_5+R_v(A_5-jB_5)]\right\},\\
&\qquad kh\neq\frac{\pi}{2},
\end{aligned}\tag{4.117}$$

or

$$\begin{aligned}
E_{\theta}^{0}&=-j30I'\frac{e^{-jkr_0}}{r_0}\left\{\frac{1}{\sin\theta}[A_6+jB_6+R_v(A_6-jB_6)]\right.\\
&+\frac{T'_U}{\sin\theta}[A_7+jB_7+R_v(A_7-jB_7)]\\
&-\frac{T'_D\sin\theta}{\frac{1}{4}-\cos^2\theta}[A_8+jB_8+R_v(A_8-jB_8)]\\
&\left.+\left(-1+\frac{\sqrt{2}}{2}T'_D\right)\tan\theta[A_9+jB_9+R_v(A_9-jB_9)]\right\},\qquad kh=\frac{\pi}{2},
\end{aligned}\tag{4.118}$$

where the superscript 0 signifies that the screen is not present,

$$A_2=\cos(kh\cos\theta)-\cos kh,\tag{4.119}$$

$$B_2=\sin(kh\cos\theta)-\cos\theta\sin kh,\tag{4.120}$$

$$A_3 = \sin kh \cos(kh\cos\theta) - \cos\theta \cos kh \sin(kh\cos\theta), \tag{4.121}$$

$$B_3 = \sin kh \sin(kh\cos\theta) + \cos\theta \cos kh \cos(kh\cos\theta) - \cos\theta, \tag{4.122}$$

$$A_4 = \tfrac{1}{2}\sin\tfrac{1}{2}kh \cos(kh\cos\theta) - \cos\theta \cos\tfrac{1}{2}kh \sin(kh\cos\theta), \tag{4.123}$$

$$B_4 = \tfrac{1}{2}\sin\tfrac{1}{2}kh \sin(kh\cos\theta) + \cos\theta \cos\tfrac{1}{2}kh \cos(kh\cos\theta) - \cos\theta, \tag{4.124}$$

$$A_5 = \sin(kh\cos\theta), \qquad B_5 = 1 - \cos(kh\cos\theta), \tag{4.125}$$

$$A_6 = 1 - \cos\theta \sin\left(\frac{\pi}{2}\cos\theta\right), \qquad B_6 = \cos\theta \cos\left(\frac{\pi}{2}\cos\theta\right), \tag{4.126}$$

$$A_7 = \cos\left(\frac{\pi}{2}\cos\theta\right), \qquad B_7 = \sin\left(\frac{\pi}{2}\cos\theta\right) - \cos\theta, \tag{4.127}$$

$$A_8 = \frac{\sqrt{2}}{4}\cos\left(\frac{\pi}{2}\cos\theta\right) - \frac{\sqrt{2}}{2}\cos\theta \sin\left(\frac{\pi}{2}\cos\theta\right), \tag{4.128}$$

$$B_8 = \frac{\sqrt{2}}{4}\sin\left(\frac{\pi}{2}\cos\theta\right) + \frac{\sqrt{2}}{2}\cos\theta \cos\left(\frac{\pi}{2}\cos\theta\right) - \cos\theta, \tag{4.129}$$

$$A_9 = \sin\left(\frac{\pi}{2}\cos\theta\right), \qquad B_9 = 1 - \cos\left(\frac{\pi}{2}\cos\theta\right), \tag{4.130}$$

and r_0 is the distance measured from the monopole base to the far-field point.

Note that all the terms in (4.117) and (4.118) are finite, as they should be. For example, let us choose the first term in (4.117) to explain. Although the denominator ($\sin\theta$) becomes zero when $\theta = 0°$, the numerator also vanishes at $\theta = 0°$, as can be verified with A_2 and B_2 in (4.119) and (4.120). In fact,

$$\lim_{\theta\to 0}\left(\frac{A_2 + jB_2}{\sin\theta}\right) = \lim_{\theta\to 0}\left(\frac{dA_2}{d\theta} + j\frac{dB_2}{d\theta}\right) \Big/ \cos\theta = 0.$$

The same explanation applies to other terms involving $(\frac{1}{4} - cos^2\theta)$ or $\cos\theta$ in the denominator.

When the ground is perfectly conducting ($\sigma=\infty$), the corresponding field can simply be obtained from (4.117) or (4.118) by taking the limit $R_v \to 1$:

$$E_\theta^\infty = E_\theta^0\,(R_v = 1), \tag{4.131}$$

where the superscript ∞ denotes the special condition when $\sigma=\infty$.

When the screen is present, it is convenient to express the electric field in the form

$$E_\theta = E_\theta^0 + \Delta E_\theta, \tag{4.132}$$

where ΔE_θ represents the difference between the actual field and the field that would exist in the absence of the screen. According to the work by Wait and Pope,[28,29] the ratio of $\Delta E_\theta / E_\theta^0$ may be approximately given by

$$\frac{\Delta E_\theta}{E_\theta^0} \cong -k\eta \frac{e^{-jkr_0}}{r_0} \cdot \frac{1}{E_\theta^\infty} \int_{\rho=0}^{a_s} H_\varphi^\infty(\rho,0) J_1(k\rho \sin\theta)\,\rho\, d\rho, \tag{4.133}$$

where σ, μ, ϵ are the ground constants, $\eta=[j\omega\mu/(\sigma+j\omega\epsilon)]^{1/2}$ (surface impedance of the earth), ω is the angular frequency, a_s the radius of the circular screen, and J_1 the Bessel function of the first kind. The quantity $H_\varphi^\infty(\rho,0)$ is the associated φ component of the magnetic field expressed in a cylindrical coordinate system (ρ,φ,z) and calculated at the surface $(z=0)$ of a perfectly conducting ground. Of course, the reason that $H_\varphi^\infty(\rho,z)$ is a function of ρ and z only is that it is independent of φ. Note that $\rho^2+z^2=r_0^2$ and $\rho/z=\tan\theta$.

The field $H_\varphi^\infty(\rho,0)$ is related to the current $I(z)$ on the antenna by[28,29]

$$H_\varphi^\infty(\rho,0) = -\frac{1}{2\pi}\frac{\partial}{\partial\rho}\int_0^h \frac{\exp\left[-jk(z^2+\rho^2)^{1/2}\right]}{(z^2+\rho^2)^{1/2}} I(z)\,dz, \tag{4.134}$$

which becomes, after using (4.115) and (4.116), respectively,

$$H_\varphi^\infty(\rho,0) = \frac{jI}{2\pi}\left\{\frac{\exp\left[-jk(\rho^2+h^2)^{1/2}\right]}{\rho} - \frac{e^{-jk\rho}}{\rho}\cos kh + T_U A_{10} - j(T_U \cos kh + T_D \cos\tfrac{1}{2}kh)A_{11} - jT_D A_{12}\right\}, \quad kh \neq \pi/2, \tag{4.135}$$

or

$$H_\varphi^\infty(\rho,0)=\frac{I'}{2\pi}\left\{\frac{h\exp\left[-jk(\rho^2+h^2)^{1/2}\right]}{\rho(\rho^2+h^2)^{1/2}}+\frac{j}{\rho}e^{-jk\rho}+T'_UA_{10}\right.$$

$$\left.+\left(-1+\cos\frac{\pi}{4}T'_D\right)A_{11}-T'_DA_{12}\right\},\qquad kh=\frac{\pi}{2},\tag{4.136}$$

where

$$A_{10}=\frac{\exp\left[-jk(\rho^2+h^2)^{1/2}\right]}{\rho}\left[\sin kh-\frac{jh}{(\rho^2+h^2)^{1/2}}\cos kh\right],$$

$$A_{11}=-\int_0^h\left[\rho(z^2+\rho^2)^{-3/2}+jk\rho(z^2+\rho^2)^{-1}\right]\exp\left[-jk(z^2+\rho^2)^{1/2}\right]dz,$$

$$\tag{4.137}$$

and

$$A_{12}=\int_0^h\left[\rho(z^2+\rho^2)^{-3/2}+jk\rho(z^2+\rho^2)^{-1}\right]\exp\left[-jk(z^2+\rho^2)^{1/2}\right]\cos\tfrac{1}{2}kz\,dz.$$

$$\tag{4.138}$$

Note that (4.137) and (4.138) must be integrated numerically. When a situation arises such that the assumption of a sinusoidal current distribution is justified, (4.136) becomes unnecessary and (4.135) simplifies to

$$H_\varphi^\infty(\rho,0)=\frac{jI}{2\pi\rho}\left\{\exp\left[-jk(\rho^2+h^2)^{1/2}\right]-e^{-jk\rho}\cos kh\right\},\qquad T_U=T_D=0.$$

$$\tag{4.139}$$

The above equations are sufficient for calculating the far-field radiated from the base-fed monopole above a flat lossy earth with a possible additional screen. By a similar formulation, the input impedance of the same antenna shown in Fig. 4.28 when the ground is not perfectly conducting can be written as[28,30]

$$Z_{11}=Z_{11}^\infty+\Delta Z,\tag{4.140}$$

where the input impedance of the monopole above a perfect ground, Z_{11}^{∞}, should be only one-half of that for the center-fed dipole of length $2h$ in free space. That is,

$$Z_{11}^{\infty}=\frac{-j30\psi_{dR}\cos kh}{\sin kh+T_U(1-\cos kh)+T_D(1-\cos\frac{1}{2}kh)},\qquad kh\neq\frac{\pi}{2},\qquad (4.141)$$

or

$$Z_{11}^{\infty}=\frac{j30\psi_{dR}}{-1+T'_U-T'_D[1-\cos(\pi/4)]},\qquad kh=\frac{\pi}{2}.\qquad (4.142)$$

The change in impedance, ΔZ, may again be expressed in terms of cylindrical coordinates by[28,30]

$$\Delta Z=-\frac{1}{I^2(0)}\int_0^{\infty}H_{\varphi}^{\infty}(\rho,0)E_{\rho}(\rho,0)2\pi\rho\,d\rho,\qquad (4.143)$$

where $I(0)$ is the base current, $H_{\varphi}^{\infty}(\rho,0)$ is the same magnetic field given by (4.135) or (4.136), and $E_{\rho}(\rho,0)$ is the tangential electrical field on the actual imperfect ground.

To calculate (4.143), we still need an expression for $E_{\rho}(\rho,0)$. Unfortunately, an exact solution for it is unknown. It is, therefore, necessary to make further simplifications. In particular, when $\sigma\gg\omega\epsilon$, we can have the following approximation[28,31]:

$$E_{\rho}(\rho,0)\cong-\eta_e H_{\varphi}^{\infty}(\rho,0),\qquad (4.144)$$

where η_e can be considered as an equivalent surface impedance of the air-ground interface. This impedance should be

$\eta_e=\eta$ (surface impedance of the imperfect earth alone) for $\rho>a_s$,

(4.145)

or

$\eta_e=\dfrac{\eta\eta_s}{\eta+\eta_s}$ (parallel combination of η and the intrinsic impedance of the ground screen, η_s) for $0\leqslant\rho\leqslant a_s$, (4.146)

where

$$\eta_s = j60kd\ln(d/2\pi c), \tag{4.147}$$

c = radius of the screen wire,

and

d = spacing between radial wires

$= 2\pi\rho/N$ (see Fig. 4.28).

Note that $\eta_s \to 0$ as $N \to \infty$, corresponding to the case when the ground screen of radial wires is replaced by a solid circular disk.

Substituting (4.144) through (4.146) into (4.143), we have

$$\Delta Z = \Delta Z_1 + \Delta Z_2, \tag{4.148}$$

with

$$\Delta Z_1 \cong \frac{\eta}{I^2(0)} \int_{a_s}^{\infty} \left[H_\varphi^\infty(\rho, 0)\right]^2 2\pi\rho\, d\rho \tag{4.149}$$

and

$$\Delta Z_2 \cong \frac{1}{I^2(0)} \int_0^{a_s} \frac{\eta\eta_s}{\eta+\eta_s} \left[H_\varphi^2(\rho, 0)\right]^2 2\pi\rho\, d\rho. \tag{4.150}$$

Physically, it is clear that the expression ΔZ_1 is the contribution by the monopole over a perfectly conducting discoid, and that ΔZ_2 accounts for the finite surface impedance of the radial screen over a lossy earth.

Once again, both (4.149) and (4.150) must be calculated numerically if the expression given in (4.135) or (4.136) is used for $H_\varphi^\infty(\rho, 0)$. On the other hand, if a simpler form in (4.139) is used instead, both (4.149) and (4.150) can be expressed in terms of the ordinary sine and cosine integrals, $S(0, x)$ and $C(0, x)$, defined in (4.6) and (4.7).

Based on the formulation outlined above, the power gain of the subject antenna discussed in this section will then be

$$G = \frac{r_0^2 |E_\theta|^2}{30 |I(0)|^2 R_{\text{in}}}, \tag{4.151}$$

where E_θ is given by (4.132), R_{in} is the real part of the input impedance Z_{11} in (4.140), and I(0) is the base current which can be obtained by putting $z = 0$ in (4.115) or (4.116).

A typical example, for $h = a_s = \lambda/4$ at $f = 10$ MHz with a (radius of the monopole) $= 0.0016$ m, c (radius of the screen wire) $= 0.0015$ m, and $N = 120$, is shown in Fig. 4.29, where the case of sea water ($\sigma = 5$ mho/m, $\epsilon_r = 80$) as the ground is presented as curve (a) and that with "good ground" ($\sigma = 0.01$ mho/m, $\epsilon_r = 10$) by curve (b). In both cases, $Z_{11}^{\infty} = 38.1966 + j20.6777\ \Omega$. On the other hand, $\Delta Z = -0.5149 - j0.2142\ \Omega$ and $0.8567 + j3.0246\ \Omega$, respectively, for curves (a) and (b). It is again evident that the value and position of the maximum power gain are dependent on the type of ground.

Since the calculations for the special case when $\sigma = \infty$, $T_U = T_D = 0$, and $a_s = \infty$ can be easily made, it is found that the accuracy for the data presented in curve (a) is indeed very satisfactory by comparison. On the other hand, there is no simple way to check the accuracy for curve (b).

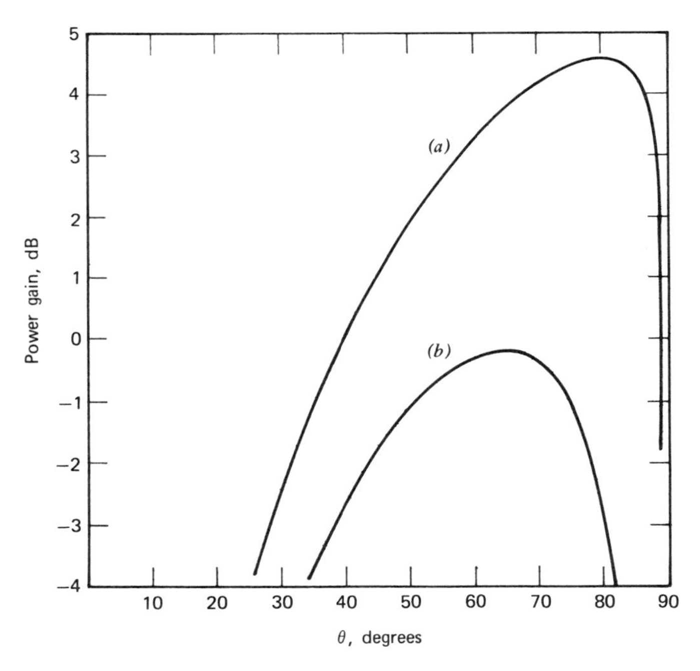

Fig. 4.29 Power gain of a vertical monopole above a flat lossy ground with $h = a_s = \lambda/4$, $a = 0.0016$ m, $c = 0.0015$ m, $N = 120$, and $f = 10$ MHz: (a) sea water ($\sigma = 5$ mho/m, $\epsilon_r = 80$), (b) good ground ($\sigma = 0.01$ mho/m, $\epsilon_r = 10$).

Furthermore, the condition, $\sigma \gg \omega\epsilon$, required for making approximations in the formulation is only marginally satisfied for curve (*b*) in this particular example. It should help to explain why the difference betweeen the maximum values in curves (*a*) and (*b*) is larger than expected.

In general, the results in terms of power gain are improved somewhat by having a larger h or a_s although the general shape of G versus θ remains about the same. To be specific, variations of G_{max} with h while all other parameters are kept the same as in Fig. 4.29 are given in Fig. 4.30, where sea water is chosen as the ground. The position of G_{max} remains between $\theta = 80°$ and $82°$. Since the improvement in G_{max} by increasing a_s is insignificant, the results are omitted here.

To demonstrate the variation in power gain when the frequency is changed, another example, with $h = a_s = \lambda/4$ at $f = 30$ MHz and with the same values for a, c, N as those used in Fig. 4.29 and sea water, is shown in Fig. 4.31. Since the magnitude of the surface impedance of the earth is larger (because of higher frequency) and h/a is smaller in this case, the maximum power gain obtained is lower and the position of G_{max} is farther away from the ground. These facts are clearly seen by comparing Figs. 4.29 and 4.31.

Thus far we have only discussed the monopole as a single antenna whose pattern, because of its vertical polarization, has only θ dependence. In order to have the azimuthal coverage at the same time, which may be desirable in some application such as direction finding, we can arrange, for example, a circular array made of this type of monopoles. Indeed, this is

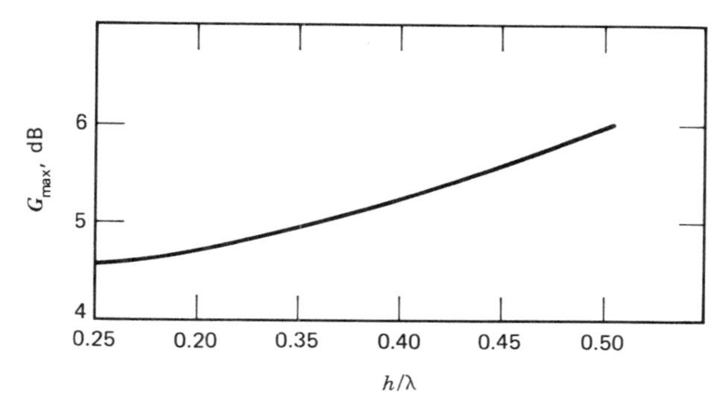

Fig. 4.30 Variations of the maximum power gain with respect to the monopole length for the same antenna (specified in Fig. 4.29) above sea water.

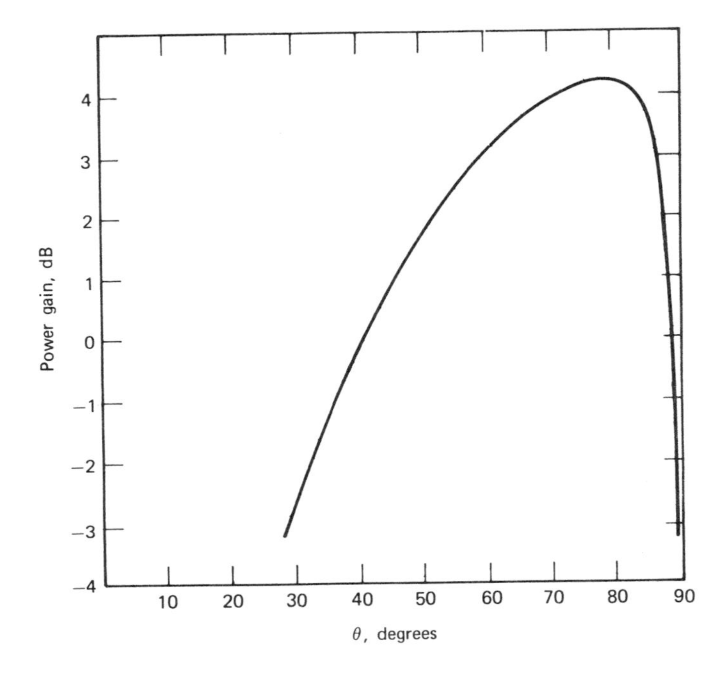

Fig. 4.31 Power gain of a vertical monopole above sea water with $h = a_s = \lambda/4$, $a = 0.0016$ m, $c = 0.0015$ m, $N = 120$, and $f = 30$ MHz.

the kind of array (with perhaps some minor modifications depending on special situations involved) which has been built as a part of the Wullenweber antenna.[32] The analytic method presented in Section 3.2 will then apply. Actual calculations can be made with a known set of excitations to individual elements if the array is used as a transmitting antenna or with a measured set of signal levels if the array is used for receiving.

4.7 Cylindrical Sleeve Antenna

Another kind of radiator also with vertical polarization is known as the cylindrical sleeve antenna, designed for broad-band application.[33] It consists of a vertically extended inner conductor of height h and radius a_i, and an outer conductor of coaxial line of height s and radius a_0 over a horizontal conducting plane [see Fig. 4.32(a)]. It differs from the base-fed monopole discussed in the previous section in that the sheath of the coaxial line does not end at the conducting plane but extends above it a distance s to form a sleeve. This essentially moves the feed point upward from $z=0$ to $z=s$. However, the discontinuity at $z=s$ does cause considerable difficulty in the rigorous solution of the problem. To avoid this trouble, an approximation is usually made by considering the antenna with a uniform height h and an equivalent radius a_e ($a_i < a_e \leqslant a_0$) and with the excitation at $z=s$, as shown in Fig. 4.32(*b*). In this approximation we have implicitly assumed that the conducting plane in Fig. 4.32(*a*) is infinitely large.

With this approximation, we can then derive the input impedance at $z=s$ through the conventional induced emf method with appropriate boundary conditions[33]:

$$Z_{\text{in}} = \frac{-\mu}{4\pi I^2(s)} \int_0^h I(z)\,dz\, L_z \int_{-h}^{h} I(z')K(z,z')\,dz', \tag{4.152}$$

where

$$L_z = -\frac{j\omega}{k^2}\left(\frac{\partial^2}{\partial z^2} + k^2\right), \qquad k = \frac{2\pi}{\lambda}, \tag{4.153}$$

$$K(z,z') = \frac{\exp\left[-jk\sqrt{(z-z')^2 + a_e^2}\,\right]}{\sqrt{(z-z')^2 + a_e^2}}, \tag{4.154}$$

$$\mu = \text{permeability},$$

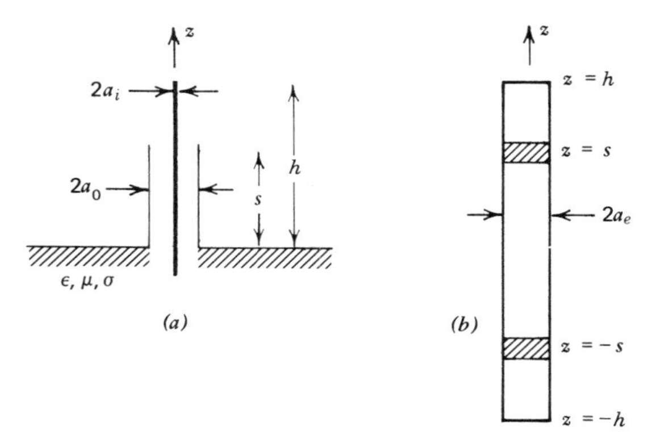

Fig. 4.32 (*a*) A sketch of sleeve antenna, (*b*) approximate equivalent of (*a*).

and $I(z)$, the current distribution over the antenna, is still unknown.

An approximate Z_{in} can usually be determined by assuming a suitable form for $I(z)$. When considering the conditions that (i) $I(z)$ must be continuous in $0 \leqslant z \leqslant h$, (ii) $I(h)=0$, and (iii) $\partial I(z)/\partial z$ must be continuous except at the driving point $z=s$, Taylor concluded that the following form may approximate $I(z)$ very well[33]:

$$I_1(z)=1+C_1(-\cos ks+\cos kz), \qquad 0 \leqslant z \leqslant s,$$

and (4.155)

$$I_2(z)=\delta_1[1-\cos k(h-z)] + C_2\{\sin k(h-z)+\delta_2[1-\cos k(h-z)]\}, \qquad s \leqslant z \leqslant h,$$

where

$$\delta_1=\frac{1}{1-\cos k(h-s)},$$

$$\delta_2=-\frac{\sin k(h-s)}{1-\cos k(h-s)}, \tag{4.156}$$

and C_1 and C_2 are complex constants to be determined. Note that the current has been normalized to unity at $z=s$. Current forms assumed in (4.155) have also be verified experimentally and have proved satisfactory provided that

$$h-s<\lambda \qquad \text{and} \qquad s<\frac{\lambda}{2}. \tag{4.157}$$

Substituting (4.155) into (4.152) and simplifying, we obtain an approximate expression for the input impedance of the sleeve antenna:

$$\bar{Z}_{in}=j30(\gamma_0+\gamma_1 C_1+\gamma_2 C_2+\gamma_{11}C_1^2+\gamma_{22}C_2^2+\gamma_{12}C_1 C_2), \tag{4.158}$$

where

$$\gamma_0=\delta_1^2 M_0 \equiv \gamma_1^2(M_{0r}+jM_{0i}),$$

$$\gamma_1=\delta_1 M_1 \equiv \delta_1(M_{1r}+jM_{1i}),$$

$$\gamma_2=2\delta_1^2 M_2 \equiv 2\delta_1^2(M_{2r}+jM_{2i}), \tag{4.159}$$

$$\gamma_{11}=M_3 \equiv M_{3r}+jM_{3i},$$

$$\gamma_{22}=2\delta_1 M_4 \equiv 2\delta_1(M_{4r}+jM_{4i}),$$

and

$$\gamma_{12}=2\delta_1 M_5 \equiv 2\delta_1(M_{5r}+jM_{5i}).$$

The detailed derivations of (4.158) and (4.159), which can be found elsewhere,[33,34] are omitted here. Instead, the real and imaginary parts of M_0 through M_5 in (4.159) are summarized as follows:

$$\begin{aligned}
M_{0r}=&[1+\cos^2 k(h-s)]\sin ka_e-\cos^2 k(h-s)\sin k\sqrt{4s^2+a_e^2}\\
&-2\cos k(h-s)\sin k\sqrt{(h-s)^2+a_e^2}+2\cos k(h-s)\sin k\sqrt{(h+s)^2+a_e^2}\\
&-\sin k\sqrt{4h^2+a_e^2}+[2ks\cos^2 k(h-s)+\sin 2k(h-s)]\overline{C}(ka_e,2ks)\\
&+[2k(h-s)\cos k(h-s)-2\sin k(h-s)]\overline{C}(ka_e,k(h-s))\\
&-[2k(h+s)\cos k(h-s)+2\sin k(h-s)]\overline{C}(ka_e,k(h+s))\\
&+2kh\overline{C}(ka_e,2kh)+2\sin 2kh\overline{C}_c(ka_e,k(h+s))\\
&-\sin 2kh\overline{C}_c(ka_e,2kh)-\sin 2kh\overline{C}_c(ka_e,2ks)\\
&-2\cos 2kh C_s(ka_e,k(h+s))+\cos 2kh C_s(ka_e,2kh)\\
&+\cos 2kh C_s(ka_e,2ks)+2C_s(ka_e,k(h-s)),
\end{aligned}\tag{4.160}$$

$$\begin{aligned}
M_{0i}=&[1+\cos^2 k(h-s)]\cos ka_e-\cos^2 k(h-s)\cos k\sqrt{4s^2+a_e^2}\\
&-2\cos k(h-s)\cos k\sqrt{(h-s)^2+a_e^2}+2\cos k(h-s)\cos k\sqrt{(h+s)^2+a_e^2}\\
&-\cos k\sqrt{4h^2+a_e^2}-[2ks\cos^2 k(h-s)+\sin 2k(h-s)]S(ka_e,2ks)\\
&-[2k(h-s)\cos k(h-s)-2\sin k(h-s)]S(ka_e,k(h-s))\\
&+[2k(h+s)\cos k(h-s)+2\sin k(h-s)]S(ka_e,k(h+s))
\end{aligned}$$

$$-2khS(ka_e,2kh)-2S_s(ka_e,k(h-s))$$

$$-2\sin 2khS_c(ka_e,k(h+s))+\sin 2khS_c(ka_e,2kh)$$

$$+\sin 2khS_c(ka_e,2ks)+2\cos 2khS_s(ka_e,k(h+s))$$

$$-\cos 2khS_s(ka_e,2kh)-\cos 2khS_s(ka_e,2ks), \tag{4.161}$$

$$M_{1r}=2\cos ks\cos k(h-s)\left[\sin ka_e-\sin k\sqrt{4s^2+a_e^2}\right]$$

$$-2\cos ks\left[\sin k\sqrt{(h-s)^2+a_e^2}-\sin k\sqrt{(h+s)^2+a_e^2}\right]$$

$$+[4ks\cos ks\cos k(h-s)+2\sin k(h-2s)]\overline{C}(ka_e,2ks)$$

$$+2[\sin ks+k(h-s)\cos ks]\overline{C}(ka_e,k(h-s))$$

$$+2[\sin ks-k(h+s)\cos ks]\overline{C}(ka_e,k(h+s))$$

$$+2\sin kh\left[\overline{C}_c(ka_e,k(h+s))-\overline{C}_c(ka_e,k(h-s))-\overline{C}_c(ka_e,2ks)\right]$$

$$-2\cos kh[C_s(ka_e,k(h+s))-C_s(ka_e,k(h-s))-C_s(ka_e,2ks)], \tag{4.162}$$

$$M_{1i}=2\cos ks\cos k(h-s)\left[\cos ka_e-\cos k\sqrt{4s^2+a_e^2}\right]$$

$$-2\cos ks\left[\cos k\sqrt{(h-s)^2+a_e^2}-\cos k\sqrt{(h+s)^2+a_e^2}\right]$$

$$-[4ks\cos ks\cos k(h-s)+2\sin k(h-2s)]S(ka_e,2ks)$$

$$-2[\sin ks+k(h-s)\cos ks]S(ka_e,k(h-s))$$

$$-2[\sin ks-k(h+s)\cos ks]S(ka_e,k(h+s))$$

$$-2\sin kh[S_c(ka_e,k(h+s))-S_c(ka_e,k(h-s))-S_c(ka_e,2ks)]$$

$$+2\cos kh[S_s(ka_e,k(h+s))-S_s(ka_e,k(h-s))-S_s(ka_e,2ks)], \tag{4.163}$$

$$M_{2r} = \tfrac{1}{2}\sin 2k(h-s)\sin k\sqrt{4s^2+a_e^2} + \sin k(h-s)\sin k\sqrt{4h^2+a_e^2}$$

$$+ [\sin k(h-s) + \tfrac{1}{2}\sin 2k(h-s)]\Big[\sin k\sqrt{(h-s)^2+a_e^2}$$

$$- \sin k\sqrt{(h+s)^2+a_e^2} - \sin ka_e\Big]$$

$$- [2kh\sin k(h-s) + \cos 2k(h-s) - 1]\bar{C}(ka_e, 2kh)$$

$$- 2\sin k(h-s) C_s(ka_e, k(h-s))$$

$$+ [-ks\sin 2k(h-s) + \cos 2k(h-s) - \cos k(h-s)]\bar{C}(ka_e, 2ks)$$

$$+ \{k(h+s)[\sin k(h-s) + \tfrac{1}{2}\sin 2k(h-s)]$$

$$+ 2 - 2\cos k(h-s)\}\bar{C}(ka_e, k(h+s))$$

$$+ \{-k(h-s)[\sin k(h-s) + \tfrac{1}{2}\sin 2k(h-s)]$$

$$+ 2\sin^2 k(h-s)\}\bar{C}(ka_e, k(h-s))$$

$$- [\cos k(h+s) - \cos 2kh]$$

$$\times \left[2\bar{C}_c(ka_e, k(h+s)) - \bar{C}_c(ka_e, 2kh) - \bar{C}_c(ka_e, 2ks)\right]$$

$$- [\sin k(h+s) - \sin 2kh]$$

$$\times [2C_s(ka_e, k(h+s)) - C_s(ka_e, 2kh) - C_s(ka_e, 2ks)], \tag{4.164}$$

$$M_{2i} = \tfrac{1}{2}\sin 2k(h-s)\cos k\sqrt{4s^2+a_e^2} + \sin k(h-s)\cos k\sqrt{4h^2+a_e^2}$$

$$+ [\sin k(h-s) + \tfrac{1}{2}\sin 2k(h-s)]\Big[\cos k\sqrt{(h-s)^2+a_e^2}$$

$$- \cos k\sqrt{(h+s)^2+a_e^2} - \cos ka_e\Big]$$

$$+ [2kh\sin k(h-s) + \cos k(h-s) - 1]S(ka_e, 2kh)$$

$$+ 2\sin k(h-s)S_s(ka_e, k(h-s))$$

$$+ [ks\sin 2k(h-s) - \cos 2k(h-s) + \cos k(h-s)]S(ka_e, 2ks)$$

$$- \{k(h+s)[\sin k(h-s) + \tfrac{1}{2}\sin 2k(h-s)]$$

$$+ 2 - 2\cos k(h-s)\}S(ka_e, k(h+s))$$

$$+ \{k(h-s)[\sin k(h-s) + \tfrac{1}{2}\sin 2k(h-s)]$$

$$- 2\sin^2 k(h-s)\}S(ka_e, k(h-s))$$

$$+ [\cos k(h+s) - \cos 2kh]$$

$$\times\ [2S_c(ka_e, k(h+s)) - S_c(ka_e, 2kh) - S_c(ka_e, 2ks)]$$

$$+ [\sin k(h+s) - \sin 2kh]$$

$$\times\ [2S_s(ka_e, k(h+s)) - S_s(ka_e, 2kh) - S_s(ka_e, 2ks)], \tag{4.165}$$

$$M_{3r} = \cos^2 ks\Big[\sin ka_e - \sin k\sqrt{4s^2+a_e^2}\Big] + C_s(ka_e, 2ks)$$

$$+ 2\cos ks(ks\cos ks - \sin ks)\bar{C}(ka_e, 2ks), \tag{4.166}$$

$$M_{3i} = \cos^2 ks\left[\cos ka_e - \cos k\sqrt{4s^2 + a_e^2}\right] - S_s(ka_e, 2ks)$$

$$-2\cos ks(ks\cos ks - \sin ks)S(ka_e, 2ks), \tag{4.167}$$

$$M_{4r} = [1 + \cos k(h-s)]\Big[\sin ka_e - \sin k\sqrt{(h-s)^2 + a_e^2} + \sin k\sqrt{(h+s)^2 + a_e^2}$$

$$-\tfrac{1}{2}\sin k\sqrt{4h^2 + a_e^2} - \tfrac{1}{2}\sin k\sqrt{4s^2 + a_e^2}\Big]$$

$$+2C_s(ka_e, k(h-s)) + [ks(1 + \cos k(h-s)) + \sin k(h-s)]\bar{C}(ka_e, 2ks)$$

$$+[kh(1 + \cos k(h-s)) - \sin k(h-s)]\bar{C}(ka_e, 2kh)$$

$$-k(h+s)[1 + \cos k(h-s)]\bar{C}(ka_e, k(h+s))$$

$$+[k(h-s)(1 + \cos k(h-s)) - 2\sin k(h-s)]\bar{C}(ka_e, k(h-s))$$

$$+\sin k(h+s)\left[2\bar{C}_c(ka_e, k(h+s)) - \bar{C}_c(ka_e, 2kh) - \bar{C}_c(ka_e, 2ks)\right]$$

$$-\cos k(h+s)[2C_s(ka_e, k(h+s)) - C_s(ka_e, 2kh) - C_s(ka_e, 2ks)],$$

$$\tag{4.168}$$

$$M_{4i} = [1 + \cos k(h-s)]\Big[\cos ka_e + \cos k\sqrt{(h+s)^2 + a_e^2} - \cos k\sqrt{(h-s)^2 + a_e^2}$$

$$-\tfrac{1}{2}\cos k\sqrt{4h^2 + a_e^2} - \tfrac{1}{2}\cos k\sqrt{4s^2 + a_e^2}\Big]$$

$$-2S_s(ka_e, k(h-s)) - [ks(1 + \cos k(h-s)) + \sin k(h-s)]S(ka_e, 2ks)$$

$$-[kh(1 + \cos k(h-s)) - \sin k(h-s)]S(ka_e, 2kh)$$

$$+k(h+s)(1 + \cos k(h-s))S(ka_e, k(h+s))$$

$$-[k(h-s)(1 + \cos k(h-s)) - 2\sin k(h-s)]S(ka_e, k(h-s))$$

$$-\sin k(h+s)[2S_c(ka_e, k(h+s)) - S_c(ka_e, 2kh) - S_c(ka_e, 2ks)]$$

$$+\cos k(h+s)[2S_s(ka_e, k(h+s)) - S_s(ka_e, 2kh) - S_s(ka_e, 2ks)],$$

$$\tag{4.169}$$

$$
\begin{aligned}
M_{5r} = {} & \cos ks \sin k(h-s)\Big[\sin k\sqrt{4s^2+a_e^2} - \sin ka_e \\
& + \sin k\sqrt{(h-s)^2+a_e^2} - \sin k\sqrt{(h+s)^2+a_e^2}\Big] \\
& - [2ks\cos ks \sin k(h-s) + \cos ks - \cos k(h-2s)]\bar{C}(ka_e, 2ks) \\
& + [k(h+s)\cos ks \sin k(h-s) + \cos kh - \cos ks]\bar{C}(ka_e, k(h+s)) \\
& - [k(h-s)\cos ks \sin k(h-s) - \cos ks + \cos k(h-2s)]\bar{C}(ka_e, k(h-s)) \\
& + (\cos kh - \cos ks)\Big[\bar{C}_c(ka_e, k(h+s)) \\
& \qquad - \bar{C}_c(ka_e, k(h-s)) - \bar{C}_c(ka_e, 2ks)\Big] \\
& + (\sin kh - \sin ks)[C_s(ka_e, k(h+s)) \\
& \qquad - C_s(ka_e, k(h-s)) - C_s(ka_e, 2ks)],
\end{aligned}
\tag{4.170}
$$

and

$$
\begin{aligned}
M_{5i} = {} & \cos ks \sin k(h-s)\Big[\cos k\sqrt{4s^2+a_e^2} - \cos ka_e \\
& + \cos k\sqrt{(h-s)^2+a_e^2} - \cos k\sqrt{(h+s)^2+a_e^2}\Big] \\
& + [2ks\cos ks \sin k(h-s) + \cos ks - \cos k(h-2s)]S(ka_e, 2ks) \\
& - [k(h+s)\cos ks \sin k(h-s) + \cos kh - \cos ks]S(ka_e, k(h+s)) \\
& + [k(h-s)\cos ks \sin k(h-s) - \cos ks + \cos k(h-2s)]S(ka_e, k(h-s)) \\
& - (\cos kh - \cos ks)[S_c(ka_e, k(h+s)) - S_c(ka_e, k(h-s)) - S_c(ka_e, 2ks)] \\
& - (\sin kh - \sin ks)[S_s(ka_e, k(h+s)) - S_s(ka_e, k(h-s)) - S_s(ka_e, 2ks)].
\end{aligned}
\tag{4.171}
$$

All of the generalized sine integrals and cosine integrals in (4.160) through (4.171) have been defined in (4.6) through (4.13).

Note that the complex constants C_1 and C_2 in (4.158) are still unknown. Since $\overline{Z}_{in}$ is an analytic function of C_1 and C_2, we can determine C_1 and C_2 by setting

$$\frac{\partial \overline{Z}_{in}}{\partial C_1}=0 \quad \text{and} \quad \frac{\partial \overline{Z}_{in}}{\partial C_2}=0, \tag{4.172}$$

yielding

$$C_1=\frac{\gamma_2\gamma_{12}-2\gamma_1\gamma_{22}}{4\gamma_{11}\gamma_{22}-\gamma_{12}^2},$$

$$C_2=\frac{\gamma_1\gamma_{12}-2\gamma_2\gamma_{11}}{4\gamma_{11}\gamma_{22}-\gamma_{12}^2}. \tag{4.173}$$

The final result of $\overline{Z}_{in}$ obtained by putting (4.173) into (4.158) is usually considered the best approximation to the true Z_{in} defined in (4.152). Using (4.173) we can also calculate the current distribution in (4.155). Numerical results of C_1 and C_2 for a typical sleeve antenna with $h=7.2390$ m, $s=2.5146$ m, $a_i=0.0056$ m, and $a_0=0.1054$ m$\cong a_e$ are given below as a function of frequency:

Frequency (MHz)	C_1	C_2
8	0.60418–j0.36310	1.61858–j0.12486
10	1.28039–j0.52402	1.52778–j0.21449
12	1.75302–j0.75330	1.52759–j0.34905
14	2.14839–j1.09585	1.58267–j0.55420
16	2.49371–j1.62849	1.65846–j0.87396
18	2.70938–j2.46835	1.67859–j1.37410
20	2.49742–j3.69701	1.44559–j2.09210
22	1.32618–j4.95753	0.63594–j2.78933
24	−0.64426–j5.15836	−0.63101–j2.78725
26	−2.04155–j4.16334	−1.50666–j1.98585
28	−2.54785–j3.16106	−1.66750–j0.97116
30	−3.10023–j2.15180	−0.94180–j0.46891
32	−2.63807–j1.07107	−0.95281–j0.69870

Examples of $\bar{Z}_{in}$ and $I(z)$ for the same antenna dimensions are shown in Figs. 4.33 through 4.37. These results are in fair agreement with existing measured data for a close (but not identical) model discussed elsewhere.[6]

The technique adopted here is actually an extension of the variational formulation.[35] Equation (4.158) is also good for the mutual impedance between two parallel identical sleeve antennas when the equivalent radius a_e is replaced by d, the center-to-center separation of the two antennas.

Strictly speaking, the above derivation is based on the assumption that the conducting plane of the sleeve antenna shown in Fig. 4.32(a) is infinitely large. In reality, of course, the said plane cannot be infinite in extent. The actual input impedance of the antenna should then be modified in a manner similar to that following (4.140). Since this modification is not significant when the radius of the finite conducting plane is not less than $\lambda/4$, as shown by the example given in the previous section, we are content here to ignore this additional consideration.

After taking care of current and input impedance, we are now ready to derive the field radiated by the sleeve antenna above a finitely conducting earth. Here again we follow the same approach as that for the vertical monopole (Section 4.6) without first considering the presence of the

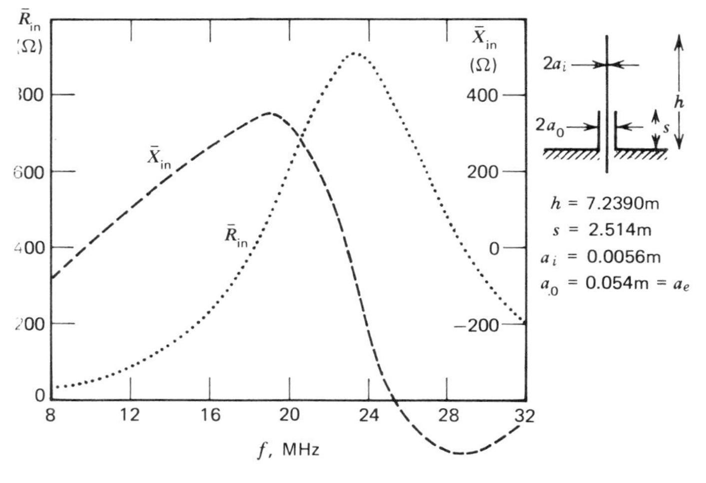

Fig. 4.33 Approximate input impedance of a sleeve antenna above a perfect ground.

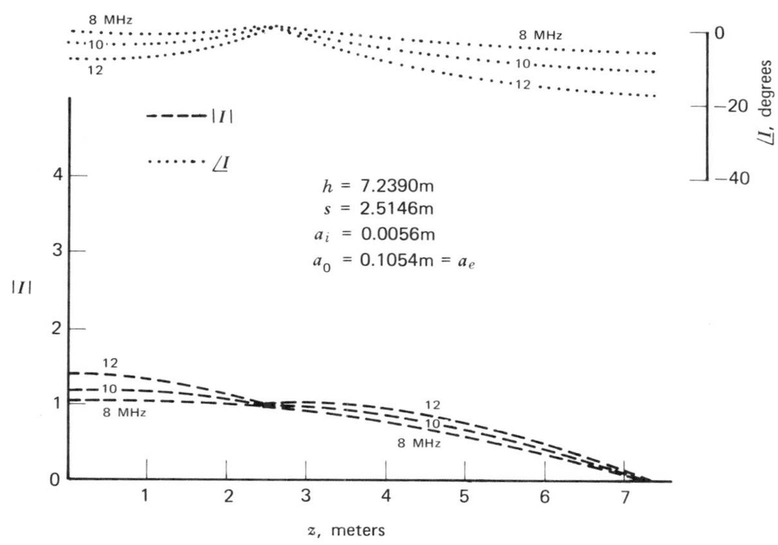

Fig. 4.34 Assumed current distribution on a sleeve antenna for $f = 8$, 10, and 12 MHz.

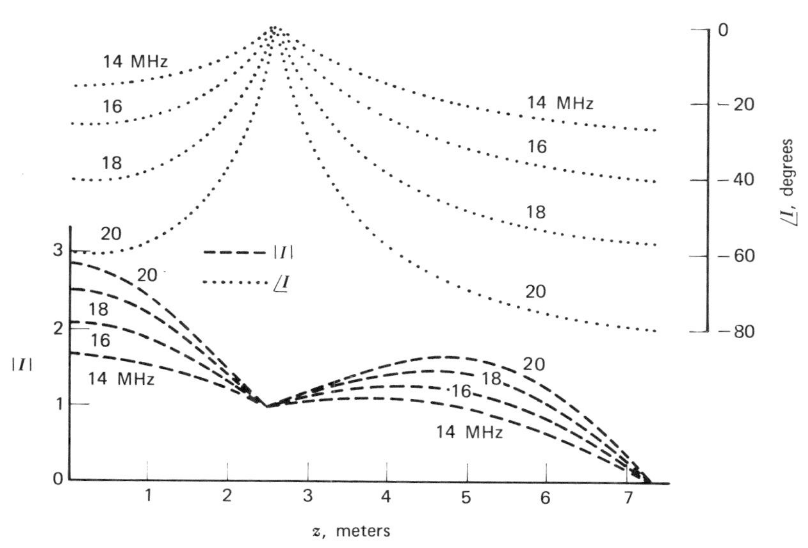

Fig. 4.35 Assumed current distribution on a sleeve antenna for $f = 14$, 16, 18, and 20 MHz.

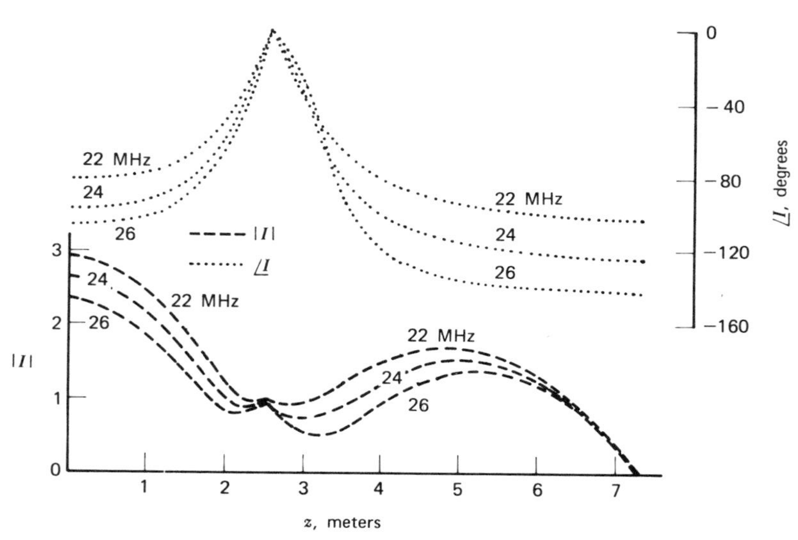

Fig. 4.36 Assumed current distribution on a sleeve antenna for $f = 22$, 24, and 26 MHz.

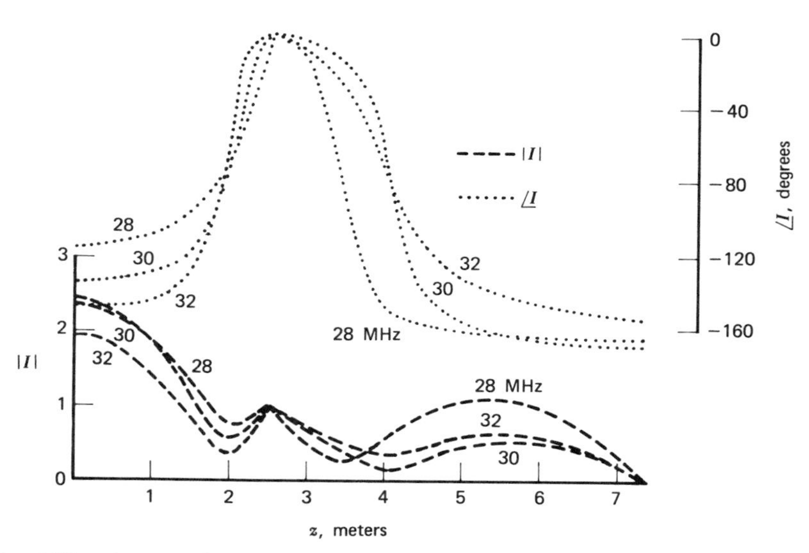

Fig. 4.37 Assumed current distribution on a sleeve antenna for $f = 28$, 30, and 32 MHz.

conducting plane. We then have

$$E_{\varphi}^{0}=0,$$

$$E_{\theta}^{0}=-j30k\frac{e^{-jkr_0}}{r_0}\sin\theta\int_0^h I(z)\exp(jkz\cos\theta)[1+R_v\exp(-j2kz\cos\theta)]\,dz$$

$$=-j30\frac{e^{-jkr_0}}{r_0}\sin\theta(E_{\theta 1}+E_{\theta 2}+E_{\theta 3}+E_{\theta 4}+E_{\theta 5}), \tag{4.174}$$

where

$$E_{\theta 1}=\int_0^s(1-C_1\cos ks)[\exp(jkz\cos\theta)+R_v\exp(-jkz\cos\theta)]\,dz$$

$$=\frac{1-C_1\cos ks}{j\cos\theta}[\exp(jks\cos\theta)-R_v\exp(-jks\cos\theta)-1+R_v], \tag{4.175}$$

$$E_{\theta 2}=C_1\int_0^s\cos kz[\exp(jkz\cos\theta)+R_v\exp(-jkz\cos\theta)]\,dz$$

$$=\frac{C_1}{\sin^2\theta}\{[\exp(jks\cos\theta)](j\cos\theta\cos ks+\sin ks)-j\cos\theta$$

$$+R_v[(-j\cos\theta\cos ks+\sin ks)\exp(-jks\cos\theta)+j\cos\theta]\}, \tag{4.176}$$

$$E_{\theta 3}=(\delta_1+C_2\delta_2)\int_s^h[\exp(jkz\cos\theta)+R_v\exp(-jkz\cos\theta)]\,dz$$

$$=\frac{(\delta_1+C_2\delta_2)}{j\cos\theta}[\exp(jkh\cos\theta)-R_v\exp(-jkh\cos\theta)$$

$$-\exp(jks\cos\theta)+R_v\exp(-jks\cos\theta)], \tag{4.177}$$

$$E_{\theta 4}=[C_2 \sin kh-(\delta_1+C_2\delta_2)\cos kh] \times \int_s^h \cos kz[\exp(jkz\cos\theta)+R_v\exp(-jkz\cos\theta)]\,dz$$

$$=\frac{C_2\sin kh-(\delta_1+C_2\delta_2)\cos kh}{\sin^2\theta}[(j\cos\theta\cos kh+\sin kh)\exp(jkh\cos\theta)$$

$$-(j\cos\theta\cos ks+\sin ks)\exp(jks\cos\theta)$$

$$+R_v(-j\cos\theta\cos kh+\sin kh)\exp(-jkh\cos\theta)$$

$$-R_v(-j\cos\theta\cos ks+\sin ks)],\exp(-jks\cos\theta) \tag{4.178}$$

and

$$E_{\theta 5}=-[C_2\cos kh+(\delta_1+C_2\delta_2)\sin kh]$$

$$\times \int_s^h \sin kz[\exp(jkz\cos\theta)+R_v\exp(-jkz\cos\theta)]\,dz$$

$$=-\frac{C_2\cos kh+(\delta_1+C_2\delta_2)\sin kh}{\sin^2\theta}$$

$$\times[(j\cos\theta\sin kh-\cos kh)\exp(jkh\cos\theta)$$

$$-(j\cos\theta\sin ks-\cos ks)\exp(jks\cos\theta)$$

$$-R_v(j\cos\theta\sin kh+\cos kh)\exp(-jkh\cos\theta)$$

$$+R_v(j\cos\theta\sin ks+\cos ks)]\exp(-jks\cos\theta). \tag{4.179}$$

To account for the finite conducting plane, we can also calculate the difference quantity ΔE_θ in a fashion similar to that in (4.133). However, considering that $\Delta E_\theta \ll E_\theta^0$ in general and that the mathematics involved here is already complicated, we choose not to pursue this modification any longer.

Typical numerical results for $|E_\theta^0|$ are given in Fig. 4.38, where the normalized free-space pattern (when $R_v=0$) of a sleeve antenna with the same dimensions as those in Fig. 4.33 and with $f=10$ MHz is shown as curve (*a*), while the corresponding pattern of the same antenna above sea water at the same frequency is shown as curve (*b*). Note that this antenna also works at other frequencies in the range 8–32 MHz. Their corresponding patterns do not differ appreciably from those in Fig. 4.38.

Because the sleeve antenna being considered is a vertical element, its radiated field is independent of the azimuthal angle φ. If, however, there are N elements arranged equally spaced on a ring circumference such as part of the Wullenweber antenna,[32] the total field will then be a function of both θ and φ depending on the excitation. For the practical purpose of having a limited azimuthal coverage, only a small number of pairs of elements, say n ($n \ll N$), are actually excited at a time. The array factor contributed by these n pairs can be written as

$$S_1 = \sum_{i=-n}^{n} I_i \exp\left[jka_1 \sin\theta \cos(\varphi - \varphi_i)\right] \qquad (i \neq 0), \qquad (4.180)$$

where a_1 is the radius of the ring array, φ is measured from the axis of symmetry (see Fig. 4.39), and φ_i and I_i are the position and feed-point current of the ith element.

To simplify the derivation, let us refer to Fig. 4.39, where eight of N elements are actually excited. Since the effect of mutual coupling will inevitably induce some currents on the neighboring unexcited elements, we choose to consider only two additional parasitic pairs closest to the excited

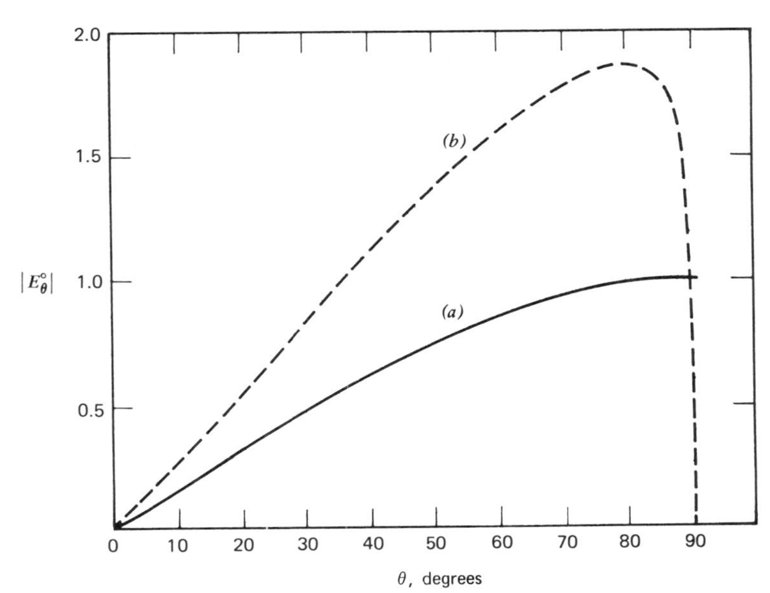

Fig. 4.38 (a) Normalized free-space vertical pattern of the sleeve antenna specified in Fig. 4.33 at $f = 10$ MHz; (b) Vertical pattern of the same antenna and frequency above sea water (normalized with respect to the free-space maximum).

elements and neglect the more distant unexcited pairs. For this reason, we use $n=6$ (four excited pairs plus two parasitic pairs) for (4.180). Of course, the elements are excited by voltage sources. Although the beam maximum from part of this ring array can be made to point in an arbitrary direction (θ_0, φ_0) in principle, the special way of choosing azimuthal symmetry as shown in Fig. 4.39 demands, however, $\varphi_0=0$. Thus, the phase of the voltage excitation must be $e^{-j\alpha_i}$ with

$$\alpha_i = ka_1 \sin\theta_0 \cos(\varphi_0 - \varphi_i) = ka_1 \sin\theta_0 \cos\varphi_i \qquad \text{with} \qquad \varphi_0 = 0. \quad (4.181)$$

If a single-sleeve antenna were considered, the voltage excitation in volts must be numerically equal to $\bar{Z}_{\text{in}}$ to produce a unit current (1 A) at the feed point [which has been assumed in (4.155)]. For this reason, the voltage excitations (both amplitudes and phases) for the array being considered must be

$$V_i = \bar{Z}_{\text{in}} e^{-j\alpha_i}. \quad (4.182)$$

Because of this kind of excitations and the geometry displayed in Fig. 4.39, we always have

$$I_i = I_{-1}, \qquad i = 1, 2, \ldots, n. \quad (4.183)$$

This means that only I_i (for positive i) are needed for calculating S_1 in (4.180).

For another practical reason, a concentric vertical screen is always placed behind the real ring array, resulting in an image ring array (assuming a perfect image) whose radius is designated as a_2. The contribution from the corresponding portion of the image ring should then be

$$S_2 = \sum_{i=-n}^{n} (-I_i) \exp[jka_2 \sin\theta \cos(\varphi - \varphi_i)] \qquad (i \neq 0). \quad (4.184)$$

The currents I_i required in (4.180) and (4.184) can be determined by a circuit consideration:

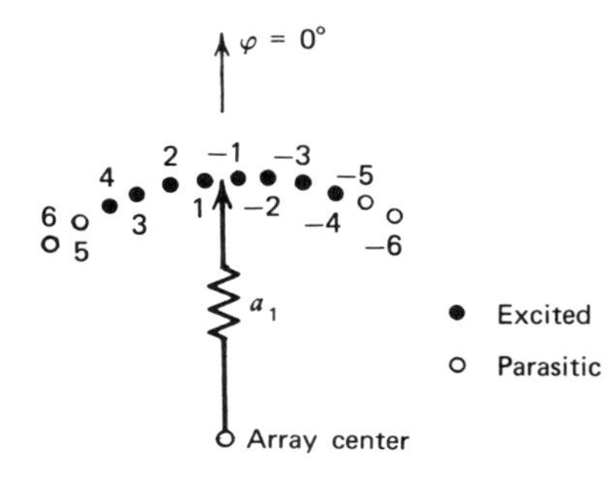

Fig. 4.39 Top view of a part of the ring array of sleeve antennas.

$$
\begin{bmatrix}
\bar{Z}_{\text{in}}-Z_{11'}+Z_{12}-Z_{12'} & Z_{12}-Z_{12'}+Z_{13} & Z_{13} & 0 & 0 & 0 \\
Z_{12}-Z_{12'}+Z_{13} & \bar{Z}_{\text{in}}-Z_{11'} & Z_{12}-Z_{12'} & Z_{13} & 0 & 0 \\
Z_{13} & Z_{12}-Z_{12'} & \bar{Z}_{\text{in}}-Z_{11'} & Z_{12}-Z_{12'} & Z_{13} & 0 \\
0 & Z_{13} & Z_{12}-Z_{12'} & \bar{Z}_{\text{in}}-Z_{11'} & Z_{12}-Z_{12'} & Z_{13} \\
0 & 0 & Z_{13} & Z_{12}-Z_{12'} & \bar{Z}_{\text{in}}+R_t-Z_{11'} & Z_{12}-Z_{12'} \\
0 & 0 & 0 & Z_{13} & Z_{12}-Z_{12'} & \bar{Z}_{\text{in}}+R_t-Z_{11'}
\end{bmatrix}
$$

$$
\times \begin{bmatrix} I_1 \\ I_2 \\ I_3 \\ I_4 \\ I_5 \\ I_6 \end{bmatrix} = \begin{bmatrix} \bar{Z}_{\text{in}}e^{-j\alpha_1} \\ \bar{Z}_{\text{in}}e^{-j\alpha_2} \\ \bar{Z}_{\text{in}}e^{-j\alpha_3} \\ \bar{Z}_{\text{in}}e^{-j\alpha_4} \\ 0 \\ 0 \end{bmatrix}, \tag{4.185}
$$

where

$\bar{Z}_{in}$ = open-circuit self-impedance of a single-sleeve antenna, given by (4.158),

Z_{12} = open-circuit mutual impedance between No. 1 and No. 2 antennas with $d=2a_1\sin(\pi/N)$,

Z_{13} = open-circuit mutual impedance between No. 1 and No. 3 antennas with $d=2a_1\sin(2\pi/N)$,

Z'_{11} = open-circuit mutual impedance between No. 1 antenna and its own image with $d=a_1-a_2$,

and

Z'_{12} = open-circuit mutual impedance between No. 1 antenna and the image of No. 2 antenna with

$$d=\left(a_1^2+a_2^2-2a_1a_2\cos\frac{2\pi}{N}\right)^{1/2}.$$

Here again we have only considered mutual interactions between relatively close pairs of elements. A resistance of R_t ohms has been added to the parasitic elements (No. 5 and No. 6), since they both are unexcited but terminated with a resistor of R_t ohms.

After obtaining the I_i from (4.185) and substituting them into (4.180) and (4.184), we have the total field radiated from the part of the array under considertion:

$$E_t=E_\theta^0(S_1+S_2). \tag{4.186}$$

The power gain will then be

$$G(\theta,\varphi)=\frac{4\pi r_0^2\times(1/120\pi)|E_t|^2}{P_{in}}$$

$$=\frac{30\sin^2\theta\left|\sum_{i=1}^{5}E_{\theta i}(S_1+S_2)\right|^2}{P_{in}}, \tag{4.187}$$

where $E_{\theta i}$, $i=1$, 2, 3, 4, and 5, are given in (4.175) through (4.179), and P_{in} representing the total power input should be

$$P_{\text{in}} = 2\,\text{Re} \sum_{i=1}^{4} V_i I^*_i$$

$$= 2\,\text{Re}\left(\bar{Z}_{\text{in}} \sum_{i=1}^{4} e^{-j\alpha_i} I^*_i\right), \tag{4.188}$$

with $\bar{Z}_{\text{in}}$ given in (4.158), α_i defined in (4.181), I^*_i meaning the complex conjugate of I_i, and Re denoting the real part of.

Based on the above derivation, we are now ready to present numerical results for the power gain as a function of frequency and ground constants. In all of these examples, we have chosen $a_1 = 133.121$ m (436.75 ft), $a_2 = 125.044$ m (410.25 ft), and dimensions of the sleeve antenna itself as those in Fig. 4.33. Since we have also followed the general practice to choose ($\theta_0 = 75°$, $\varphi_0 = 0°$) as the designed direction of the main beam, a series of $G(75°, \varphi)$ and $G(\theta, 0°)$ for $N = 120$ and $R_t = 50$ ohms are presented in Figs. 4.40 through 4.43. From these figures we can have a clear understanding about (i) the broadband property of the sleeve antenna, and (ii) overall characteristics expected from the array of this antenna.

A final note is in order before we conclude this section. Although four pairs of excited elements and specific dimensions for the sleeve antenna and array geometry were used for illustrative purposes, the formulation presented here should apply equally well to other configurations provided, of course, that the conditions in (4.157) are met.

4.8 Concluding Remarks

In this chapter we have analyzed various arrays of horizontal dipoles, vertical monopoles, or sleeve antennas. The procedures and principles involved in this analytic work are (i) to use an assumed special form for the current distribution on the antenna, (ii) to derive the self- and mutual impedances by the induced emf method, (iii) to calculate the radiation fields based on the assumed current, (iv) to employ the reflection coefficients to take care of the imperfect grounds, and (v) to formulate the power gain according to (1.7). Throughout this chapter we have been concerned only with the far fields without payng due attention to the surface wave near the grazing angle ($\theta = 90°$). Admittedly, the formulation is not as rigorous as we would like, but the emphasis here is to devise an approach, simplify it by some approximations, and obtain some useful application-oriented quantitative information. Modern computer methods may be developed in the future to help secure more direct and accurate

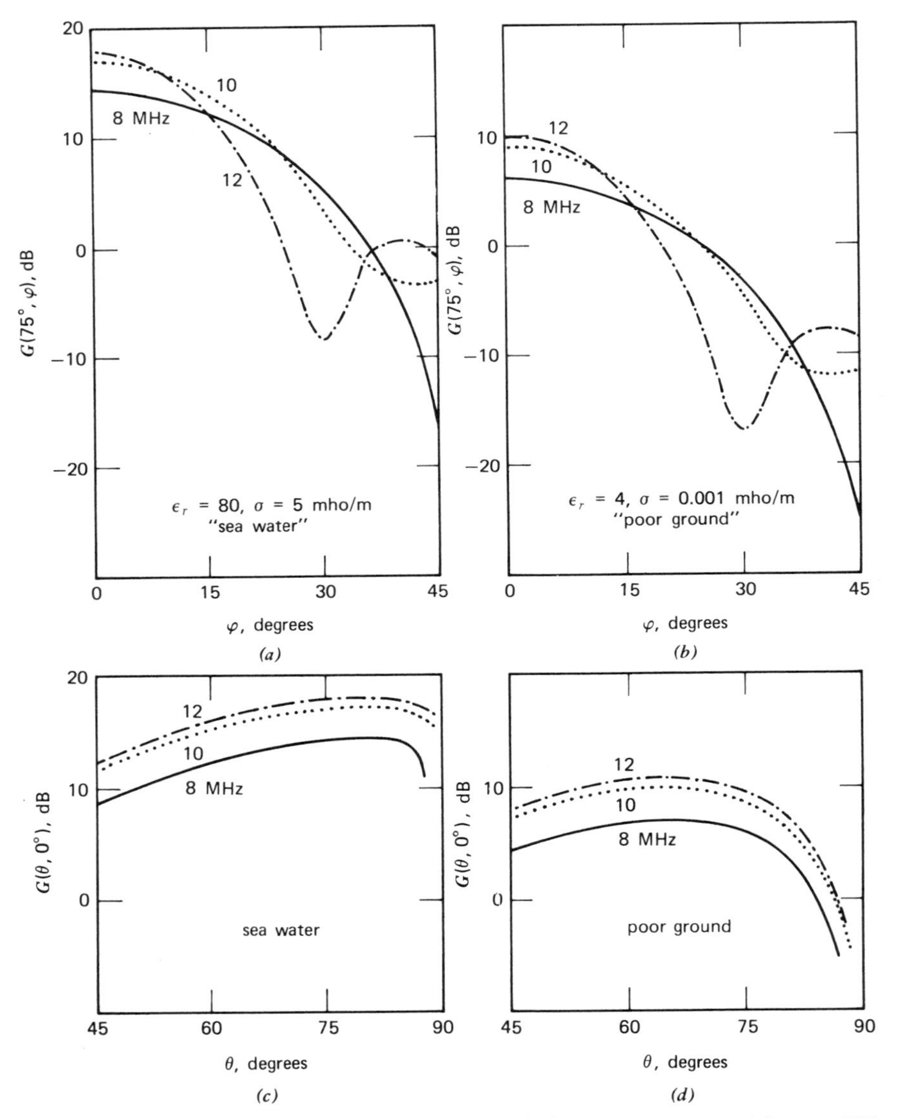

Fig. 4.40 Power gain of a ring array of sleeve antennas with $a_1 = 133.121$ m, $a_2 = 125.044$ m, $n = 6$, $N = 120$, two grounds, and $f = 8$, 10, and 12 MHz: (a) and (b) as a function of φ; (c) and (d) as a function of θ.

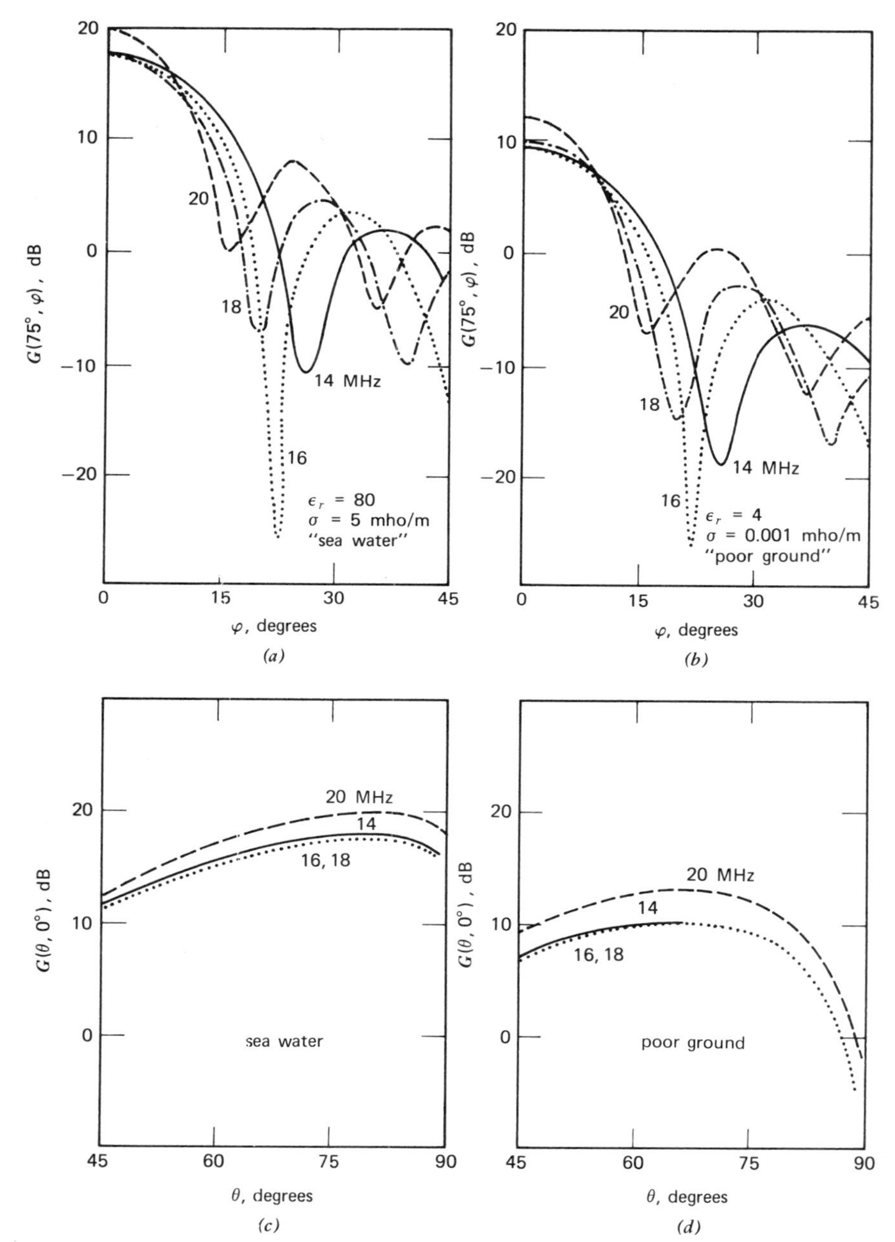

Fig. 4.41 Continuation of Fig. 4.40 with $f = 14$, 16, 18, and 20 MHz.

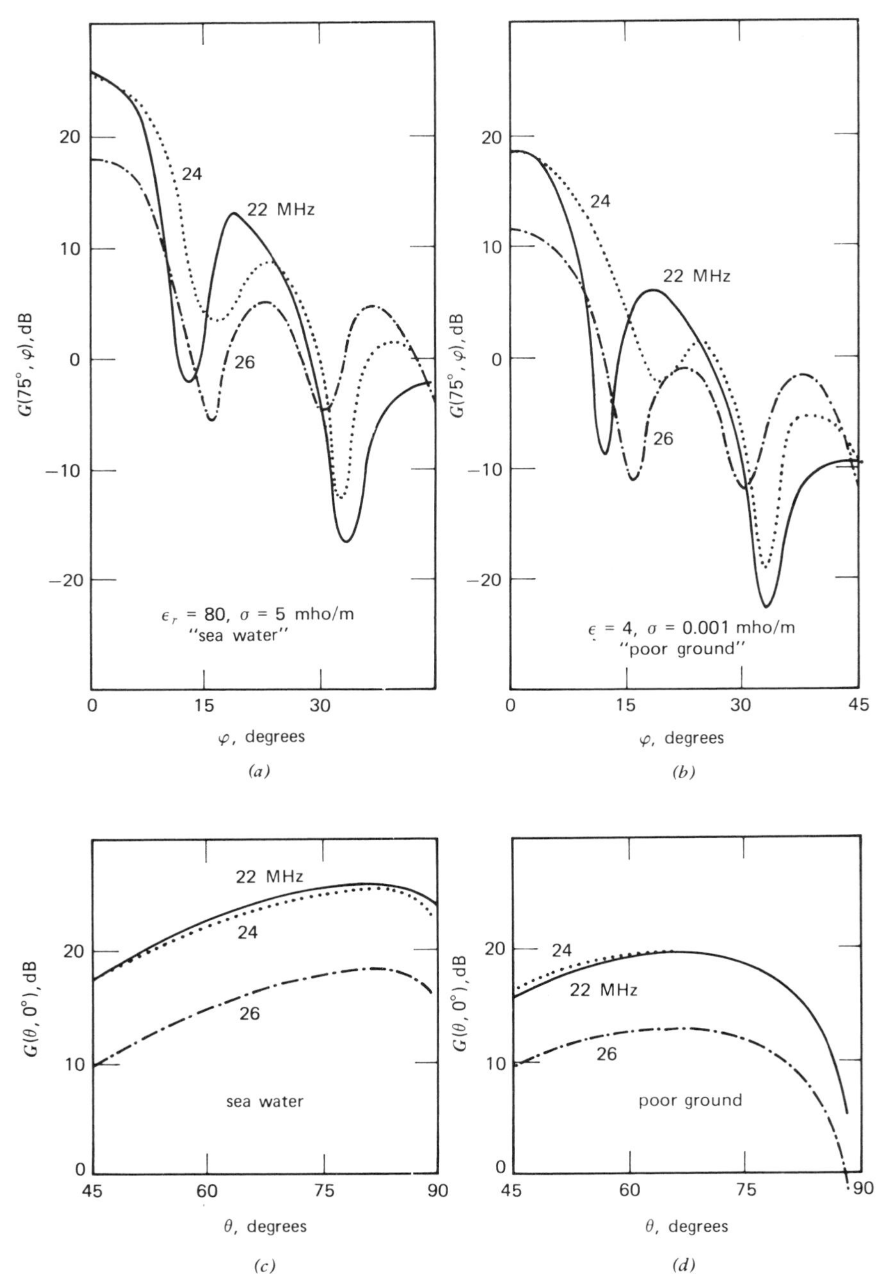

Fig. 4.42 Continuation of Fig. 4.40 with $f = 22$, 24, and 26 MHz.

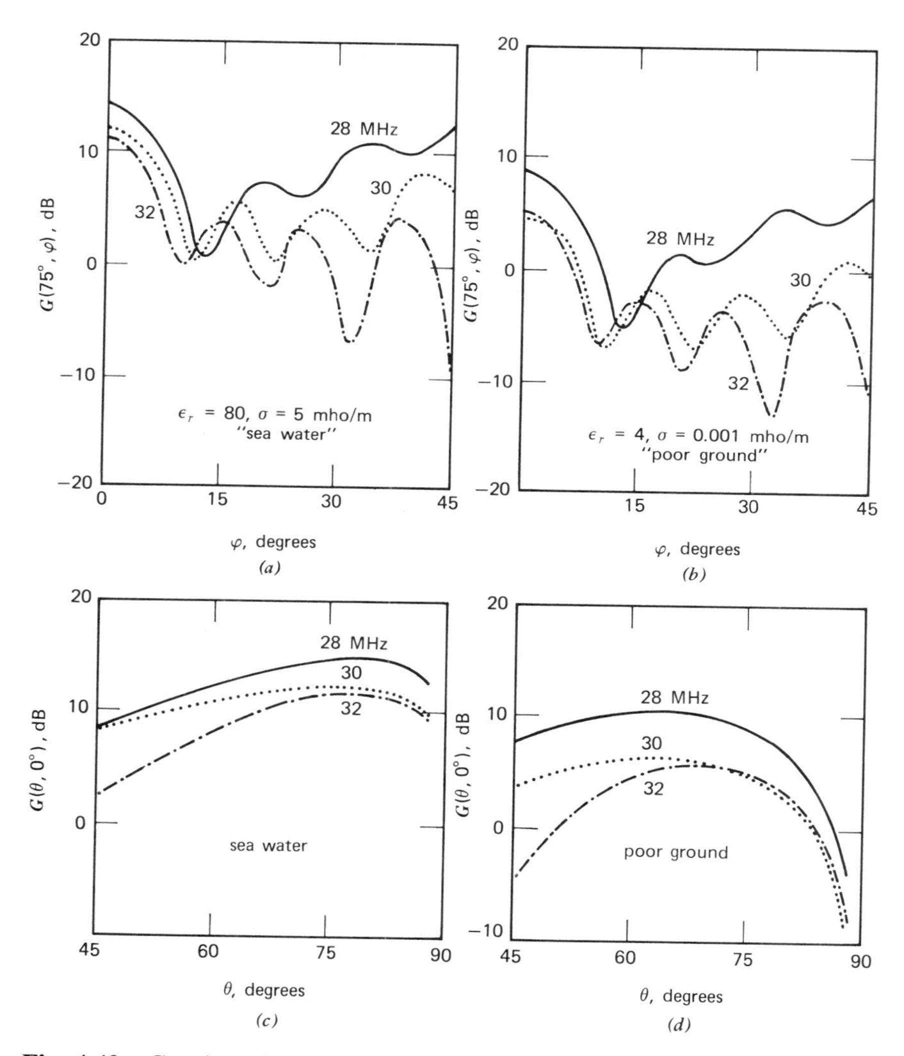

Fig. 4.43 Continuation of Fig. 4.40 with $f = 28$, 30, and 32 MHz.

results. Until then we must still try to make the practical problems simpler to handle and easier to understand.

REFERENCES

1. **King, R. W. P. and T. T. Wu.** Currents, charges, and near fields of cylindrical antennas, *Radio Sci.,* Vol. 69D, No. 3, pp. 429–446, March, 1965.

2. **Jordan, E. C.** *Electromagnetic Waves and Radiating Systems,* Prentice-Hall, Inc., Englewood Cliffs, N. J., 1950.

3. **Ma, M. T. and L. C. Walters.** Power gains for antennas over lossy plane ground, *Technical Report No. ERL 104-ITS 74,* ESSA, Boulder, Colo., April, 1969.

4. **Sommerfield, A.** *Partial Differential Equations,* Academic Press, New York, 1949.

5. **Banõs, A.** *Dipole Radiation in the Presence of a Conducting Half-Space,* Pergamon Press, Oxford, England, 1966.

6. **King, R. W. P.** *Theory of Linear Antennas,* Harvard University Press, Cambridge, Mass., 1956.

7. **Wait, J. R.** Characteristics of antennas over lossy earth, Chap. 23 in *Antenna Theory,* Part II, edited by R. E. Collin and F. J. Zucker, McGraw-Hill Book Co., New York, 1969.

8. **Bhattacharyya, B. K.** Input resistances of horizontal electric and vertical magnetic dipoles over a homogeneous ground, *IEEE Trans. Antennas and Propagation,* Vol. AP-11, No. 3, pp. 261–266, May, 1963.

9. **Chang, D. C. and J. R. Wait.** Theory of a vertical tubular antenna located above a conducting half-space, *IEEE Trans. Antennas and Propagation,* Vol. AP-18, No. 2, pp. 182–188, March, 1970.

10. **Olsen, R. G. and D. C. Chang.** Input impedance change of a half-wave vertical antenna over a dissipative earth, *Technical Report No. 2,* EE Department, University of Colorado, Boulder, September, 1970.

11. **Taylor, C. D., E. A. Aronson, and C. W. Harrison, Jr.** Theory of coupled monopoles, *IEEE Trans. Antenas and Propagation,* Vol. AP–18, No. 3, pp. 360–366, May, 1970.

12. **Carter, P. S.** Circuit relations in radiating systems and application to antenna problems, *Proc. IRE,* Vol. 20, No. 6, pp. 1004–1041, June, 1932.

13. **Mack, R. B.** A study of circular arrays, *Harvard University Technical Report No. 382,* part 2, Cruft Laboratory, Cambridge, Mass., May, 1963.

14. **Chang, V. W. H. and R. W. P. King.** Theoretical study of dipole array of *N* parallel elements, *Radio Sci.,* Vol. 3, No. 5, pp. 411–424, May, 1968.

15. **Chang, V. W. H. and R. W. P. King.** On two arbitrarily located identical parallel antennas, *IEEE Trans. Antennas and Propagation,* Vol. AP-16, No. 3, pp. 309–317, May, 1968.

16. *Tables of Generalized Sine and Cosine Integral Functions,* Vols. I and II, Harvard University Press, Cambridge, Mass., 1949.

17. **Norton, K. A.** The propagation of radio waves over the surface of the earth and in the upper atmosphere, *Proc. IRE,* Vol. 24, No. 10, pp. 1367–1387, October, 1936; Vol. 25, No. 9. pp. 1203–1236, September, 1937.

18. **Barghausen, A. F., J. W. Finney, L. L. Proctor, and L. D. Schultz.** Predicting long-term operational parameters of HF sky-wave telecommunication systems, *Technical Report No. ERL 110-ITS 78,* ESSA, Boulder, Colo., May, 1969.

19. **Harrington, R. F.** Antenna excitation for maximum gain, *IEEEE Trans. Antennas and Propagation,* Vol. AP-13, No. 6, pp. 896–903, November, 1965.

20. Test Procedure for Antennas, No. 149 (revision of 48 IRE 252), *IEEE Trans. Antennas and Propagation,* Vol. AP-13, No. 3, pp. 453–459, May, 1965.

21. **FitzGerrell, R. G.** Gain measurements of vertically polarized antennas over imperfect ground, *IEEE Trans. Antennas and Propagation,* Vol. AP-15, No. 2, pp. 211–216, March, 1967.

22. **Surtees, W. J.** Antenna performance as affected by inhomogeneous imperfectly-conducting ground, Ph.D. dissertation, University of Toronto, Ontario, Canada, 1952.

23. **Burke, G. J., S. Gee, E. K. Miller, A. J. Poggio, and E. S. Selden,** Analysis of antennas over lossy half-space, *MB Associates Report,* San Ramon, Calif.

24. **Miller, E. K., A. J. Poggio, G. J. Burke, and E. S. Selfon.** Analysis of wire antennas in the presence of a conducting half-space, Part I. The vertical antenna in free space, *Can. J. Phys.,* Vol. 50, No. 9, pp. 879–888, May, 1972.

25. **Southward, G. C.** Arrays of linear elements, Chap. 5 in *Antenna Engineering Handbook,* edited by H. Jasik, McGraw-Hill Book Co., New York, 1961.

26. **Stewart, A. C., M. E. Chrisman, and C. O. Stearns.** Computed antenna patterns for Voice of America curtain arrays, *Report 8220,* NBS, Boulder, Colo., February, 1964.

27. **King, H. E.** Mutual impedance of unequal length antennas in echelon, *IRE Trans. Antennas and Propagation,* Vol. AP-5, No. 3, pp. 306–313, July, 1957.

28. **Wait, J. R. and W. A. Pope.** The characteristics of a vertical antenna with a radial conductor ground system, *Appl. Sci. Res.,* Vol. B4, pp. 177–195, March, 1954.

29. **Wait, J. R.** Effect of the ground screen on the field radiated from a monopole, *IRE Trans. Antennas and Propagation,* Vol, AP-4, No. 2, pp. 179–181, April, 1956.

30. **Maley, S. W. and R. J. King.** The impedance of a monopole antenna with a circular conducting-disk ground system on the surface of a lossy half space, *J. Res. NBS,* Vol. 65D, No. 2, pp. 183–188, March-April, 1961.

31 **Wait, J. R. and W. J. Surtees.**, Impedance of a top-loaded antenna of arbitrary length over a circular grounded screen, *J. Appl. Phys.,* Vol. 25, No. 5, pp. 553–555, May, 1954.

32. **Ma, M. T. and L. C. Walters.** Theoretical methods for computing characteristics of Wullenweber antennas, *Proc. IEE* (London), Vol. 117, No. 11, pp. 2095–2101, November, 1970.

33. **Taylor, J.** The sleeve antenna, *Harvard University Technical Report No. 128,* Cruft Laboratory, Cambridge, Mass., 1951.

34. **Ma, M. T. and L. C. Walters.** Theoretical methods for computing characteristics of Wullenweber antennas, *ESSA Technical Report No. ERL 159-ITS 102,* Boulder, Colo., March, 1970.

35. **Storer, J. E.** Variational solution to the problem of the symmetrical antenna, *Harvard University Technical Report No. 101,* Cruft Laboratory, Cambridge, Mass., 1950.

ADDITIONAL REFERENCES

Allen, J. L. Gain and impedance variation in scanned dipole arrays, *IRE Trans. Antennas and Propagation,* Vol. AP-10, No. 5, pp. 566–572, September, 1962.

Allen, J. L. On array element impedance variation with spacing, *IEEE Trans. Antennas and Propagation,* Vol. AP-12, No. 3, pp. 371–372, May, 1964.

Biggs, A. W. Radiation fields from a horizontal electric dipole in a semi-infinite conducting medium, *IRE Trans Antennas and Propagation,* Vol. AP-10, No. 4, pp. 358–362, July, 1962.

Blair, W. E. Experimental verification of dipole radiation in a conducting half-space, *IEEE Trans. Antennas and Propagation,* Vol. AP-11, No. 3, pp. 269–275, May, 1963.

Bohn, E. V. The current distribution and imput impedance of cylindrical antennas, *IRE Trans. Antennas and Propagation,* Vol. AP-5, No. 4, pp. 343–348, October, 1957.

Collin, R. E. and F. J. Zuker, editors. *Antenna Theory,* Parts I and II, McGraw-Hill Book Company, New York, 1969.

Ehrenspeck, H, W. and H. Poehler. A new method for obtaining maximum gain from Yagi antennas, *IRE Trans. Antennas and Propagation,* Vol. AP-7, No. 4, pp. 379–386, October, 1959.

Fishenden, R. M. and E. R. Wiblin. Design of Yagi aerials, *Proc. IEE* (London), Part III, Vol. 96, pp. 5–12, January, 1949.

Harrison, C. W., Jr., and R. W. P. King. On the impedance of a base-driven vertical antenna with a radial ground system, *IRE Trans. Antennas and Propagation,* Vol. AP-10, No. 5, pp. 640–642, September, 1962.

Krause, L. O. Enhancing HF received fields with large planar and cylindrical ground screens,*IEEE Trans. Antennas and Propagation,* Vol. AP-15, No. 6, pp. 785–795, November, 1967.

Mailloux, R. J. Antenna and wave theories of infinite Yagi-Uda arrays, *IEEE Trans. Antennas and Propagation,* Vol. AP-13, No. 4, pp. 499–509, July, 1965.

Mailloux, R. J. The long Yagi-Uda array, *IEEE Trans. Antennas and Propagation,* Vol. AP-14, No. 2, pp. 128–137, March, 1966.

Poggio, A. J. and P. E. Mayes Pattern bandwidth optimization of the sleeve monopole antenna, *IEEE Trans. Antennas and Propagation,* Vol. AP-14, No. 5, pp. 643–645, September, 1966.

Reid, D. G. The gain of an idealized Yagi array, *J. IEE* (London), Part IIIA, Vol. 93, pp. 564–566, 1946.

Sengupta, D. L. On the phase velocity of wave propagation along an infinite Yagi structure, *IRE Trans. Antennas and Propagation,* Vol. AP-7, No. 4, pp. 234–239, July, 1959.

Sivaprasad, K. U and R. W. P. King. A study of arrays of dipoles in a semi-infinite dissipative medium, *IEEE Trans. Antenna and Propagation,* Vol. AP-11, No. 3, pp. 240–256, May, 1963.

Southworth, G. C. Arrays of linear elements, Chap. 5 in *Antenna Engineering Handbook*, edited by H. Jasik, McGraw-Hill Book Company, New York, 1961.

Walkinshaw, W. Theoretical treatment of short Yagi aerials, *J. IEE* (London), part III, Vol. 93, pp. 598–614, 1946.

Wolff, E. A. *Antenna Analysis*, Chap. 3, John Wiley and Sons, Inc., New York, 1966.

Yagi, H. Beam transmission of ultra-short waves, *Proc. IRE*, Vol. 16, No. 6, pp. 715–741, June, 1928.

CHAPTER 5

LOG-PERIODIC DIPOLE ARRAYS ABOVE LOSSY GROUND

In Chapter 4, we analyzed arrays of standing-wave antennas above lossy grounds. Specifically, the antennas considered therein were dipoles, monopoles, and sleeve antennas. The array considered in this chapter, the *log-periodic dipole array,* also consists essentially of dipoles as its elements, and could, in principle, be included as part of Chapter 4. On the other hand, it could also, as some have suggested,[1] be included with the traveling-wave antennas to be discussed in Chapter 6, in view of the special manner of excitation involved. However, because of the importance of unusual characteristics in the sense of the broad range of frequency operation associated with this array, we choose to treat the subject as a separate chapter.

Since the term frequency-independent antenna was introduced by Rumsey, employing the so-called *angle concept*,[2] many types of broadband antennas have been developed with different design approaches. One of these implies that the input impedance of an antenna whose physical dimension is identical to its complement should be frequency independent.[3] The other makes the input impedance and radiation pattern of an antenna vary periodically with the logarithm of frequency and requires that the change of characteristics with frequency over a logarithmic period be negligible.[3] Both of these approaches result essentially from one basic principle, that the performance of a lossless antenna should remain unchanged if its dimensions in terms of the operating wavelength are held constant.[4]

Strictly speaking, some dimension of the antennas designed from these ideas should be infinite, extending from an infinitesimal feed point at one end to infinity at the other end. A practical antenna can only be obtained by taking a portion from this ideal infinite structure and by requiring that truncation at both ends cause only minor deviations in the final performance. The actual finite physical dimension near the feed point (one end) and that of the other end determine, respectively, the limits of the highest and lowest frequency of operation. If, by proper design, these two frequency limits are far apart (say 3 : 1 or 5 : 1) and the electrical performance of the antenna within this frequency range changes only slightly, it is said that a broadband antenna has been developed.

Antennas of this nature take different physical shapes such as equiangular spiral,[5] wire trapezoidal tooth or sheet circular tooth,[4] zigzag wire,[6] log-periodic dipole array,[7,8] log-periodic monopole array,[9] and so on. The log-periodic dipole array is the main subject discussed in this chapter.

Two general approaches for studying log-periodic dipole arrays have been suggested in the past. One requires a detailed analysis of the propagation properties of the wave along the structure,[10] and the other treats the topic from a circuit-and-array point of view.[11] Since our main concern in this book is arrays and the method adopted for calculating impedances in Chapter 4 is based on the circuit viewpoint, we are following the circuit-and-array approach originally proposed by Carrel[11] for analyzing this array. Moreover, since the array is more complicated, we choose to analyze it in free space first in order to have a better understanding of its behavior.

5.1 Log-Periodic Dipole Arrays in Free Space

A representative log-periodic dipole array is shown in Fig. 5.1(a). It consists of a number of parallel center-fed dipoles arranged side by side in a plane. The lengths and radii of dipole elements form a geometric progression with a common ratio τ, called the scale factor:

$$\frac{h_n}{h_{n+1}} = \frac{a_n}{a_{n+1}} = \tau, \qquad n = 1, 2, \ldots, N-1, \qquad 0 < \tau < 1 \tag{5.1}$$

where h_n and a_n are, respectively, the half-length and radius of the nth dipole and N is total number of dipoles in the array. The ratio of half length to radius is supposed to be the same for all the dipoles in a given array. A line through the ends of dipole elements makes an angle α with the array axis at the virtual apex 0. The spacing factor σ' is defined as the ratio of the distance between two adjacent dipoles to twice the full length of the longer dipole. According to the geometry of Fig. 5.1(a), we have

$$\sigma' = \frac{d_n}{4h_{n+1}} = \frac{y_{n+1} - y_n}{4h_{n+1}} = \frac{(h_{n+1} - h_n)\cot\alpha}{4h_{n+1}} = \tfrac{1}{4}(1-\tau)\cot\alpha. \tag{5.2}$$

The dipoles are energized from a balanced constant-impedance feeder with adjacent dipoles connected to the feeder in a "phase-reversal" fashion, as shown in Fig. 5.1(b).

According to Carrel's analysis from the circuit viewpoint,[11] we can consider the entire structure as a parallel combination of two parts, one of them being the dipole elements and the other being the feeder circuit. This concept is clearly represented in Fig. 5.2. Note that Z_T in Fig. 5.2(*b*) is the termination impedance connected to the last (longest) dipole at a usual distance of $d_N = h_N/2$ from it.[7]

The current-voltage relations for the element circuit [Fig. 5.2(*a*)] can be written as

$$[V_a] = [Z_a][I_a] \qquad \text{or} \qquad [I_a] = [Z_a]^{-1}[V_a], \tag{5.3}$$

where

$$[I_a] = \begin{bmatrix} I_{1a} \\ I_{2a} \\ \vdots \\ I_{Na} \end{bmatrix} \qquad \text{and} \qquad [V_a] = \begin{bmatrix} V_{1a} \\ V_{2a} \\ \vdots \\ V_{Na} \end{bmatrix} \tag{5.4}$$

are $N \times 1$ column matrices representing, respectively, the driving base currents and response voltages for the dipole elements, and

$$[Z_a] = \begin{bmatrix} Z_{11a} & Z_{12a} & \cdots & Z_{1Na} \\ Z_{21a} & Z_{22a} & \cdots & Z_{2Na} \\ \cdots & \cdots & \cdots & \cdots \\ Z_{N1a} & Z_{N2a} & \cdots & Z_{NNa} \end{bmatrix} \tag{5.5}$$

is the associated $N \times N$ open-circuit impedance matrix.

Clearly, the matrix elements on the main diagonal of $[Z_a]$ in (5.5) should represent the self-impedances of the dipoles, which can be calculated according to (4.32) or (4.33), depending on the values of h_n. The off-diagonal elements in (5.5) will then represent the mutual impedances between dipoles indicated by the indices. These mutual impedances can be calculated by (4.38) or (4.44).

Similarly, the current-voltage relations for the feeder circuit shown in Fig. 5.2(*b*) can be expressed by

$$[I_f] = [Y_f][V_f] = [Y_f][V_a], \tag{5.6}$$

where

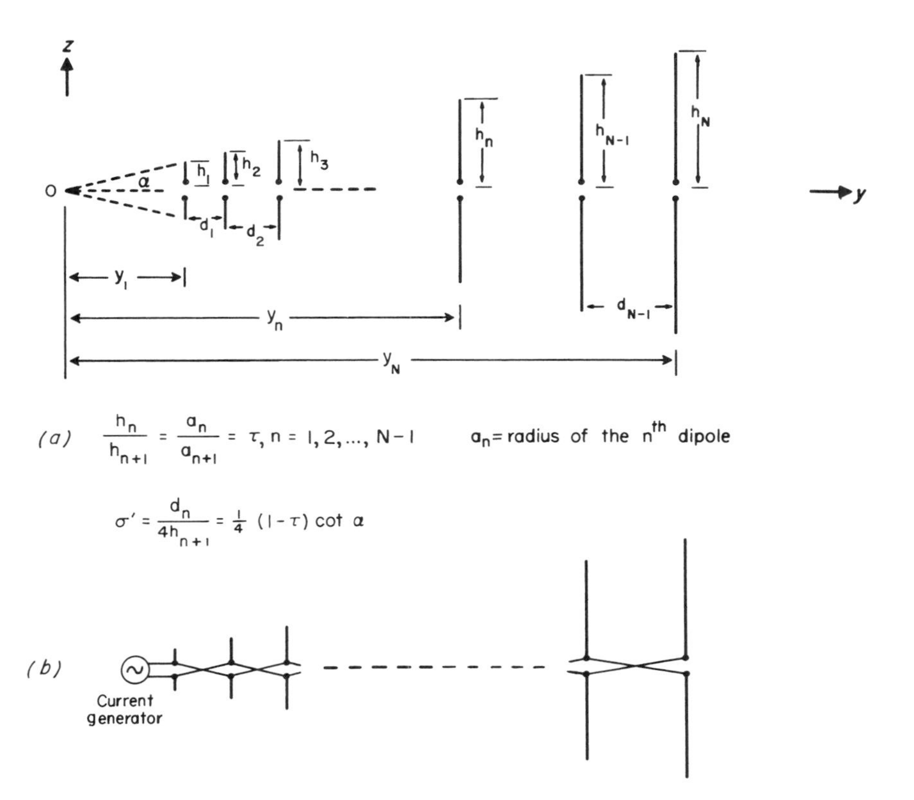

Fig. 5.1 A sketch of the log-periodic dipole array: (*a*) symbols and definitions, (*b*) method of feeding.

$$[I_f] = \begin{bmatrix} I_{1f} \\ I_{2f} \\ \vdots \\ I_{Nf} \end{bmatrix} \quad \text{and} \quad [V_f] = \begin{bmatrix} I_{1f} \\ V_{2f} \\ \vdots \\ V_{Nf} \end{bmatrix} = \begin{bmatrix} V_{1a} \\ V_{2a} \\ \vdots \\ V_{Na} \end{bmatrix} \tag{5.7}$$

are, respectively, the driving currents and response voltages for each section of the transmission line constituting a complete feeder circuit, and

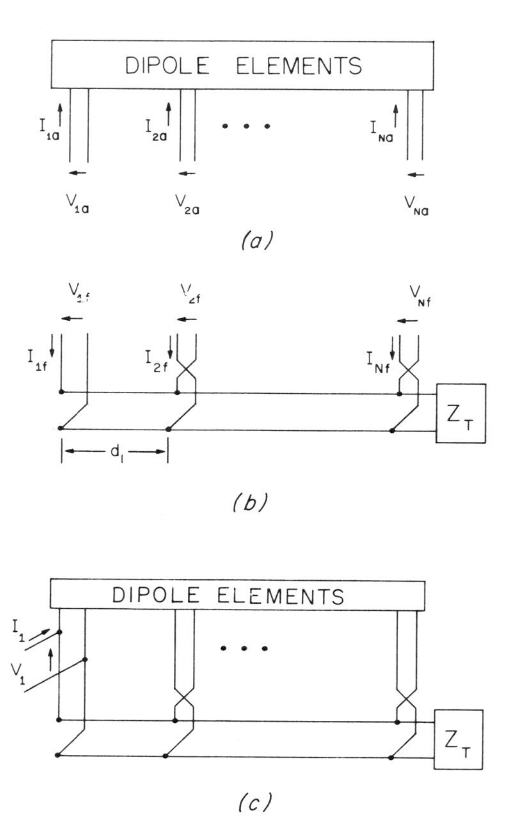

Fig. 5.2 Equivalent circuits of the log-periodic dipole array: (*a*) element circuit, (*b*) feeder circuit, (*c*) complete circuit.

$$[Y_f] = \begin{bmatrix} Y_{11f} & Y_{12f} & \cdots & Y_{1Nf} \\ Y_{21f} & Y_{22f} & \cdots & Y_{2Nf} \\ \cdots & \cdots & \cdots & \cdots \\ Y_{N1f} & Y_{N2f} & \cdots & Y_{NNf} \end{bmatrix} \tag{5.8}$$

is the associated $N \times N$ short-circuit admittance matrix of the feeder. The matrix elements of $[Y_f]$ in (5.8) depend naturally on the lengths of the transmission line in each section and the characteristic admittance Y_0. The parameter Y_0 will be known once a design or a choice of the transmission line is made.

According to the classical formulation for a two-port network,[12] we obtain the matrix elements for (5.8) as follows:

$$
\begin{aligned}
Y_{11f} &= -jY_0 \cot kd_1, \\
Y_{22f} &= -jY_0 (\cot kd_1 + \cot kd_2), \\
&\cdots\cdots\cdots\cdots\cdots\cdots \\
Y_{(N-1)(N-1)f} &= -jY_0 (\cot kd_{N-2} + \cot kd_{N-1}), \\
Y_{NNf} &= -jY_0 \cot kd_{N-1} + Y'_T, \qquad Y'_T = Y_0 \frac{\cos kd_N + jY_0 Z_T \sin kd_N}{Y_0 Z_T \cos kd_N + j \sin kd_N}, \\
Y_{12f} &= Y_{21f} = -jY_0 \csc kd_1, \\
Y_{23f} &= Y_{32f} = -jY_0 \csc kd_2, \\
&\cdots\cdots\cdots\cdots\cdots\cdots \\
Y_{(N-1)Nf} &= Y_{N(N-1)f} = -jY_0 \csc kd_{N-1},
\end{aligned}
\tag{5.9}
$$

where d_i, $i = 1, 2, \ldots, N-1$, are clearly indicated in Fig. 5.1(*a*). Note that all the expressions for the short-circuit admittances in (5.9) have a negative sign, which is the result of the phase-reversal connection evidenced in Fig. 5.1(*b*). Note also that the relation $[V_f] = [V_a]$ is used in (5.6). The relation is true because the two circuits are connected in parallel at the dipole bases. Since there is no difference between the feeder voltage matrix $[V_f]$ and the antenna voltage matrix $[V_a]$, we use $[V]$ to represent both, as specifically shown in Fig. 5.2(*c*).

Adding (5.3) and (5.6), we obtain the total input current matrix:

$$
\begin{aligned}
[I] &= [I_a] + [I_f] = [I_a] + [Y_f][V] \\
&= [I_a] + [Y_f][Z_a][I_a] \\
&= \{[U] + [Y_f][Z_a]\}[I_a],
\end{aligned}
\tag{5.10}
$$

where $[U]$ is the $N \times N$ unit matrix.

The matrix elements of $[I]$ in (5.10) should represent the input currents to each node point (dipole base) where the antenna and feeder circuits are combined. In the actual model considered here [see Fig. 5.1(b)], all the matrix elements in $[I]$ are zero except I_1, which is the only current source [at the base of the shortest dipole] for the entire array. Without loss of generality, we assume $I_1 = 1$ A. Thus, the dipole base current matrix $[I_a]$ can be determined from (5.10) by matrix inversion:

$$[I_a] = \{[U] + [Y_f][Z_a]\}^{-1}[I]$$

$$= \{[U] + [Y_f][Z_a]\}^{-1} \begin{bmatrix} 1 \\ 0 \\ 0 \\ \vdots \\ 0 \end{bmatrix}. \tag{5.11}$$

The response voltages appearing at the dipole bases can then be determined by substituting (5.11) into (5.3). The input impedance Z_{in} of the entire array, which is numerically equal to the voltage V_1 across the feed point since the input current has been assumed to be unity, should be

$$Z_{\text{in}} = V_1 = Z_{11a}I_{1a} + Z_{12a}I_{2a} + \cdots + Z_{1Na}I_{Na}$$

$$= \sum_{i=1}^{N} Z_{1ia}I_{ia} \equiv R_{\text{in}} + jX_{\text{in}}. \tag{5.12}$$

Before we proceed to the task of deriving expressions for fields, power gain, and ground effect, it should be instructive to give a numerical example at this point and to examine some of the special properties associated with this array. Let us study the following example:

$$N = 12, \qquad \tau = 0.87, \qquad \alpha = 12.50°, \qquad Z_T = 0, \qquad Z_0 = \frac{1}{Y_0} = 450 \text{ ohms},$$

$$\frac{h}{a} = 500, \text{ and } h_{12} = 7.50 \text{ m} \qquad \text{(half-length of the longest dipole)}.$$

When $f = 10$ MHz, we list the important parameters as follows:

$$h_1=0.0541\lambda, \quad h_2=0.0622\lambda, \quad h_3=0.0714\lambda, \quad h_4=0.0821\lambda,$$

$$h_5=0.0944\lambda, \quad h_6=0.1085\lambda, \quad h_7=0.1247\lambda, \quad h_8=0.1433\lambda,$$

$$h_9=0.1648\lambda, \quad h_{10}=0.1894\lambda, \quad h_{11}=0.2177\lambda, \quad h_{12}=0=0.2502\lambda,$$

$$Z_{11a}=2.2758-j1767.0139\ \Omega, \qquad Z_{22a}=3.0278-j1515.8728\ \Omega,$$

$$Z_{33a}=4.0374-j1293.9828\ \Omega, \qquad Z_{44a}=5.3996-j1096.9476\ \Omega,$$

$$Z_{55a}=7.2503-j920.7765\ \Omega, \qquad Z_{66a}=9.7866-j761.8090\ \Omega,$$

$$Z_{77a}=13.3038-j616.5711\ \Omega, \qquad Z_{88a}=18.2572-j481.6185\ \Omega,$$

$$Z_{99a}=25.3801-j353.3126\ \Omega, \qquad Z_{1010a}=35.9173-j227.4312\ \Omega,$$

$$Z_{1111a}=52.1358-j98.4674\ \Omega, \qquad Z_{1212a}=78.5824+j41.8443\ \Omega.$$

The dipole base currents obtained from the matrix inversion (5.11) are

$$\begin{aligned}
&I_{1a}=0.2855\ \underline{/95.33^\circ}, && I_{2a}=0.3555\ \underline{/-88.72^\circ},\\
&I_{3a}=0.3385\ \underline{/59.49^\circ}, && I_{4a}=0.2798\ \underline{/-149.54^\circ},\\
&I_{5a}=0.4991\ \underline{/-14.41^\circ}, && I_{6a}=0.7346\ \underline{/120.53^\circ},\\
&I_{7a}=0.7758\ \underline{/-74.97^\circ}, && I_{8a}=1.0764\ \underline{/64.24^\circ},\\
&I_{9a}=1.8787\ \underline{/147.97^\circ}, && I_{10a}=2.3343\ \underline{/-114.76^\circ},\\
&I_{11a}=2.3556\ \underline{/-65.07^\circ}, && I_{12a}=1.1395\ \underline{/-12.60^\circ}.
\end{aligned} \tag{5.13}$$

The input impedance of the array calculated from (5.12) is

$$Z_{\text{in}}=464.3673+j49.4957\ \text{ohms}. \tag{5.14}$$

In view of the assumption made in (5.11) that the input current is $I_1=1\ \underline{/0^\circ}$ A, we may conclude that the currents in (5.13) are actually values relative to the input current. From there we see that the eleventh dipole with half-length 0.2177λ (with respect to the operating frequency) has the largest current amplitude. If we normalize the currents further with

I_{11a}, we should be able to see clearly the relative distribution of currents over the dipole bases. By doing so, we have

$$\begin{aligned}
\frac{I_{1a}}{I_{11a}} &= 0.1212 \underline{/160.39^\circ}, & \frac{I_{2a}}{I_{11a}} &= 0.1509 \underline{/-23.65^\circ}, \\
\frac{I_{3a}}{I_{11a}} &= 0.1437 \underline{/124.56^\circ}, & \frac{I_{4a}}{I_{11a}} &= 0.1188 \underline{/-84.47^\circ}, \\
\frac{I_{5a}}{I_{11a}} &= 0.2119 \underline{/50.66^\circ}, & \frac{I_{6a}}{I_{11a}} &= 0.3118 \underline{/-174.41^\circ}, \\
\frac{I_{7a}}{I_{11a}} &= 0.3293 \underline{/-9.90^\circ}, & \frac{I_{8a}}{I_{11a}} &= 0.4569 \underline{/129.30^\circ}, \\
\frac{I_{9a}}{I_{11a}} &= 0.7975 \underline{/-146.96^\circ}, & \frac{I_{10a}}{I_{11a}} &= 0.9909 \underline{/-49.69^\circ}, \\
\frac{I_{11a}}{I_{11a}} &= 1.0000 \underline{/0^\circ}, & \frac{I_{12a}}{I_{11a}} &= 0.4837 \underline{/52.47^\circ}.
\end{aligned} \tag{5.15}$$

A few important conclusions may now be drawn. First, although there are N dipoles in the array, the actual number of dipoles which contribute significantly to the radiated field is less than N in general. In the above example, $N=12$, there are perhaps only five contributing elements (eighth through twelfth dipole) whose normalized current amplitudes are relatively large (>0.40). The remaining elements (first through seventh) serve only the purpose of making this array operable over a broader frequency range. For this reason, Carrel divided the entire array structure into three different regions.[11] The first region is called the transmission region, which includes the feed point and the first few short dipoles whose current amplitudes are relatively small. This region is followed by an active region, which consists of a number of dipoles whose full lengths are close to one-half wavelength. It is this active region which serves the main objective of radiating. The last portion of the array, called the unexcited region, contains the longer elements toward the far end. This unexcited region generally helps predict the end effect due to truncation discussed at the beginning of this chapter. For the currents in (5.15), the first seven dipoles may be considered members of the transmission region, while the next five dipoles can be classified to constitute the active region. There is practically no element in the unexcited region for this particular frequency (10 MHz).

This unexicted region will eventually appear when the operating frequency increases, as will be seen later. On the other hand, if we had started with two more longer dipoles, they would fall into the unexcited region even when $f=10$ MHz.

The second conclusion we may draw from (5.15) is that, because the current phases associated with the tenth and twelfth dipoles are, respectively, lagging and leading, relative to the most active element (eleventh dipole), we can predict, by the knowledge learned from Chapter 1, an endfire type of pattern radiating toward the feed point. This result will be clear when we discuss the pattern in the next section. The third conclusion is that the most active element will occur more or less at one of those elements whose half-lengths are slightly shorter than $\lambda/4$. In the above example, it happens to be the eleventh dipole. It could be the tenth dipole, as evidenced by the fact that $|I_{10a}/I_{11a}|=0.9909$. All of these three properties will remain when the array is operated with higher frequencies.

If we change the frequency to 12 MHz for the same array, we obtain

$$h_7=0.1497\lambda, \qquad h_8=0.1720\lambda, \qquad h_9=0.1977\lambda,$$

$$h_{10}=0.2273\lambda, \qquad h_{11}=0.2612\lambda, \qquad h_{12}=0.3003\lambda;$$

$$\begin{aligned}
\frac{I_{7a}}{I_{9a}}&=0.6113\,\underline{/-168.82^\circ}, & \frac{I_{8a}}{I_{9a}}&=0.7192\,\underline{/-84.11^\circ}, \\
\frac{I_{9a}}{I_{9a}}&=1.0000\,\underline{/0^\circ}, & \frac{I_{10a}}{I_{9a}}&=0.7347\,\underline{/45.79^\circ}, \\
\frac{I_{11a}}{I_{9a}}&=0.3174\,\underline{/96.89^\circ}, & \frac{I_{12a}}{I_{9a}}&=0.0612\,\underline{/130.44^\circ};
\end{aligned} \tag{5.16}$$

and

$$Z_{\text{in}}=372.5437-j172.3898\,\text{ohms}.$$

From the above, we see that the most active element has shifted to the ninth dipole, whose half-length is 0.1977λ. Obviously, only the seventh through tenth dipoles are contributing significantly this time, therefore constituting the active region. The first six dipoles are in the transmission region. The last two elements may be considered as belonging to the unexcited region. The current phases of the eighth and tenth dipoles again satisfy the basic requirement for an endfire radiation. Note how drastically the current amptitude on the longest dipole has been attenuated. Also note that, because the first six elements are relatively unimportant as far as our discussion here is concerned, the numerical data pertaining to them are omitted in (5.16).

The unique properties outlined above still hold true even when the frequency is increased to 32 MHz. Under this condition, the most active element will shift to the second dipole, leaving practically no element in the transmission region and most of the elements in the unexcited region. This situation is just the opposite of that when the lowest frequency of 10 MHz was considered. Therefore, as far as the particular array studied in this section is concerned, it should be able to operate within the range of $10 \leqslant f \leqslant 32$ MHz, indeed a significantly broad band. Alternatively speaking, the computed operating bandwidth for this array may be termed as

$$B_0 = \frac{32}{10} = 3.2 \tag{5.17}$$

If the active region were very narrow, consisting of one or two dipoles, the theoretical operating bandwidth should be determined by the ratio of h_N/h_1. This ratio is generally called the structure bandwidth.[11] That is

$$B_s = \frac{h_N}{h_1} = \tau^{1-N}. \tag{5.18}$$

For the above example, $B_s = 4.6248$.

Since the active region always has some width, it is apparent that the actual operating bandwidth, B_0, is always less than B_s. According to Carrel,[11] it may be defined as

$$B_0 = \frac{B_s}{B_a}, \tag{5.19}$$

where B_a is the average width of the active region. In view of the fact that there is no clear-cut boundary between the active region and the other two regions, it is rather difficult to define B_a precisely. Instead, an empirical formula for B_a has been proposed:[11]

$$B_a = 1.1 + 30.7\sigma'(1-\tau). \tag{5.20}$$

Using $\tau = 0.87$ and $\sigma' = 0.1466$, we obtain, for the example studied here, $B_a = 1.6851$ and $B_0 = 2.7445$, which may be compared with that in (5.17) to gain a feeling for the deviation between the computed and empirical results.

Instead of tabulating all the base currents and input impedances for different frequencies as we did in (5.13) through (5.16), we summarize them in Figs. 5.3 through 5.9 for every 2 MHz in the entire range $10 \leqslant f \leqslant 32$ MHz. Ideally, the input impedance should remain frequency inde-

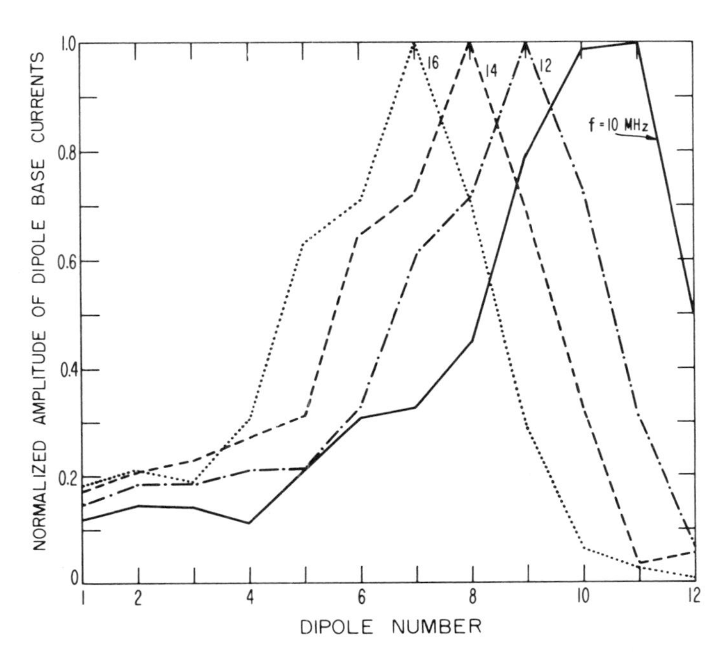

Fig. 5.3 Normalized amplitude of dipole base currents of the log-periodic dipole array in free space with $N=12$, $\tau=0.87$, $\alpha=12.5°$, $Z_T=0, Z_0=450\ \Omega$, $h/a=500$, $h_{12}=7.50$ m, $f=10$, 12, 14, and 16 MHz.

pendent. However, in view of many imperfections in the formulation such as the finite size of the array (truncation at both ends), the assumed form for current distribution on dipoles, and the assumption of no interaction between dipoles and the transmission line, the input impedance does vary with frequencies although the variation is restrained within a relatively small area shown in Fig. 5.9. This variation is indeed not substantial when it is compared with that shown in the table preceding (5.13) for a single dipole.

The degree of variation of Z_{in} is generally measured by quantities known as the mean resistance level R_0 and the corresponding standing-wave ratio (SWR), which are, respectively, defined as

$$R_0 = \sqrt{R_{min} R_{max}} \tag{5.21}$$

and

$$\mathrm{SWR} = \sqrt{\frac{R_{max}}{R_{min}}}, \tag{5.22}$$

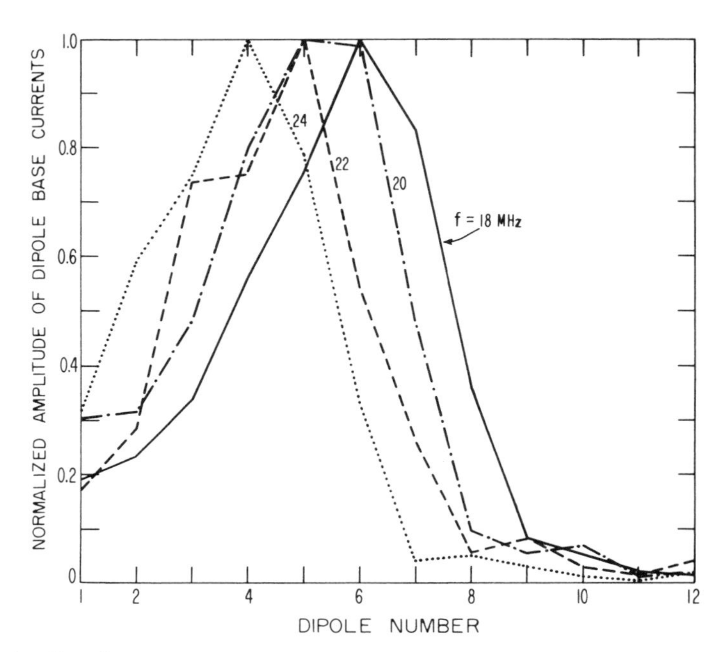

Fig. 5.4 Continuation of Fig. 5.3 with $f = 18$, 20, 22, and 24 MHz.

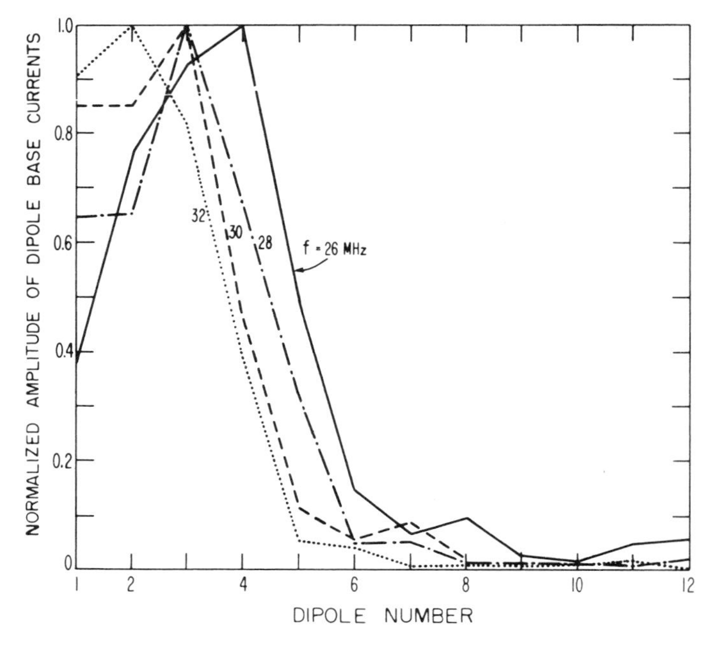

Fig. 5.5 Continuation of Fig. 5.3 with $f = 26$, 28, 30, and 32 MHz.

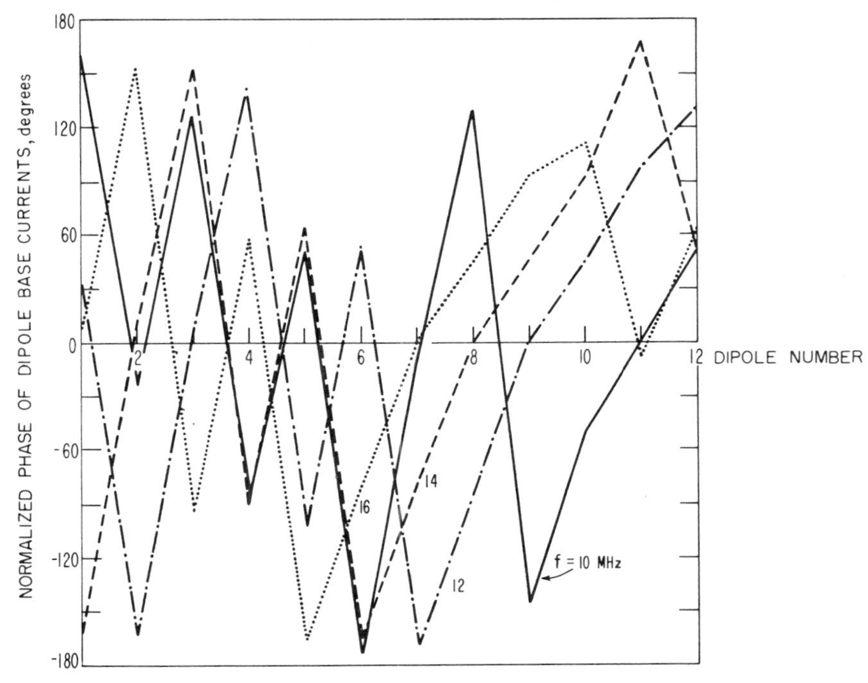

Fig. 5.6 Normalized phase of dipole base currents of the same array as in Fig. 5.3, also in free space, with $f=10$, 12, 14, and 16 MHz.

where R_{min} and R_{max} are, respectively, the minimum and maximum real parts of the input impedance calculated within the entire operating bandwidth. From the set of actually computed values presented in Fig. 5.9, we have

$$R_{\text{min}}=145.50 \text{ ohms} \qquad \text{occurring at} \qquad f=32 \text{ MHz},$$

$$R_{\text{max}}=464.37 \text{ ohms} \qquad \text{occurring at} \qquad f=10 \text{ MHz},$$

$$R_0=259.94 \text{ ohms}, \tag{5.23}$$

and

$$\text{SWR}=1.79.$$

If we operate the same array for the frequency range $14 \ll f \ll 28$ MHz instead (still $B_0=2$), the corresponding set will be

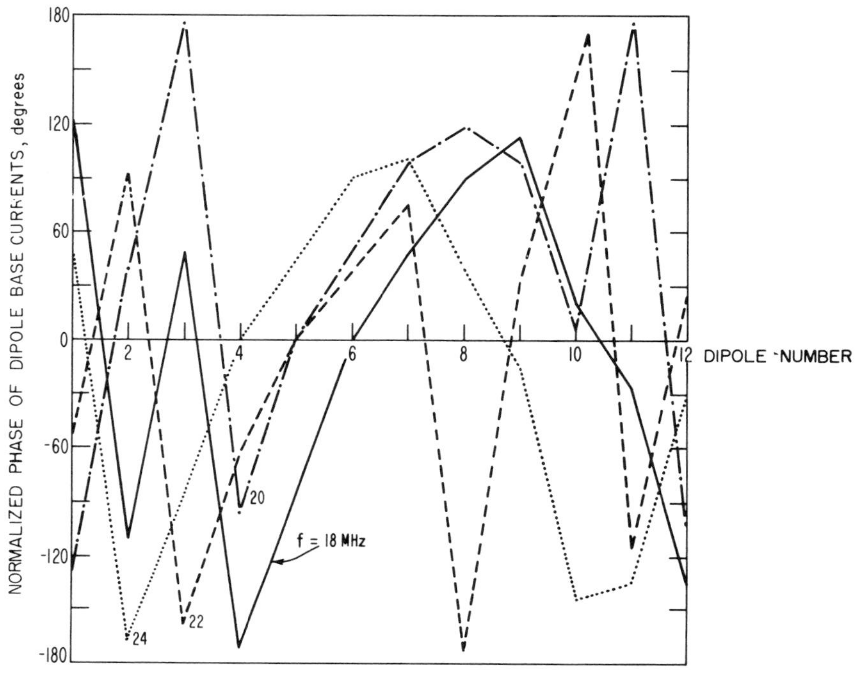

Fig. 5.7 Continuation of Fig. 5.6 with $f = 18$, 20, 22, and 24 MHz.

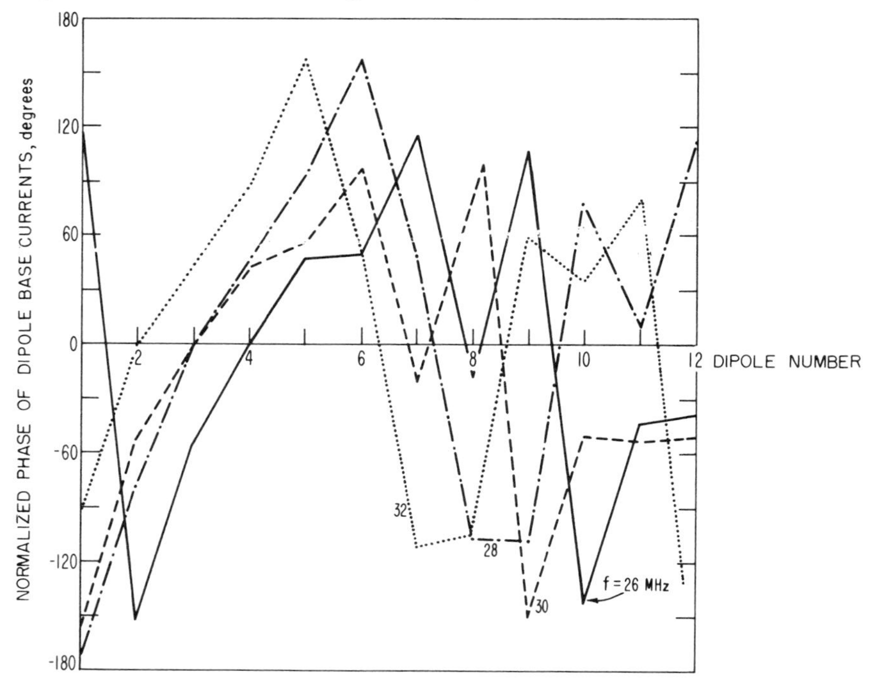

Fig. 5.8 Continuation of Fig. 5.6 with $f = 26$, 28, 30, and 32 MHz.

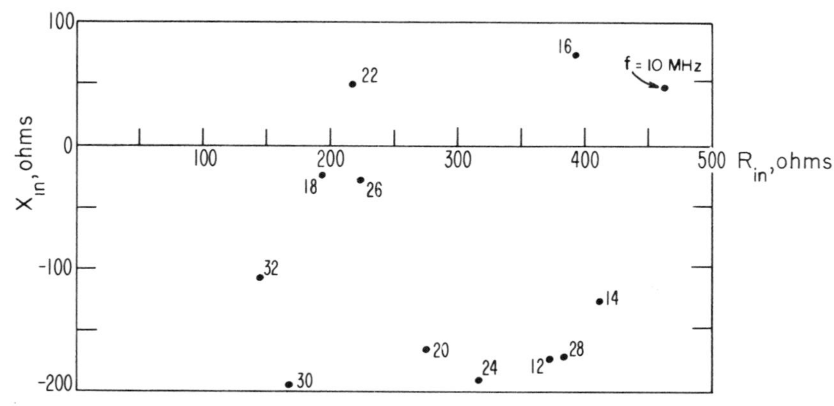

Fig. 5.9 Input impedance of the same array as in Fig. 5.3, in free space.

$$R_{\min}=193.82 \text{ ohms} \quad \text{at} \quad f=18\ MHz,$$

$$R_{\max}=410.90 \text{ ohms} \quad \text{at} \quad f=14 \text{ MHz},$$

$$R_0=282.21 \text{ ohms}, \tag{5.24}$$

and

$$\text{SWR}=1.46.$$

Note that all of the values in (5.23) and (5.24) are computed results, based on the fomulation given in this section. Obviously, the mean resistance level and the standing wave ratio should depend on $Z_0, \tau, \sigma'\,(\text{or}\,\alpha)$ for a given array of N dipoles within an operating bandwidth B_0. Specific results showing some of these effects are presented in Figs. 5.10 through 5.13, where only the solid dots are actually computed. From Figs. 5.11 and 5.13, we see clearly that there are a few favorable values of σ' (or α) to minimize the standing-wave ratio.

An approximate formula for the mean resistance level has also been proposed by Carrel:[11]

$$R'_0=\frac{Z_0}{\sqrt{1+(Z_0/Z_a)(\sqrt{\tau}\,/4\sigma')}}, \tag{5.25}$$

where Z_a, called the average characteristic impedance of a dipole antenna in free space, is determined mainly by h/a. According to some of the original derivations, treating the dipole as an opened-out transmission line,[13,14] Z_a also contains another frequency-dependent term involving

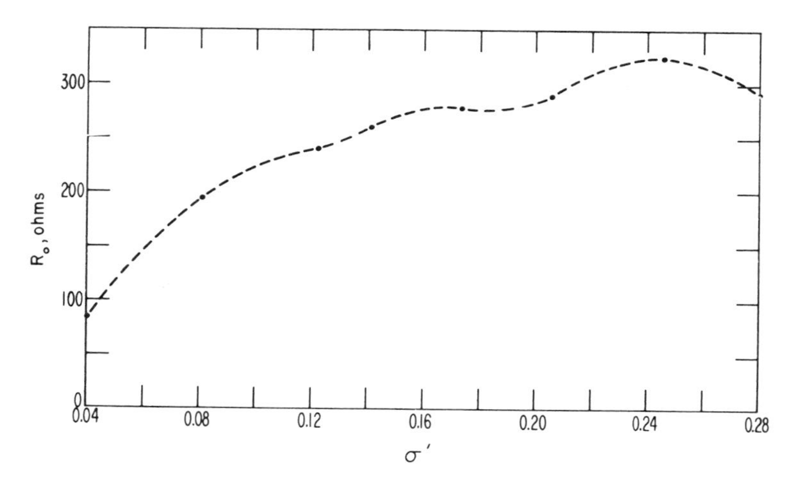

Fig. 5.10 Variation of R_0 with σ' for a log-periodic dipole array in free space with $N=12$, $\tau=0.87$, $Z_T=0$, $Z_0=450\ \Omega$, $h/a=500$, and $h_{12}=7.50$ m in 10 MHz $\leqslant f \leqslant$ 32 MHz.

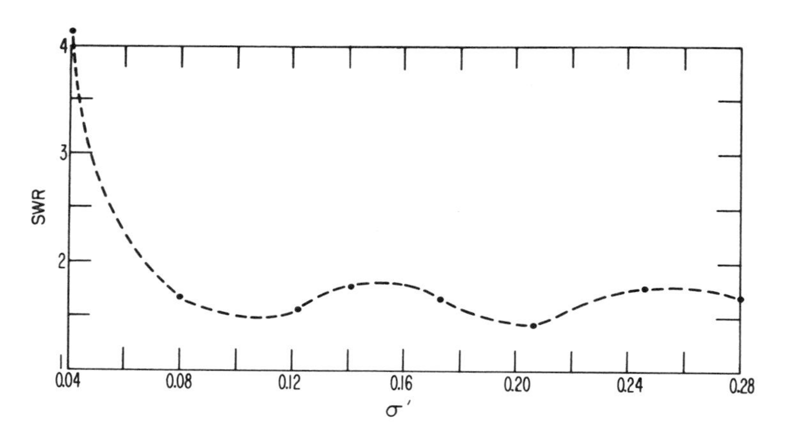

Fig. 5.11 Variation of SWR with σ' for the same array described in Fig. 5.10.

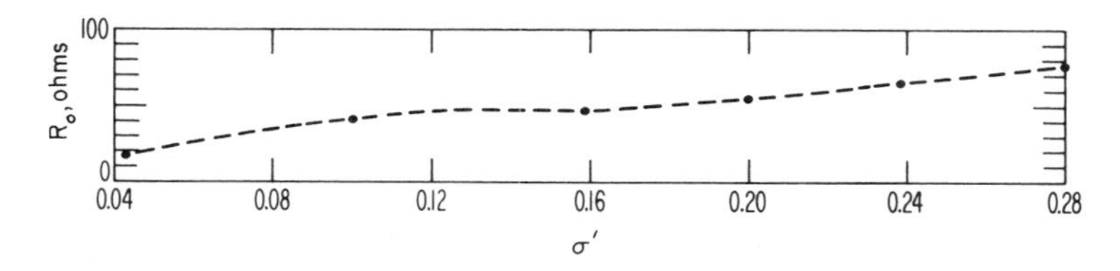

Fig. 5.12 Variation of R_0 with σ' for a log-periodic dipole array in free space with $N=12$, $\tau=0.95$, $Z_T=0$, $Z_0=100\ \Omega$, $h/a=100$, and $h_{12}=7.50$ m in 10 MHz $\leqslant f \leqslant$ 16 MHz.

h/λ. Since there are many dipoles with different values of h in the log-periodic dipole array considered here, Carrel replaced this frequency-dependent term by an average numerical factor.[11] His result is

$$Z_a \cong 120\left(\ln\frac{h}{a} - 2.25\right). \tag{5.26}$$

A set of examples for R'_0 is given in Fig. 5.14, which may be compared with the corresponding cases in Figs. 5.10 and 5.12.

5.2 Vertical Log-Periodic Dipole Array—Power Gain

A In Free Space. Having outlined the procedures for calculating $[I_a]$ and Z_{in} and given some discussion through a numerical example, we are now ready to derive the radiation pattern and power gain. The exact expression for the far field radiated from the array depends on the actual coordinates of the array. The configuration shown in Fig. 5.1(a), where the

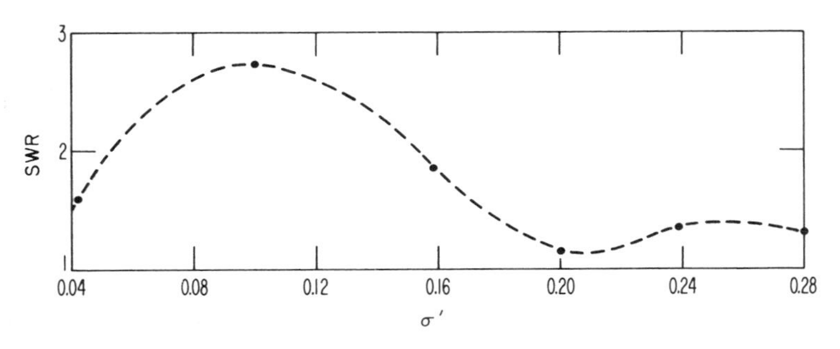

Fig. 5.13 Variation of SWR with σ' for the same array described in Fig. 5.12.

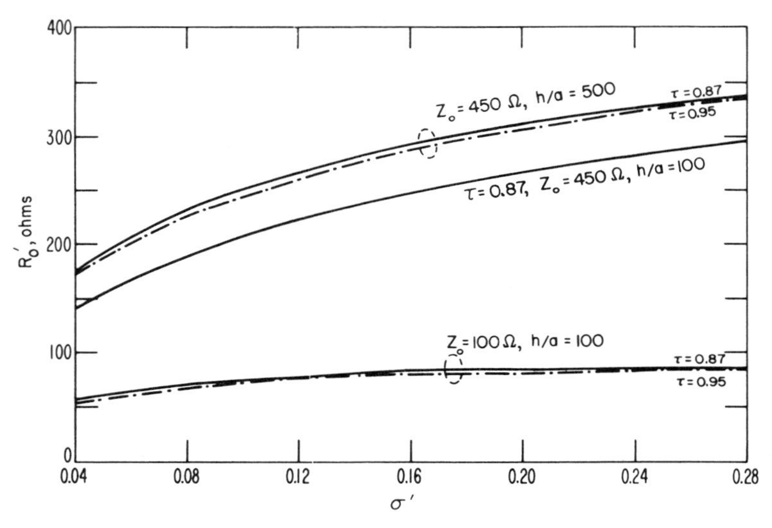

Fig. 5.14 Approximate mean resistance level of log-periodic dipole array as a function of σ'.

array axis is arbitrarily chosen as the y axis and the entire array is in the yz plane, may be called the vertical log-periodic dipole array. We still can follow the same procedure as in Section 4.2 to determine the component field expressions. Using the same notation as those shown in Fig. 4.2, we have for the present case [Fig. 5.1(a)]

$$\theta' = 0, \qquad \psi = \theta, \qquad \alpha' = \frac{\pi}{2}, \qquad s = H_s = z,$$

and the current distribution on the ith dipole is

$$I_i(z) = I_{mi}\left[\sin k(h_i - |z|) + T_{Ui}(\cos kz - \cos kh_i) + T_{Di}\left(\cos\frac{kz}{2} - \cos\frac{kh_i}{2}\right)\right], \qquad kh_i \neq \frac{\pi}{2},$$

$$= I'_{mi}\left[\sin k|z| - 1 + T'_{Ui}\cos kz - T'_{Di}\left(\cos\frac{kz}{2} - \cos\frac{\pi}{4}\right)\right], \qquad kh_i = \frac{\pi}{2}, \tag{5.27}$$

where we have, for the purpose of avoiding confusion with the notation $[I]$ in (5.10), used I_{mi} or I'_{mi} to denote the current maximum on the ith dipole. The quantities T_{Ui}, T_{Di}, T'_{Ui}, and T'_{Di} in (5.27) are calculable according to (4.14) through (4.17). To clarify the notation further, we also note that the base current of the ith dipole determined from the assumed current form in (5.27) can be easily obtained by setting $z=0$.This base current should be equal to I_{ia} in (5.11). That is,

$$I_{ia}=I_i(0)=I_{mi}\left[\sin kh_i+T_{Ui}(1-\cos kh_i)+T_{Di}\left(1-\cos\frac{kh_i}{2}\right)\right],\qquad kh_i\neq\frac{\pi}{2}$$

$$=I'_{mi}\left[-1+T'_{Ui}-T'_{Di}\left(1-\cos\frac{\pi}{4}\right)\right],\qquad kh_i=\frac{\pi}{2},\tag{5.28}$$

or

$$I_{mi}=\frac{I_{ia}}{\sin kh_i+T_{Ui}(1-\cos kh_i)+T_{Di}\,[1-\cos(kh_i/2)]},\qquad kh_i\neq\frac{\pi}{2},$$

$$I'_{mi}=\frac{I_{ia}}{-1+T'_{Ui}-T'_{Di}[1-\cos(\pi/4)]},\qquad kh_i=\frac{\pi}{2}.\tag{5.29}$$

The component far field radiated from the first dipole (the shortest) will then be, according to (4.59) and (4.60),

$$E_{1\theta}=-j30k\frac{e^{-jkr_1}}{r_1}\sin\theta\int_{-h_1}^{h_1}I_1(z)\exp(jkz\cos\theta)\,dz$$

$$=-j60I_{m1}\frac{e^{-jkr_1}}{r_1}F_1,\qquad kh_1\neq\frac{\pi}{2},$$

$$=-j60I'_{m1}\frac{e^{-jkr_1}}{r_1}F'_1,\qquad kh_1=\frac{\pi}{2},\tag{5.30}$$

and

$$E_{1\varphi}=0,$$

where

$$F_1=\frac{1}{\sin\theta}[\cos(kh_1\cos\theta)-\cos kh_1]$$

$$+\frac{T_{U1}}{\sin\theta}[\sin kh_1\cos(kh_1\cos\theta)-\cos\theta\cos kh_1\sin(kh_1\cos\theta)]$$

$$+\frac{T_{D1}\sin\theta}{\frac{1}{4}-\cos^2\theta}[\tfrac{1}{2}\sin\tfrac{1}{2}kh_1\cos(kh_1\cos\theta)-\cos\theta\cos\tfrac{1}{2}kh_1\sin(kh_1\cos\theta)]$$

$$-(T_{U1}\cos kh_1+T_{D1}\cos\tfrac{1}{2}kh_1)\tan\theta\sin(kh_1\cos\theta), \tag{5.31}$$

$$F'_i=\frac{1}{\sin\theta}\left[1-\cos\theta\sin\left(\frac{\pi}{2}\cos\theta\right)\right]+T'_{U1}\frac{\cos[(\pi/2)\cos\theta]}{\sin\theta}$$

$$-\frac{T'_{D1}\sin\theta}{\frac{1}{4}-\cos^2\theta}\left[\frac{\sqrt{2}}{4}\cos\left(\frac{\pi}{2}\cos\theta\right)-\frac{\sqrt{2}}{2}\cos\theta\sin\left(\frac{\pi}{2}\cos\theta\right)\right]$$

$$+\left(-1+\frac{\sqrt{2}}{2}T'_{D1}\right)\tan\theta\sin\left(\frac{\pi}{2}\cos\theta\right), \tag{5.32}$$

and r_1 is the distance from the base of the first dipole to the far-field point.

Summing up the contributions from all the dipoles in the array, we obtain

$$E_\theta=-j60\sum_{i=1}^{N}I_{mi}\frac{e^{-jkr_i}}{r_i}F_i, \qquad kh_i\neq\frac{\pi}{2},$$

$$=-j60\sum_{i=1}^{N}I'_{mi}\frac{e^{-jkr_i}}{r_i}F'_i, \qquad kh_i=\frac{\pi}{2}, \tag{5.33}$$

where F_i and F'_i can, respectively, be obtained from (5.31) and (5.32) if h_1, T_{U1}, and T_{D1} there are replaced, respectively by h_i, T_{Ui}, and T_{Di}. Note that, for the mere purpose of convenient presentation, we have given two expressions in (5.33), depending on whether kh_i is $\pi/2$. In reality, since all the lengths of the dipoles in the array are different, as evidenced by (5.1), there is at most one dipole, say the nth$(1\ll n\ll N)$ whose half-length may satisfy $kh_n=\pi/2$. In this case, all the remaining N-1 terms $(i\neq n)$ in the first expression of (5.33) are valid. Only the term identified with $i=n$ should be changed according to the second expression of (5.33).

Using the far-field approximation for the distance in the phase term,

$$r_i\cong r_1-y_i\sin\theta\sin\varphi \tag{5.34}$$

where y_i is the coordinate (position) of the base of the ith dipole, we have

$$E_\theta \cong -j60\frac{e^{-jkr_1}}{r_1}\sum_{i=1}^{N} I_{mi}\exp(jky_i\sin\theta\sin\varphi)F_i. \tag{5.35}$$

The power gain for this vertical log-periodic dipole array will then become, according to the definition given in (4.64),

$$G(\theta,\varphi) = \frac{4\pi r_1^2\cdot(1/120\pi)|E_\theta|^2}{W_{\text{in}}} = \frac{r_1^2|E_\theta|^2}{30R_{\text{in}}}$$

$$= \frac{120|S|^2}{R_{\text{in}}}, \tag{5.36}$$

where R_{in} is the real part of the input impedance Z_{in} in (5.12) and

$$S = \sum_{i=1}^{N} I_{mi}\exp(jky_i\sin\theta\sin\varphi)F_i, \tag{5.37}$$

with I_{mi} given in (5.29). Here again we must keep in mind that one of the F_i and I_{mi} in (5.37) may have to be changed to F_i' and I_{mi}', respectively, in view of the note made following (5.33).

As we mentioned in the previous section when the numerical results of I_{ia} were presented, the array should radiate as an endfire array. Referring to Fig. 5.1(a), we see clearly that the endfire direction points toward the negative y axis ($\theta = 90°$ and $\varphi = -90°$). For this reason, numerical results of $G(\theta, -90°)$, $G(\theta, 90°)$, and $G(90°, \varphi)$ for the same array discussed in the previous section are given, respectively, in Figs. 5.15(a), 5.15(b), and 5.16. Note that $G(\theta, -90°)$ and $G(\theta, 90°)$ should be symmetric with respect to $\theta = 90°$. Thus, the curves in $90° \leqslant \theta \leqslant 180°$ are not shown in Fig. 5.15. Similarly, the portion of $G(90°, \varphi)$ in $90° \leqslant \varphi \leqslant 270°$ is also omitted in Fig. 5.16.

B Above lossy ground. After considering the input impedance, fields, and power gain of a vertical log-periodic dipole array in free space for the purpose of understanding how and why it works over a relatively broad band of frequencies, we are now ready to extend the formulation for this antenna above a flat, homogeneous and lossy ground. If the vertical log-periodic dipole array shown in Fig. 5.1(a) is placed H meters above the ground represented by the xy plane, the situation may be depicted in Fig. 5.17. The main problem is then how to determine the dipole base currents I_{ia}, including the imperfect ground effect. In view of the complicated mathematics encountered thus far in this chapter, we decide to determine

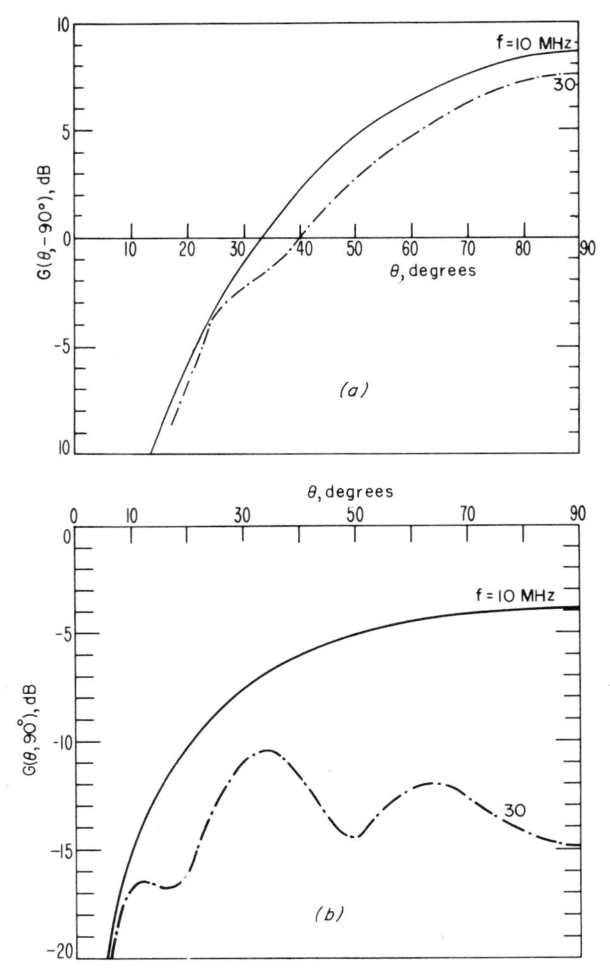

Fig. 5.15 Power gain (in the vertical plane) of the array described in Fig. 5.3 in feee space: (*a*) $\varphi = -90°$ and (*b*) $\varphi = 90°$.

I_{ia} by further approximations deduced from two viewpoints, depending on the type of ground involved.

(i) For ground with relatively low conductivity, such as "poor ground," sea ice, or polar ice cap, we propose to ignore the presence of the ground for the purpose of determining the dipole base currents. In this regard, all the procedures outlined in Section 5.1 for obtaining I_{ia} [(5.3) through (5.11)] are still considered approximately valid, because these poor grounds are not likely to have substantial, if any, effects on the current distribution.

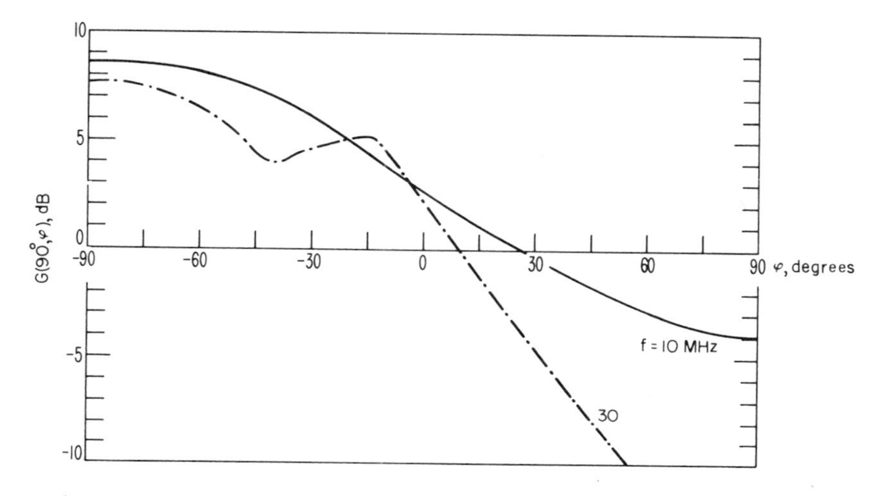

Fig. 5.16 Power gain (in the azimuthal plane) of the array described in Fig. 5.3 in free space.

When calculating the input impedance and radiated fields, we then take the imperfect ground effect into account by adding necessary terms pertaining to the vertically polarized reflection coefficient, as we did in Chapter 4. Thus, the input impedance, under the assumption made, becomes

$$Z'_{\text{in}} = Z_{\text{in}} + R'_v \sum_{i=1}^{N} Z_{1(N+i)a} I_{ia}, \tag{5.38}$$

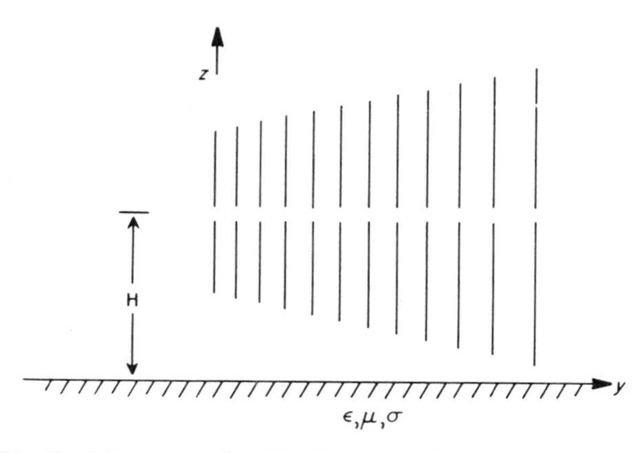

Fig. 5.17 Vertical log-periodic dipole array above flat lossy ground.

where Z_{in} is given in (5.12), R'_v in (4.73), I_{ia} $[i=1, 2,\ldots, N]$ are those obtained from (5.11), and $Z_{1(N+i)a}$ is the open-circuit mutual impedance between the first dipole (shortest) and the image of the ith dipole [which can be calculated according to (4.112) and Fig. 4.20].

Applying (4.59) and performing the necessary integration, we obtain the field radiated from this antenna as follows:

$$E'_\theta = \exp(jkH\cos\theta)[1+R_v\exp(-j2kH\cos\theta)]E_\theta, \tag{5.39}$$

where E_θ is given in (5.35) and R_v is defined in (4.58).

The power gain thus becomes

$$G'(\theta,\varphi) = \frac{r_1^2|E'_\theta|^2}{30R'_{\text{in}}} = \frac{120}{R'_{\text{in}}}|S\exp(jkH\cos\theta)[1+R_v\exp(-j2kH\cos\theta)]|^2, \tag{5.40}$$

where R'_{in} is now the real part of (5.38) and S is given in (5.37).

(ii) For ground with relatively high conductivity such as sea water or fresh water, we propose to determine the dipole base currents as if the whole array were above a perfectly conducting ground and then use R_v to take care of the imperfect ground effect. With this assumption, we can still devise a procedure similar to that outlined in Section 5.1 but with a doubled size for all the matrices involved. Referring to the geometry shown in Fig. 5.17, we see that the dipole base currents for the real array should be identical to those for the image array and that there is no physical connection between the actual transmission line and its image. Thus, we have:

$[I''_a]$ is the corresponding dipole base current matrix for the array and its image, $2N \times 1$,

$$[I''_a] = \begin{bmatrix} [I'_a] \\ [I'_a] \end{bmatrix}, \tag{5.41}$$

$[Y''_f]$ is the corresponding short-circuit admittance matrix for the transmission line and its image, $2N \times 2N$,

$$[Y''_f] = \begin{bmatrix} [Y_f] & [0] \\ [0] & [Y_f] \end{bmatrix}, \tag{5.42}$$

and $[Z_a'']$ is the corresponding open-circuit impedance matrix for the array and its image, $2N \times 2N$,

$$[Z_a''] = \begin{bmatrix} [Z_a] & [Z'_a] \\ [Z'_a] & [Z_a] \end{bmatrix}, \tag{5.43}$$

where $[I_a']$ is an $N \times 1$ submatrix of $[I_a'']$ in (5.41), representing the dipole base current matrix for the array alone when it is placed above a perfectly conducting ground; $[Y_f]$ is an $N \times N$ submatrix of $[Y_f'']$ in (5.42), representing the short-circuit admittance matrix for the real transmission line, which is given in (5.8);[0] is an $N \times N$ null matrix; $[Z_a]$ is an $N \times N$ submatrix of $[Z_a'']$ in (5.43), representing the open-circuit impedance matrix for the real array, which is given in (5.5); and $[Z_a']$ is also an $N \times N$ submatrix of $[Z_a'']$, representing the open-circuit mutual impedance matrix between the real dipoles and their images, given by

$$[Z'_a] = \begin{bmatrix} Z_{1(N+1)a} & Z_{1(N+2)a} & \cdots & Z_{1(2N)a} \\ Z_{2(N+1)a} & Z_{2(N+2)a} & \cdots & Z_{2(2N)a} \\ \cdots & \cdots & \cdots & \cdots \\ Z_{N(N+1)a} & Z_{N(N+2)a} & \cdots & Z_{N(2N)a} \end{bmatrix}. \tag{5.44}$$

In (5.44), $Z_{i(N+j)a}$ clearly represents the open-circuit mutual impedance between the ith real dipole and the image of the jth dipole. It can also be calculated according to (4.112) and Fig. 4.20. As such, we should have $Z_{i(N+j)a} = Z_{j(N+i)a}$.

Using the same derivation as that leading to (5.10), we obtain

$$[I''] = \{[U''] + [Y_f''][Z_a'']\}[I_a''], \tag{5.45}$$

where $[I'']$ is the corresponding input current matrix ($2N \times 1$) for the entire system and $[U'']$ is the $2N \times 2N$ unit matrix.

If we also partition $[I'']$ and $[U'']$ into submatrices such as

$$[I''] = \begin{bmatrix} [I] \\ [I] \end{bmatrix}, \qquad \text{with} \qquad [I] = \begin{bmatrix} 1 \\ 0 \\ 0 \\ \vdots \\ 0 \end{bmatrix}, \tag{5.46}$$

and

$$[U''] = \begin{bmatrix} [U] & [0] \\ [0] & [U] \end{bmatrix}, \tag{5.47}$$

we can simplify (5.45) to

$$\begin{bmatrix} [I] \\ [I] \end{bmatrix} = \left\{ \begin{bmatrix} [U] & [0] \\ [0] & [U] \end{bmatrix} + \begin{bmatrix} [Y_f][Z_a] & [Y_f][Z'_a] \\ [Y_f][Z'_a] & [Y_f][Z_a] \end{bmatrix} \right\} \begin{bmatrix} [I'_a] \\ [I'_a] \end{bmatrix}$$

$$= \begin{bmatrix} [U]+[Y_f][Z_a] & [Y_f][Z'_a] \\ [Y_f][Z'_a] & [U]+[Y_f][Z_a] \end{bmatrix} \begin{bmatrix} [I'_a] \\ [I'_a] \end{bmatrix}, \tag{5.48}$$

or

$$[I] = \{[U]+[Y_f][Z_a]+[Y_f][Z'_a]\}[I'_a], \tag{5.49}$$

which yields

$$[I'_a] = \{[U]+[Y_f][Z_a]+[Y_f][Z'_a]\}^{-1}[I]. \tag{5.50}$$

Comparing (5.50) with (5.11), we see that the difference between the free-space and perfect-ground approaches is identified with the extra term $[Y_f][Z'_a]$ in (5.50). With the dipole base currents so determined, the remaining task of calculating the input impedenace and power gain still follows (5.12) and (5.36) with only minor modifications. Specifically, they become, respectively,

$$Z''_{\text{in}} = \sum_{i=1}^{N} [Z_{1ia} + R'_v Z_{1(N+i)a}] I'_{ia}$$

and (5.51)

$$G''(\theta,\varphi) = \frac{120}{R''_{\text{in}}} \left| S'' \exp(jkH\cos\theta)[1 + R_v \exp(-j2kH\cos\theta)] \right|^2,$$

where R''_{in} is the real part of Z''_{in} and

$$S'' = \sum_{i=1}^{N} I''_{mi} \exp(jky_i \sin\theta \sin\varphi) F_i, \tag{5.52}$$

with I''_{mi} relating to I'_{ia} in a manner similar to that in (5.29) depending on kh_i.

To illustrate the procedures so outlined, let us study once again the same example discussed in the previous section. When the array is placed above poor ground ($\epsilon_r = 4$ and $\sigma = 0.001$ mho/m) with $H = 8$ meters, we use the free-space approximation. The normalized dipole base currents so obtained should be identical to those given in Figs. 5.3 through 5.8. The input impedances for the same frequency range ($10 \leqslant f \leqslant 32$ MHz), calculated according to (5.38), are presented in Fig. 5.18, which may be compared with Fig. 5.9 for the corresponding free-space case. More specifically, when $f = 10$ MHz, we have

$$Z'_{\text{in}} = 435.0526 + j48.6046 \text{ ohms}. \tag{5.53}$$

Clearly, the difference between Z'_{in} in (5.53) and Z_{in} in (5.14) is contributed by the second term in (5.38). The associated mean resistant level and standing wave ratio become, respectively,

$$R_0 = 267.92 \text{ ohms},$$

$$\text{SWR} = 1.62 \tag{5.54}$$

with

$$R'_{\min} = 165.00 \text{ ohms} \qquad \text{occurring at } f = 32 \text{ MHz}$$

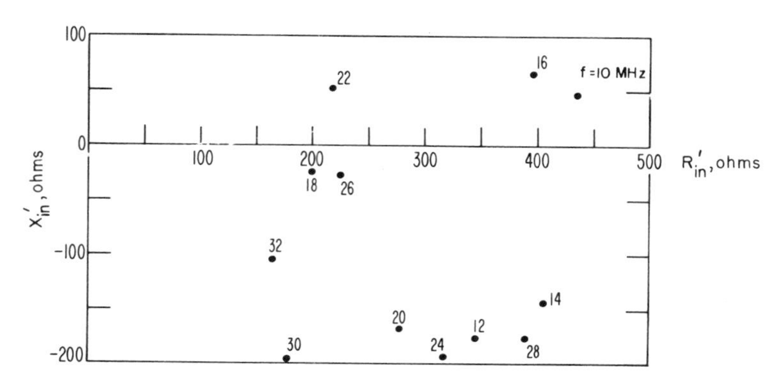

Fig. 5.18 Input impedance of the same array described in Fig. 5.3 placed 8 m above poor ground ($\epsilon_r = 4$, and $\sigma = 0.001$ mho/m).

and

$$R'_{max} = 435.00 \text{ ohms} \qquad \text{occurring at } f = 10\,\text{MHz}.$$

Although the radiation pattern or power gain in free space is relatively frequency insensitive in the entire operating bandwidth, as evidenced in Figs. 5.15 and 5.16, the shape of $G'(\theta, -90°)$ will, however, vary substantially with frequency and θ in view of the height factor in (5.40). For this reason, a set of $G'(\theta, -90°)$ is presented in Figs. 5.19 and 5.20. Furthermore, because of the imperfect ground effect represnted by R_v, the maximum value of $G'(\theta, -90°)$ no longer occurs at $\theta = 90°$, as we have experienced many times in Chapter 4. Instead, $G'(\theta, -90°)_{max}$ generally occurs around $\theta = 75°$ for the particular ground being considered. On the other hand, since the height H is not tied with the variable φ, the shape of $G'(75°, \varphi)$ will remain almost the same as that for $G(90°, \varphi)$ in Fig. 5.16, although the level is changed somewhat differently for different frequencies. Therefore, the variation of $G'(75°, \varphi)$ is not presented here.

When the same array is placed above sea water ($\epsilon_r = 80$ and $\sigma = 5$ mho/m) with the same height ($H = 8$ meters), we use the perfect-ground approximation. The normalized dipole base currents obtained according to (5.50) will be slightly different from those given in Figs. 5.3 through 5.8.

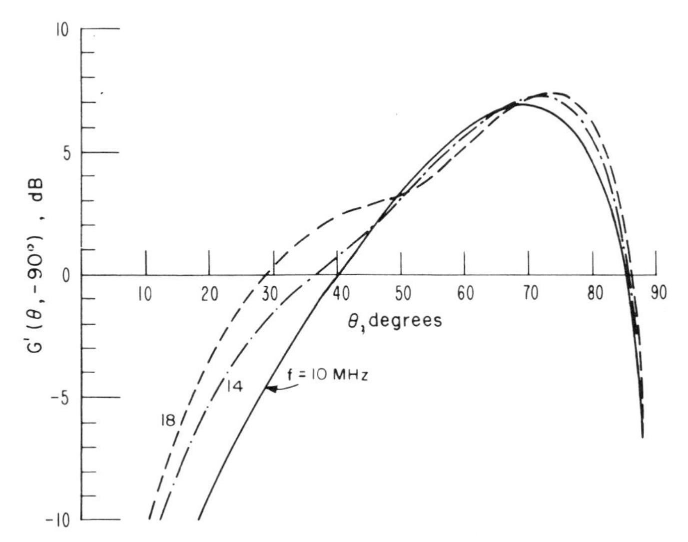

Fig. 5.19 Power gain of the same array described in Fig. 5.18 with $f = 10$, 14, and 18 MHz.

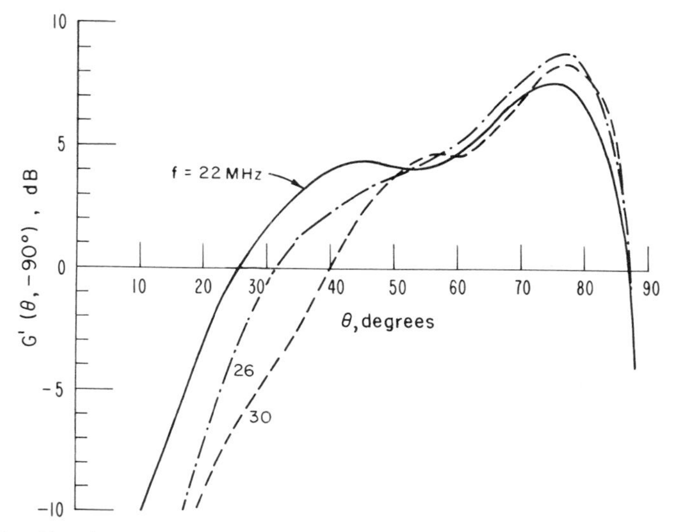

Fig. 5.20 Continuation of Fig. 5.19 with $f=22$, 26, and 30 MHz.

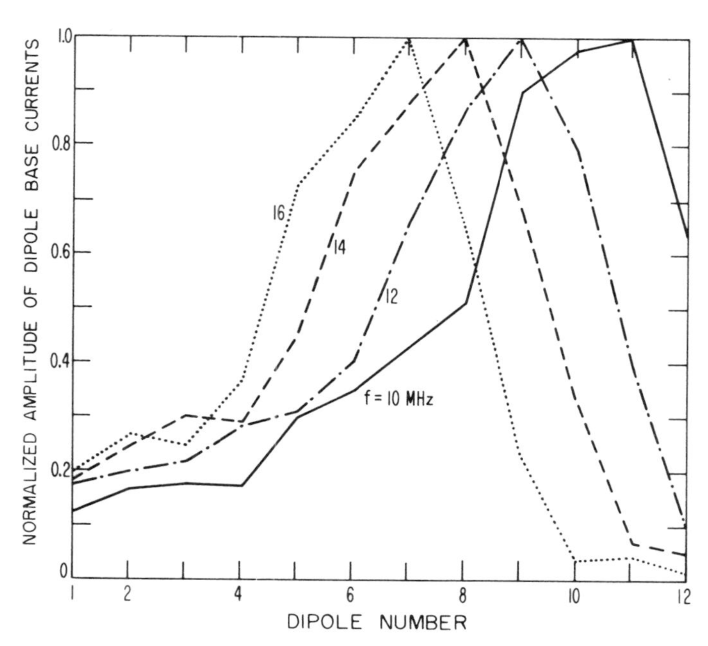

Fig. 5.21 Normalized amplitude of dipole base currents of the log-periodic dipole array ($N=12$, $\tau=0.87$, $\alpha=12.5°$, $Z_T=0$, $Z_0=450\ \Omega$, $h/a=500$, and $h_{12}=7.50$ m) placed 8 m above sea water ($\epsilon_r=80$ and $\sigma=5$ mho/m), with $f=10$, 12, 14, and 16 MHz.

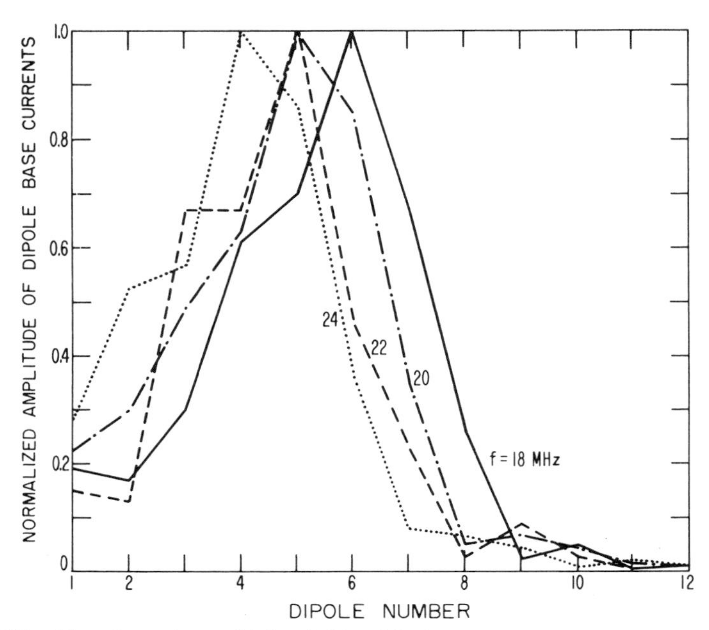

Fig. 5.22 Continuation of Fig. 5.21 with $f = 18$, 20, 22, and 24 MHz.

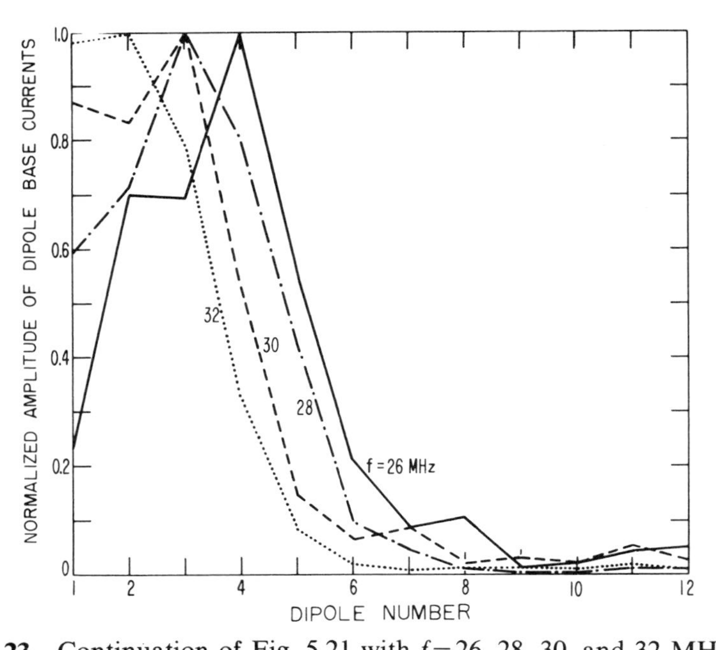

Fig. 5.23 Continuation of Fig. 5.21 with $f = 26$, 28, 30, and 32 MHz.

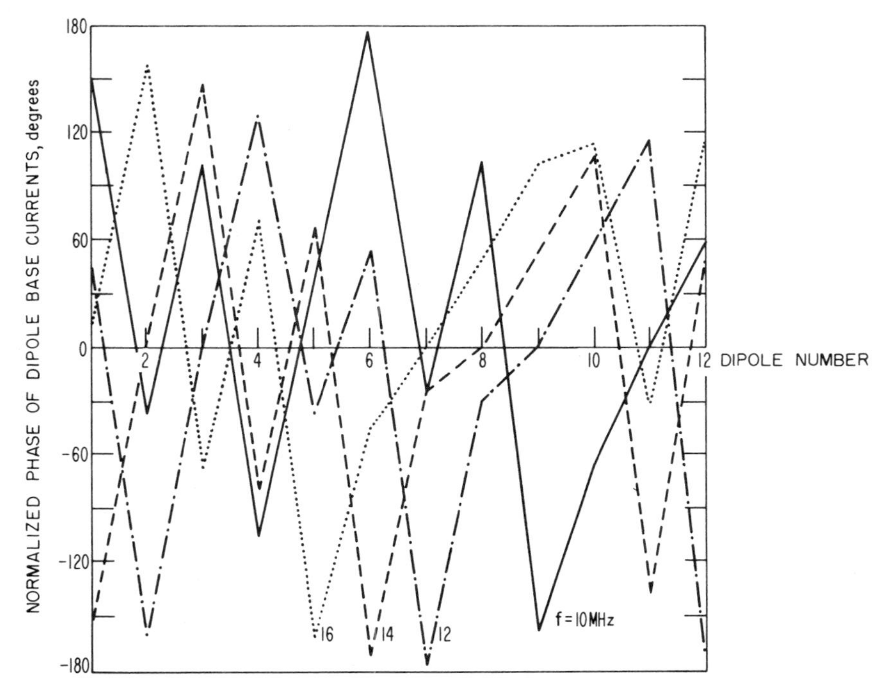

Fig. 5.24 Normalized phase of the same array described in Fig. 5.21 with $f = 10$, 12, 14, and 16 MHz.

They are presented in Figs. 5.21 through 5.26 with the same scale as that used in Figs. 5.3 and 5.6 for easy comparison. More specifically, when $f = 10$ MHz, we have

$$\begin{aligned}
\frac{I'_{1a}}{I'_{11a}} &= 0.1347 \,\underline{/152.94^\circ}, & \frac{I'_{2a}}{I'_{11a}} &= 0.1721 \,\underline{/-35.88^\circ}, \\
\frac{I'_{3a}}{I'_{11a}} &= 0.1800 \,\underline{/102.86^\circ}, & \frac{I'_{4a}}{I'_{11a}} &= 0.1771 \,\underline{/-104.00^\circ}, \\
\frac{I'_{5a}}{I'_{11a}} &= 0.3007 \,\underline{/42.80^\circ}, & \frac{I'_{6a}}{I'_{11a}} &= 0.3514 \,\underline{/178.32^\circ}, \\
\frac{I'_{7a}}{I'_{11a}} &= 0.3207 \,\underline{/-24.26^\circ}, & \frac{I'_{8a}}{I'_{11a}} &= 0.5162 \,\underline{/106.39^\circ}, \\
\frac{I'_{9a}}{I'_{11a}} &= 0.9029 \,\underline{/-156.33^\circ}, & \frac{I'_{10a}}{I'_{11a}} &= 0.9803 \,\underline{/-65.32^\circ}, \\
\frac{I'_{11a}}{I'_{11a}} &= 1.0000 \,\underline{/0^\circ}, & \frac{I'_{12a}}{I'_{11a}} &= 0.6337 \,\underline{/58.72^\circ},
\end{aligned} \tag{5.55}$$

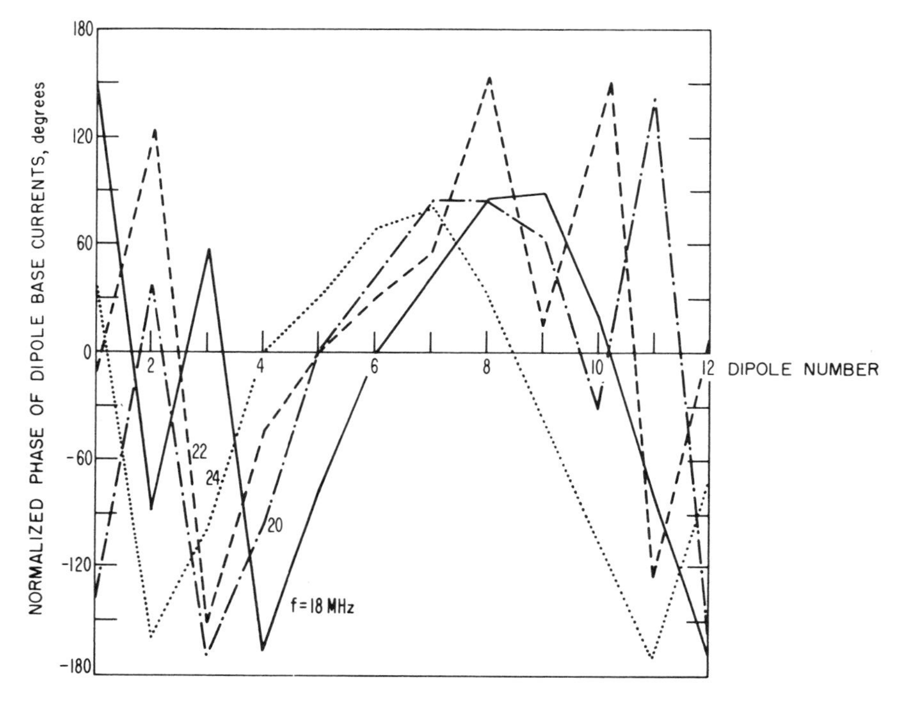

Fig. 5.25 Continuation of Fig. 5.24 with $f=18$, 20, 22, and 24 MHz.

which may be compared with those in (5.15) for the free-space case. It is clear that the currents on the eighth through twelfth dipoles in (5.55) will still produce an endfire radiation pattern with the beam maximum pointing toward the apex of the array ($\varphi=-90°$).

The input impedance Z''_{in} and the power gain in the vertical plane $G''(\theta, -90°)$, calculated according to (5.51), are given in Figs. 5.27 through 5.29. In this case with sea water as the ground, the associated mean resistance level and standing wave ratio are, respectively,

$$R_0 = 232.04 \text{ ohms}$$

and (5.56)

$$\text{SWR} = 1.60,$$

with

$$R''_{\min} = 144.93 \text{ ohms} \qquad \text{occurring at } f = 26 \text{ MHz}$$

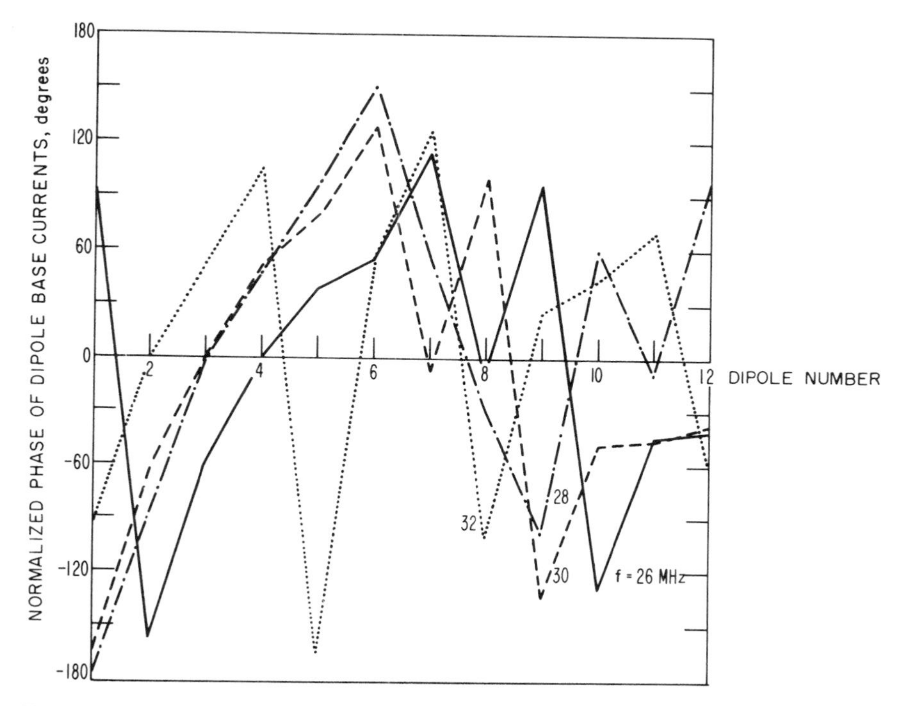

Fig. 5.26 Continuation of Fig. 5.24 with $f = 26$, 28, 30, and 32 MHz.

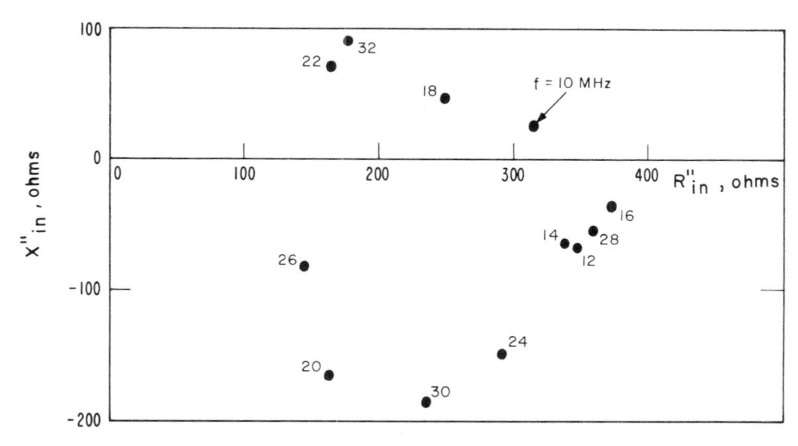

Fig. 5.27 Input impedance of the same array described in Fig. 5.21.

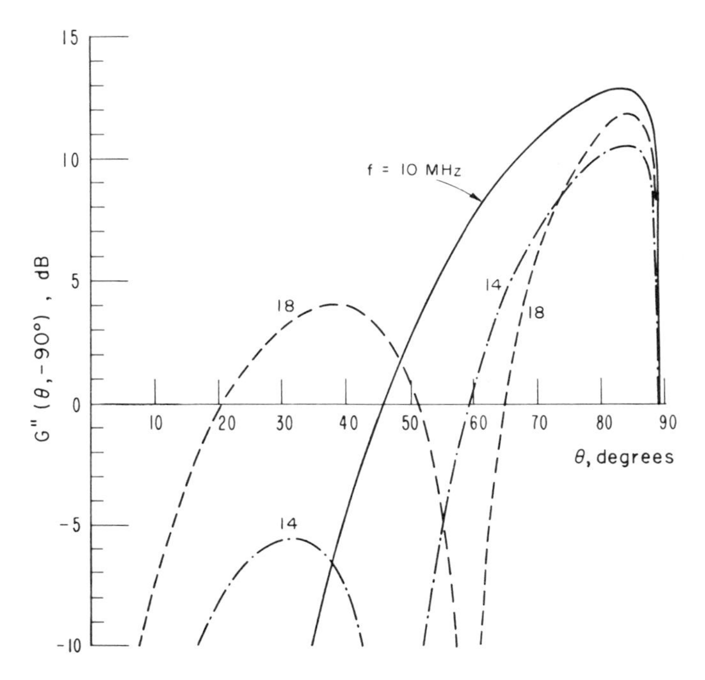

Fig. 5.28 Power gain of the same array as in Fig. 5.21 with $f = 10$, 14, and 18 MHz.

and

$$R''_{\max} = 371.51 \text{ ohms} \qquad \text{occurring at } f = 16 \text{ MHz.}$$

Note also that $G''(\theta, -90°)_{\max}$ now occurs approximately at $\theta = 82°$, and that $G''(\theta, -90°)$ generally is more frequency sensitive (via kH and R_v) than $G'(\theta, -90°)$ for the poor ground. In fact, clear sidelobes appear in Figs. 5.28 and 5.29 for frequencies above 14 MHz, resulting from the array factor contributed by the real array and its relatively good image. Again, we have omitted the graph for $G''(82°, \varphi)$ for the same reason explained before.

Before concluding this section, we should remember that the two approaches, free-space approximation and perfect-ground approximation, formulated here are rather artificial. As such, the results presented should be regarded only as limits of a practical situation which may fall between these two ideal cases.

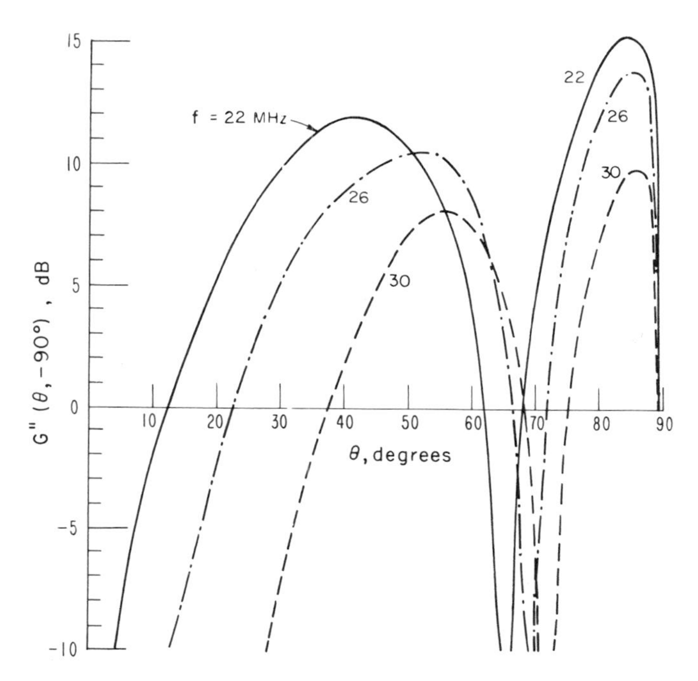

Fig. 5.29 Continuation of Fig. 5.28 with $f=22$, 26, and 30 MHz.

5.3 Modified Vertical Log-Periodic Dipole Arrays Above Lossy Ground

In the last example given in the previous section, the height of the array, H, is specified in meters. Its value must be greater than h_N (see Fig. 5.17). The equivalent electric height, kH, changes with frequency. The specific value of 8 meters used for H in the example corresponds to 0.267λ when $f=10$ MHz. It changes to 0.853λ when the frequency is increased to 32 MHz. Even though the free-space pattern or power gain of the array does not change much in the entire operating bandwidth 10 MHz$\leqslant f \leqslant$32 MHz (as the array is so designed), the final pattern or power gain of the array, including the ground and height effects, may vary substantially in view of the elementary theory presented in Chapter 1. In fact, these possible substantial variations are clearly indicated in Figs. 5.28 and 5.29. Because the array itself has broadband characteristics on the one hand and the height factor is frequency-sensitive on the other, broadband applications of the vertical log-periodic dipole array shown in Fig. 5.17 are, therefore,

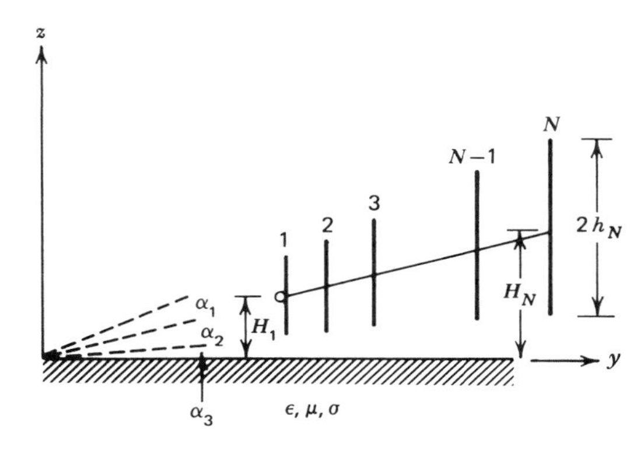

Fig. 5.30 Modified vertical log-periodic dipole array above lossy ground.

quite limited. This explains why a log-periodic monopole array built directly on the ground surface is more generally used.[9] Since the basic property of a log-periodic monopole array is almost the same as that of the array discussed in the previous section, we are not repeating the analysis here.

Alternatively, for the vertical log-periodic dipole array, the trouble caused by the fixed height can be overcome by a modified version of the array shown in Fig. 5.30.[15,16] In this case, the dipoles are still parallel to the z axis, but the array axis is no longer parallel to the y axis. Instead, it makes an angle $\alpha_2+\alpha_3$ with the ground. The lengths and radii of the dipoles still satisfy (5.1). The space between the dipoles along the array axis is, however, more complicated than (5.2). To derive an expression similar to (5.2), let us consider Fig. 5.31. Clearly, we have

$$\cos(\alpha_2+\alpha_3)=\frac{\overline{0a'}}{\overline{0a}}=\frac{\overline{0b'}}{\overline{0b}}. \tag{5.57}$$

By definition,

$$d_n \equiv \overline{0b}-\overline{0a}=(\overline{0b'}-\overline{0a'})\frac{\overline{0a}}{\overline{0a'}}. \tag{5.58}$$

Since

$$\begin{aligned}\overline{0b'} &= \overline{bb'}\cot(\alpha_2+\alpha_3)\\ &=(h_{n+1}+\overline{0b'}\tan\alpha_3)\cot(\alpha_2+\alpha_3),\end{aligned} \tag{5.59}$$

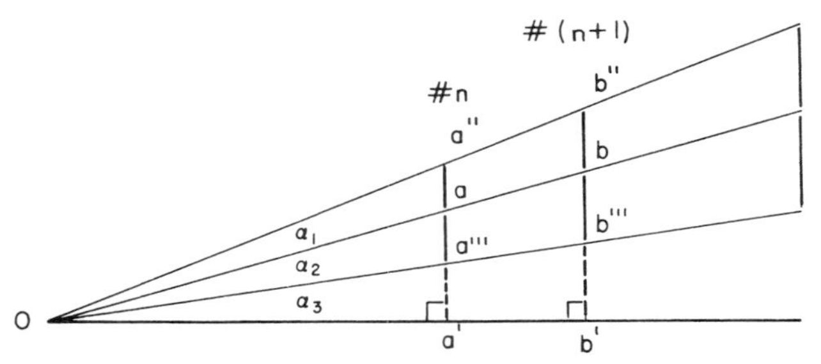

Fig. 5.31 A working diagram of Fig. 5.30 for deriving (5.62)–(5.64).

we obtain

$$\overline{0b'} = \frac{h_{n+1}\cot(\alpha_2+\alpha_3)}{1-\tan\alpha_3\cot(\alpha_2+\alpha_3)}. \tag{5.60}$$

Similarly,

$$\overline{0a'} = \frac{h_n\cot(\alpha_2+\alpha_3)}{1-\tan\alpha_3\cot(\alpha_2+\alpha_3)} = \frac{\tau h_{n+1}\cot(\alpha_2+\alpha_3)}{1-\tan\alpha_3\cot(\alpha_2+\alpha_3)}. \tag{5.61}$$

Substituting (5.60), (5.61), and (5.57) into (5.58), we have

$$d_n = (1-\tau)h_{n+1}\frac{\cot(\alpha_2+\alpha_3)}{1-\tan\alpha_3\cot(\alpha_2+\alpha_3)}\cdot\frac{1}{\cos(\alpha_2+\alpha_3)}.$$

Then,

$$\sigma' \equiv \frac{d_n}{4h_{n+1}} = \tfrac{1}{4}(1-\tau)\frac{1}{\sin(\alpha_2+\alpha_3)-\tan\alpha_3\cos(\alpha_2+\alpha_3)}. \tag{5.62}$$

Note that the three angles, α_1, α_2, and α_3, are not all independent. Applying the sine law to triangles $0aa''$ and $0aa'''$ in Fig. 5.31, we obtain

$$\frac{\sin\alpha_1}{\sin(90^\circ-\alpha_1-\alpha_2-\alpha_3)} = \frac{h_n}{\overline{0a}} = \frac{\sin\alpha_2}{\sin(90^\circ+\alpha_3)},$$

or

$$\frac{\sin\alpha_1}{\cos(\alpha_1+\alpha_2+\alpha_3)} = \frac{\sin\alpha_2}{\cos\alpha_3}. \tag{5.63}$$

Since the height of each individual dipole base above ground varies now, we can also derive a general expression for it. From Fig. 5.31, we have

$$H_n = \overline{aa'} = \overline{0a'}\tan(\alpha_2+\alpha_3) = \frac{\overline{a'a''}\tan(\alpha_2+\alpha_3)}{\tan(\alpha_1+\alpha_2+\alpha_3)}$$

$$= \frac{(h_n+H_n)\tan(\alpha_2+\alpha_3)}{\tan(\alpha_1+\alpha_2+\alpha_3)}$$

or

$$H_n = h_n \frac{\tan(\alpha_2+\alpha_3)}{\tan(\alpha_1+\alpha_2+\alpha_3)-\tan(\alpha_2+\alpha_3)}$$

$$= h_n \frac{\sin(\alpha_2+\alpha_3)\cos(\alpha_1+\alpha_2+\alpha_3)}{\sin\alpha_1} = h_n \frac{\sin(\alpha_2+\alpha_3)\cos\alpha_3}{\sin\alpha_2}$$

$$= h_n\left[1+\frac{\sin\alpha_3\cos(\alpha_2+\alpha_3)}{\sin\alpha_2}\right] > h_n. \tag{5.64}$$

Note that the last inequality is obtained because $\alpha_2>0$, $\alpha_3>0$, and $\alpha_2+\alpha_3<\pi/2$.

It is clear that when

$$\alpha_1=\alpha_2=\alpha \qquad \text{and} \qquad \alpha_3=-\alpha, \tag{5.65}$$

(5.62) will reduce to (5.2), and Fig. 5.30 will convert to Fig. 5.1(*a*) if the ground in Fig. 5.30 is ignored. On the other hand, if we still wish to consider the ground, the array apex in Fig. 5.30 must be raised by a height H, which should be no less than h_N. Then, Fig. 5.30 will convert to Fig. 5.17 under the conditions in (5.65).

After making these fundamental changes in geometric parameters for this array, we can derive the far fields in a manner similar to that leading to (5.33) or (5.39):

$$E_\varphi = 0,$$

$$E_\theta = -j60\frac{e^{-jkr_1}}{r_1}S_{mv}, \tag{5.66}$$

where

$$S_{mv} = \sum_{i=1}^{N} I_{mi} \exp\left[jky_i \sec(\alpha_2+\alpha_3)\cos\psi'\right]$$

$$\times \exp(jkH_i\cos\theta)\left[1+R_v\exp(-j2kH_i\cos\theta)\right]F_i. \tag{5.67}$$

In the above, r_1 is the distance from the base of the first dipole to the distant point where the total field is calculated, the maximum current I_{mi} is related to the base current I_{ia} [obtained by matrix inversion in (5.11) or (5.50)] by (5.29), y_i is the y coordinate of the base of the ith dipole, H_i is given in (5.64), F_i can be obtained from (5.31) if h_1 and a_1 there are replaced, respectively, by h_i and a_i, and

$$\cos\psi' = \cos\theta\sin(\alpha_2+\alpha_3) + \sin\theta\cos(\alpha_2+\alpha_3)\sin\varphi. \tag{5.68}$$

Note that the term $\exp(jkH_i\cos\theta)[1+R_v\exp(-j2kH_i\cos\theta)]$ is now inside the summation sign of (5.67). The quantity ψ' in (5.68) is the angle between the far-field point (θ,φ) and the array axis ($\theta'=\pi/2-\alpha_2-\alpha_3$, $\varphi'=\pi/2$). We should also remember that one of the I_{mi} and F_i in (5.67) may have to be replaced by I'_{mi} and F'_i if kh_i for that particular dipole happens to be $\pi/2$.

The power gain of the array is then

$$G(\theta,\varphi) = \frac{120}{R_{\text{in}}}|S_{mv}|^2, \tag{5.69}$$

where R_{in} is the real part of

$$Z_{\text{in}} = \sum_{i=1}^{N}\left[Z_{1ia} + R'_v Z_{1(N+1)a}\right]I_{ia}, \tag{5.70}$$

with the understanding that I_{ia} can be approximately obtained from (5.11) if the ground involved has a very low conductivity or from (5.50) if the ground conductivity is very high. Of course, the meanings of Z_{1ia} and $Z_{1(N+i)a}$ remain the same as those in (5.12), (5.38), and (5.51).

To illustrate the improvement made by this modified vertical log-periodic dipole array, as far as the reduction of sidelobe levels is concerned, we present as follows important characteristics of a comparable numerical example:

$$N=12, \quad \tau=0.84, \quad h_{12}=12.5 \text{ m}, \quad \alpha_1=10.18°, \quad \alpha_2=11.01°, \quad \alpha_3=1.04°,$$

$$Z_T=0, \quad Z_0=450\ \Omega, \quad \frac{h}{a}=200, \quad \epsilon_r=80, \quad \text{and} \quad \sigma=5 \text{ mho/m (sea water)}$$

Freq. (MHz)	Z_{in} (ohms)	Most Active Element	Base Height of Most Active Element in λ's
6	$303.33-j112.16$	tenth	$H_{10}=0.1930$
8	$277.24+j\ 63.77$	ninth	$H_9=0.2161$
10	$338.69+j\ 58.24$	seventh	$H_7=0.1906$
12	$247.17+j\ 17.15$	sixth	$H_6=0.1921$
14	$232.98-j\ 81.10$	fifth	$H_5=0.1883$
16	$425.51-j\ 44.30$	fifth	$H_5=0.2152$
18	$239.44-j127.13$	fourth	$H_4=0.2034$
20	$350.98-j134.15$	third	$H_3=0.1898$
22	$223.05-j222.24$	third	$H_3=0.2088$

(5.71)

From the above table, we see that, in the entire operating bandwidth $6 \leqslant f \leqslant 22$ MHz, the base height (above ground) of the most active dipole varies only between 0.1883λ and 0.2161λ. Because all of them are less than 0.25λ, we can conclude that no sidelobes with substantial levels will be present in the pattern. This result is confirmed in Fig. 5.32, where power gains with $f=6$, 12, and 18 MHz are presented. Corresponding results for other frequencies between 6 and 22 MHz are too close to those in Fig. 5.32 to be included. The associated mean resistance level and standing wave ratio can also be easily calculated with the Z_{in}'s in (5.71):

$$R_0=308.07 \text{ ohms}$$

and

$$\text{SWR}=1.38. \tag{5.72}$$

Again the power gain in the azimuthal surface such as $G(82°, \varphi)$ is omitted here because it differs only slightly from that shown in Fig. 5.16.

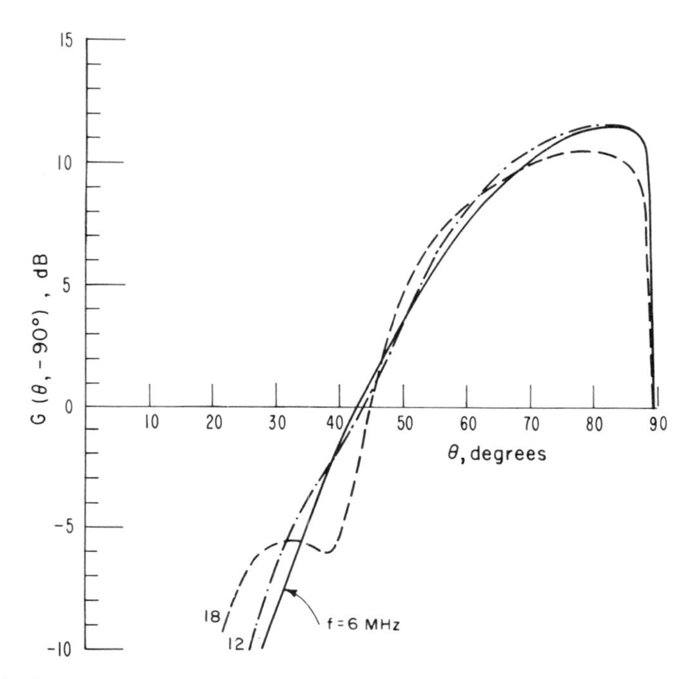

Fig. 5.32 Power gain of a modified vertical log-periodic dipole array above sea water with $N=12$, $\tau=0.84$, $h_{12}=12.5$ m, $\alpha_1=10.18°$, $\alpha_2=11.01°$, $\alpha_3=1.04°$, $Z_T=0$, $Z_0=450\ \Omega$, and $h/a=200$.

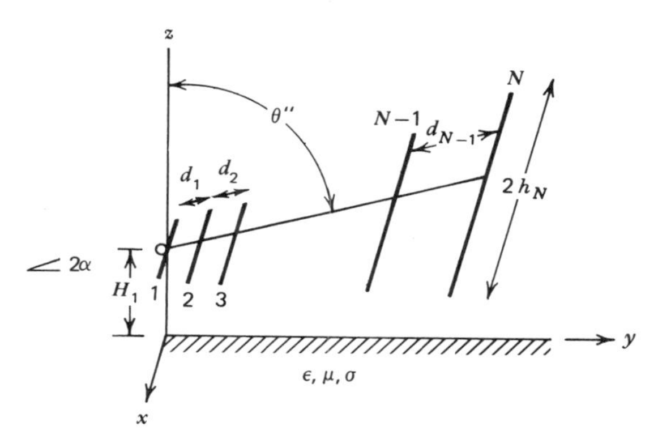

Fig. 5.33 Horizontal log-periodic dipole array above lossy ground.

5.4 Horizontal Log-Periodic Dipole Arrays Above Lossy Ground

After studying the vertical log-periodic dipole arrays in the previous two sections, we now proceed to discuss the horizontally polarized log-periodic dipole array. A typical array of this kind is pictured in Fig. 5.33. The dipole elements are now parallel to the x axis (and thus the ground) and the array axis is also in the yz plane, making an angle θ'' with the z axis. Since the basic relations (5.1) and (5.2) also hold true for h_i, a_i, and σ' here, the formulation for determining I_{ia} by (5.11) and Z'_{in} by (5.38) when the ground considered is relatively "poor" and that for obtaining I'_{ia} by (5.50) and Z''_{in} by (5.51) when the ground is relatively good are still good, except that R'_v in (5.38) and (5.51) should be replaced by R'_h, which is given in (4.72). The only other change we must make here is to rederive the far fields.

Referring to Fig. 4.2, we have

$$\alpha'=0°, \qquad \theta'=90°, \qquad \varphi'=0°, \qquad s=x, \qquad H_s=H_i,$$

$$\cos\psi=\sin\theta\cos\varphi, \tag{5.73}$$

and

$$I_i(x)=I_{mi}\left[\sin k(h_i-|x|)+T_{Ui}(\cos kx-\cos kh_i)+T_{Di}\left(\cos\frac{kx}{2}-\cos\frac{kh_i}{2}\right)\right], \qquad kh_i\neq\frac{\pi}{2}$$

$$=I'_{mi}\left[\sin k|x|-1+T'_{Ui}\cos kx-T'_{Di}\left(\cos\frac{kx}{2}-\cos\frac{\pi}{4}\right)\right], \qquad kh_i=\frac{\pi}{2}. \tag{5.74}$$

The contribution to the fields, according to (4.59) and (4.60), from the ith dipole are then

for $kh_i\neq\pi/2$,

$$E_{\theta i}=j30k\frac{e^{-jkr_i}}{r_i}\cos\theta\cos\varphi[1-R_v\exp(-j2kH_i\cos\theta)]$$
$$\times\int_{-h_i}^{h_i}I_i(x)\exp(jkx\cos\psi)\,dx$$
$$=j60I_{mi}\frac{e^{-jkr_i}}{r_i}\cos\theta\cos\varphi[1-R_v\exp(-j2kH_i\cos\theta)]F_i,$$

$$E_{\varphi i}=-j60I_{mi}\frac{e^{-jkr_i}}{r_i}\sin\varphi[1+R_h\exp(-j2kH_i\cos\theta)]F_i; \tag{5.75}$$

and for $kh_i = \pi/2$,

$$E_{\theta i} = j60 I'_{mi} \frac{e^{-jkr_i}}{r_i} \cos\theta \cos\varphi [1 - R_v \exp(-j2kH_i \cos\theta)] F'_i,$$
$$E_{\varphi i} = -j60 I'_{mi} \frac{e^{-jkr_i}}{r_i} \sin\varphi [1 + R_h \exp(-j2kH_i \cos\theta)] F'_i, \tag{5.76}$$

where

$$F_i = \frac{\cos(kh_i \cos\psi) - \cos kh_i}{\sin^2\psi} - \frac{T_{Ui} \cos kh_i \sin(kh_i \cos\psi)}{\cos\psi}$$
$$+ \frac{T_{Ui}}{\sin^2\psi} [\sin kh_i \cos(kh_i \cos\psi) - \cos\psi \cos kh_i \sin(kh_i \cos\psi)]$$
$$+ \frac{T_{Di}}{\frac{1}{4} - \cos^2\psi} [\tfrac{1}{2} \sin \tfrac{1}{2} kh_i \cos(kh_i \cos\psi) - \cos\psi \cos \tfrac{1}{2} kh_i \sin(kh_i \cos\psi)]$$
$$- \frac{T_{Di} \cos \frac{1}{2} kh_i \sin(kh_i \cos\psi)}{\cos\psi}, \tag{5.77}$$

$$F'_i = \frac{1 - \cos\psi \sin[(\pi/2)\cos\psi]}{\sin^2\psi} - \frac{\sin[(\pi/2)\cos\psi]}{\cos\psi}$$
$$+ \frac{T'_{Ui} \cos[(\pi/2)\cos\psi]}{\sin^2\psi} + \frac{T'_{Di} \cos(\pi/4) \sin[(\pi/2)\cos\psi]}{\cos\psi}$$
$$- \frac{T'_{Di}}{\frac{1}{4} - \cos^2\psi} \{\tfrac{1}{2} \sin(\pi/4) \cos[(\pi/2)\cos\psi]$$
$$- \cos\psi \cos(\pi/4) \sin[(\pi/2)\cos\psi]\}, \tag{5.78}$$

$\cos\psi$ is given in (5.73), R_h and R_v can be, respectively, found in (4.53) and (4.58), and r_i is the distance from the base of the ith dipole to the far-field point.

Summing up the contributions from all the dipoles, we obtain

$$E_\theta = j60 \frac{e^{-jkr_1}}{r_1} \cos\theta \cos\varphi \, S_\theta$$

and

$$E_\varphi = -j60 \frac{e^{-jkr_1}}{r_1} \sin\varphi \, S_\varphi, \tag{5.79}$$

where

$$S_\theta = \sum_{i=1}^{N} I_{mi} \exp\left(jky_i \csc\theta'' \cos\psi''\right)\left[1 - R_v \exp\left(-j2kH_i \cos\theta\right)\right]F_i, \tag{5.80}$$

$$S_\varphi = \sum_{i=1}^{N} I_{mi} \exp\left(jky_i \csc\theta'' \cos\psi''\right)\left[1 + R_h \exp\left(-j2kH_i \cos\theta\right)\right]F_i, \tag{5.81}$$

and

$$\cos\psi'' = \cos\theta \cos\theta'' + \sin\theta \sin\theta'' \sin\varphi. \tag{5.82}$$

Physically, the parameter ψ'' is the angle between the far-field point (θ, φ) and the array axis $(\theta'', \pi/2)$. The far-field condition,

$$r_i = r_1 - \left(\sum_{n=1}^{i-1} d_n\right) \cos\psi'' = r_1 - y_i \csc\theta'' \cos\psi'', \tag{5.83}$$

with y_i being the y coordinate of the base of the ith dipole, has been used in (5.80) and (5.81).

Note that H_i in (5.75), (5.76), (5.80), and (5.81) is related to H_1, the height of the first dipole above ground, by

$$H_i = H_1 + \left(\sum_{n=1}^{i-1} d_n\right) \cos\theta'' = H_1 + y_i \cot\theta''. \tag{5.84}$$

Note also that one of the I_{mi} and F_i in (5.80) and (5.81) should be replaced, respectively, by I'_{mi} and F'_i in (5.78) if kh_i associated with that particular dipole happens to be $\pi/2$.

The power gain of this array can then be expressed as

$$G(\theta, \varphi) = \frac{120}{R_{\text{in}}} \left[\cos^2\theta \cos^2\varphi |S_\theta|^2 + \sin^2\varphi |S_\varphi|^2\right], \tag{5.85}$$

where R_{in} is the real part of (5.38) if the dipole base currents are obtained from (5.11) or the real part of (5.51) if the dipole base currents are determined from (5.50) (with, of course, R'_h replacing R'_v).

Naturally, numerical results for this array depend on the many parameters involved. Since we have already analyzed and discussed the influence of f, τ, α, Z_0, h/a, and ground constants on dipole base currents, input impedance, or power gain in the previous sections, we will not repeat the same analyses here. Furthermore, results presented in Sections 4.3 and 4.4 concerning a single horizontal dipole and Yagi-Uda antennas should also be useful for predicting various characteristics of the array being considered. The only new parameter involved in this section is perhaps θ'', although it is related to the heights of dipoles above ground. For this reason, we present in Figs. 5.34 through 5.36 results of $G(\theta, -90°)$, $G(\theta, -45°)$, and $G(\theta, 0°)$ with $\theta''=46°$ for a horizontal log-periodic dipole array above two different grounds at a single frequency of 12 MHz only. The other parameters associated with this array are

$$N=12, \qquad \tau=0.87, \qquad \alpha=18°, \qquad Z_T=0,$$

$$Z_0=300\ \Omega, \qquad h_{12}=16.7340\ \text{m},$$

$$H_1=2.9566\ \text{m}, \qquad \text{and} \qquad h/a=250.$$

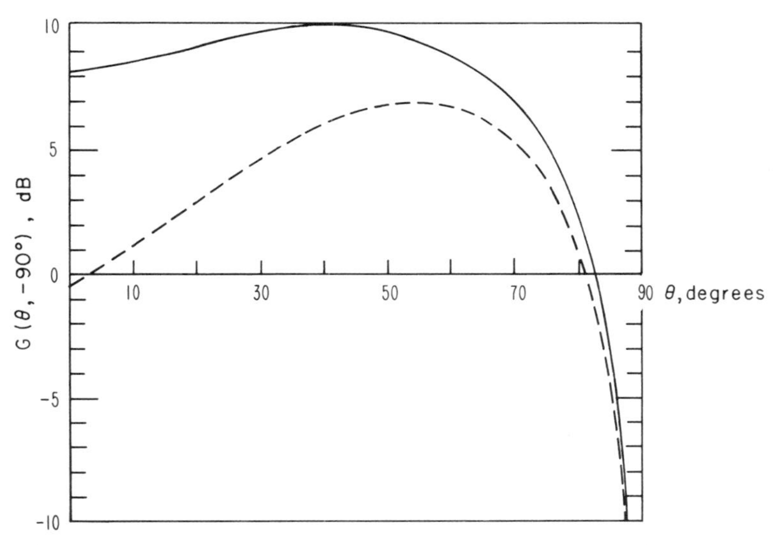

Fig. 5.34 Power gain of a horizontal log-periodic dipole array with $N=12$, $\tau=0.87$, $\alpha=18.0°$, $h_{12}=16.7340$ m, $H_1=2.9566$ m, $Z_T=0$, $Z_0=300\ \Omega$, $h/a=250$, $\theta''=46.0°$, $f=12$ MHz, and $\varphi=-90°$. Solid and dashed curves are for sea water and poor ground, respectively.

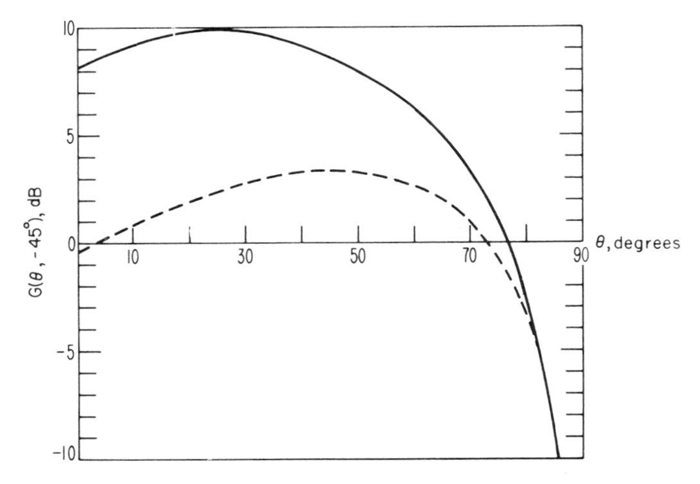

Fig. 5.35 Continuation of Fig. 5.34 with $\varphi = -45°$.

Actually, this array is operable for frequencies between 6 and 16 MHz. When it is operated at the high-frequency limit (16 MHz), the second dipole will be the most active element with $H_2 = 0.2195\lambda$, which is still under 0.25λ. This is the reason why $H_1 = 2.9566$ m is chosen in this example. For $f = 12$ MHz used in Figs. 5.34 through 5.36, dipoles No. 2 through No. 5 are actually in the active region, with No. 4 dipole as the

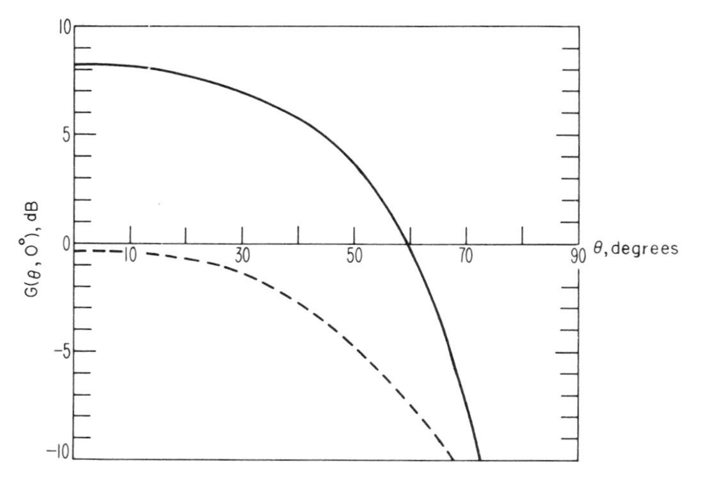

Fig. 5.36 Continuation of Fig. 5.34 with $\varphi = 0°$.

most active element. Under this condition, we have $H_2=0.1648\lambda$, $H_3=0.2178\lambda$, $H_4=0.2789\lambda$, and $H_5=0.3491\lambda$, explaining why the general shapes of curves in Fig. 5.34 are very similar to those in Fig. 4.6 for the case of single horizontal dipole with $H=3\lambda/8$. Obviously, the reason for the difference in power gain levels is that there are approximately four contributing elements for Fig. 5.34, while there is only one element for Fig. 4.6. The maximum level of power gain, $G(\theta,-90°)_{\max}$, achieved by the horizontal log-periodic dipole array in this example is also comparable with that obtained for the three-element Yagi-Uda antenna presented in Fig. 4.16. The change in positions of $G(\theta,-90°)_{\max}$ is the direct result of the change in θ'' ($\theta''=46°$ in Fig. 5.34 and $\theta''=90°$ in Fig. 4.16). Corresponding results for $\theta''=90°$ at $f=12$ MHz and $H_1=H_i=9.375$ m $(3\lambda/8)$ are presented in Fig. 5.37 for comparison purpose. Therefore, as far as the final result for power gain is concerned, the array studied here does not differ much from the Yagi-Uda antenna discussed in Section 4.4. The major advantage associated with the horizontal log-periodic dipole array is, of course, that it is operable over a much wider frequency band.

Before completing this section, we should note that the array geometry shown in Fig. 5.33 is also valid for the special cases $\theta''=0°$ and $\theta''=180°$. These two arrangements have recently found application in radio astronomy. Under these conditions, the first expression in (5.83) and (5.84), namely,

$$r_i=r_1-\left(\sum_{n=0}^{i-1} d_n\right)\cos\psi'' \qquad \text{and} \qquad H_i=H_1+\left(\sum_{n=1}^{i-1} d_n\right)\cos\theta'',$$

should be used in (5.80) and (5.81) to avoid the mathematical difficulty with $\csc\theta''$ and $\cot\theta''$ there.

For $\theta''=0°$, the maximum radiation of the array in free space points toward $\theta=180°$. When the array is above ground, an essential requirement is that the ground must be highly conductive to produce an effective reflection, because otherwise there is no reason to choose such an arrangement. Upon satisfaction of the ground condition, this array arrangement ($\theta''=0°$) does enhance the broadband property associated with the array. This is true because, when the operating frequency is increased (λ decreased), the most active element moves downward making H_{act} (the height of the most active element above ground) smaller. However, kH_{act}, under this condition, changes insignificantly so that the shape of the final pattern is relatively insensitive to the frequency change. Results of $G(\theta,-90°)$ and $G(\theta,0°)$ with $\theta''=0°$ for the same array described in Fig. 5.34 above sea water are shown in Fig. 5.38, from which we see clearly that the

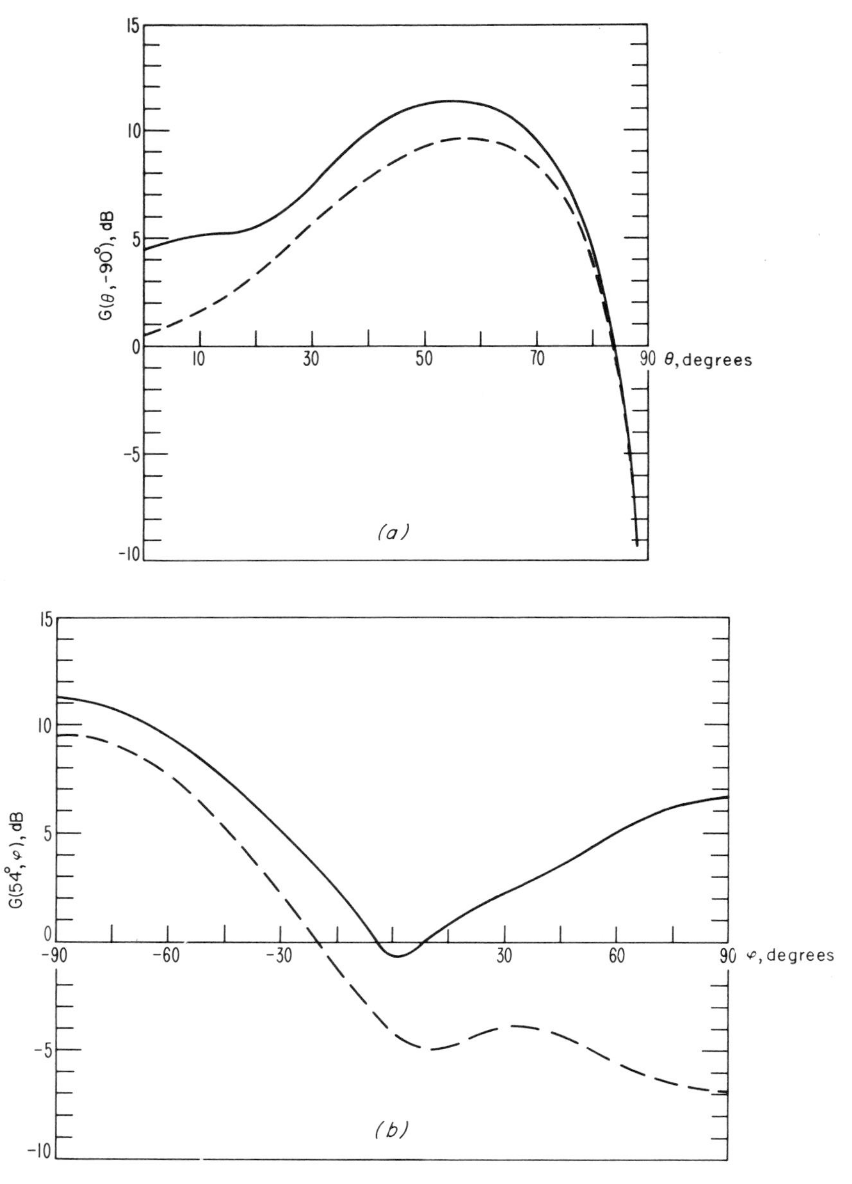

Fig. 5.37 Power gain of the array described in Fig. 5.34 but with $\theta'' = 90°$ and $H_1 = 9.375$ m: (*a*) in the vertical plane and (*b*) in the azimuthal surface. Solid and dashed curves are for sea water and poor ground, respectively.

level of the maximum power gain obtainable remains essentially the same as that when $\theta''=46°$ and that the position of G_{max} now shifts to $\theta=0°$. Of course, the disadvantage with this arrangement is that there always are a few longer elements above the active elements, which may have some blocking effect on the final result. The minor unnatural shape of the curve near $44° \leqslant \theta \leqslant 64°$ in Fig. 5.38(a) may be the consequence of this effect. For other frequencies or array dimensions, the level of power gain at $\theta=0°$ may be slightly smaller than that at, say, $\theta=10°$. In any case, the level of power gain at $\theta=0°$ is still very substantial, even though it may not be the maximum.

For $\theta''=180°$, the maximum radiation of the array in free space points toward $\theta=0°$. When the array is above ground, the first requirement is that H_1 must be adjusted to at least equal $\sum_{n=1}^{N} d_n$ in length. The second requirement is that the ground must be very poor to have an ineffective reflection. Then, when the operating frequency is increased, the most active element moves upward, making H_{act} larger. Under this condition, kH_{act} will increase quite substantially. Since the ground is poor, this variation in kH_{act} will not cause appearance of large sidelobes, as would be the case if the ground were good. Thus, the shape of the final pattern still remains very much like that in free space. An example for $\theta''=180°$ and $H_1=49.0$ m is hereby presented in Fig. 5.39.

5.5 Concluding Remarks

In this chapter, we have given detailed analyses of both horizontal and vertical log-periodic dipole arrays based, primarily, on Carrel's approach. Addition of ground effects and the assumption of three terms for the current distribution on each dipole in the array followed the basic mehtods outlined and discussed in Chapter 4. Although the overall formulation given in this chapter is straightforward, it does have a couple of weak points. First of all, there is no clear-cut boundary between the different regions along the entire array structure. Currents on the first few short dipoles in the transmission region, though relatively small in amplitudes, still contribute significantly to the input impedance. Secondly, the analytic method is not general enough to be applicable for other types of log-periodic structure.[17,18]

From the numerical results presented, we learned that only a small portion of the array (elements in the active region) actually contributes to the radiated field. This fact implies that the array does not make use of space very effectively. Definitely, a larger value of τ can be used to increase the width of the active region in the sense that more elements will fall into it, thereby yielding a better result in terms of power gain.

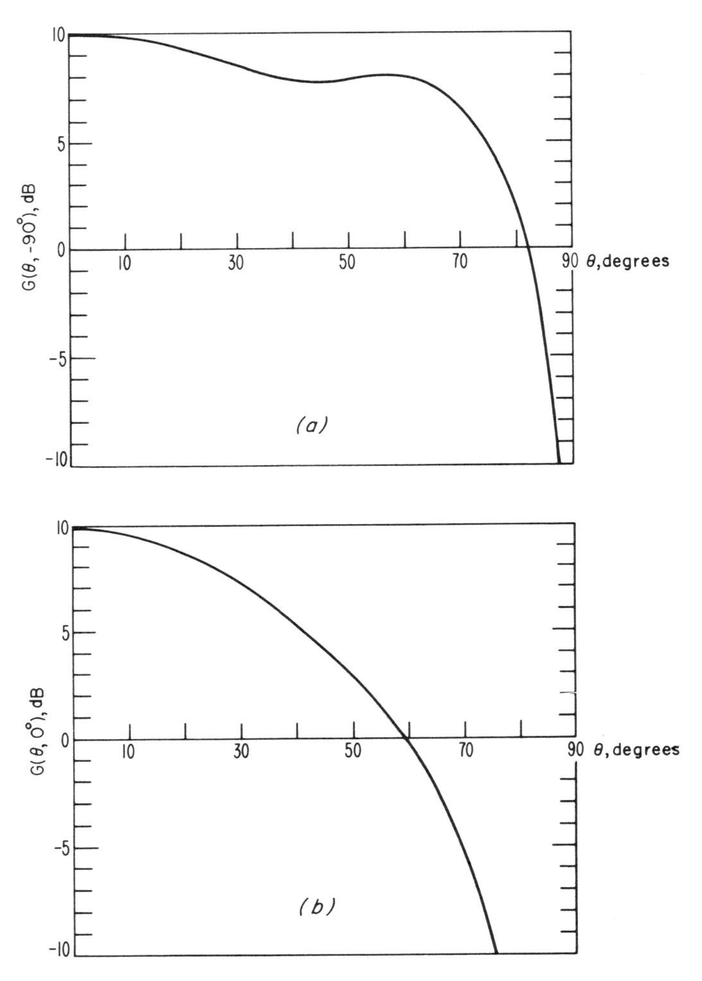

Fig. 5.38 Power gain of the array described in Fig. 5.34 above sea water but with $\theta''=0°$: (*a*) $\varphi=-90°$ and (*b*) $\varphi=0°$.

However, the operating bandwidth will then be relatively decreased for a given overall length. Apparently, the rather ineffective use of space may well be the price we must pay for the broadband operation. Although a possibility of modifying the design so that there may be multiple active regions along the array structure has been suggested,[19] no significant results by a systematic approach, theoretical or experimental, have been published.

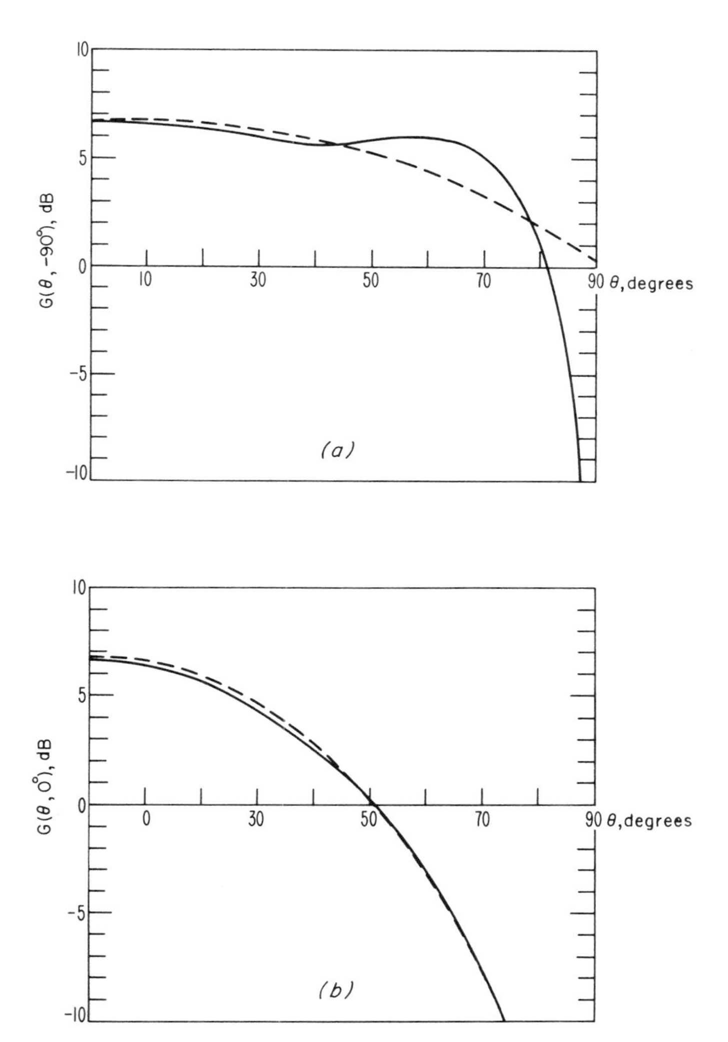

Fig. 5.39 Power gain of the array described in Fig. 5.34 but with $\theta'' = 180°$ and $H_1 = 49.0$ m: (*a*) $\varphi = -90°$ and (*b*) $\varphi = 0°$. Solid and dashed curves are for poor ground and free space, respectively.

It should be noted that some laboratory measured data were included in Carrel's original report.[11] Those data did appear in good agreement with his computed results. Although our theoretical results were not formally compared with any experimental data, part of them have, indeed, been confirmed by many private sources. This should help establish some degree of confidence about the curves and tables presented in the chapter.

As evidenced by the dipole base currents given in Section 5.2, we recognize that the radiation pattern produced by a log-periodic dipole array in free space is ordinarily endfire (backfire). By adding parasitic elements and changing the feed system shown in Fig. 5.1(*b*) to a new system called "snaking strip-transmission," a broadside log-periodic antenna can be realized.[20] The parasitic elements are used to suppress the radiation in the backfire direction. The broadside radiation pattern so produced has a narrower beamwidth.

Of course, the log-periodic dipole array discussed in this chapter can be regarded as a single unit. Identical units can then be arranged to form an array according to the basic theory presented in Chapters 1–3. The final radiation pattern of such a system will be the product of that radiated by a single log-periodic dipole array and the related array factor. Since the array factor is frequently-sensitive in general, the use of such an array is, therefore, very limited. However, by some special arrangements, a more directive broadband array consisting of log-periodic dipole arrays can be achieved.[21]

REFERENCES

1. **Hessel, A.** General characteristics of traveling-wave antennas, Chap. 19 in *Antenna Theory*, Part II, edited by R. E. Collin and F. J. Zucker, McGraw-Hill Book Company, New York, 1969.

2. **Rumsey, V. H.** Frequency-independent antennas, *IRE National Convention Record,* part 1, Vol. 5, pp. 114–118, 1957.

3. **DuHamel, R. H. and D. E. Isbell.** Broadband log-periodic antenna structures, *IRE National Convention Record,* part 1, Vol. 5, pp. 119–128, 1957.

4. **Deschamps, G. A. and R. H. DuHamel.** Frequency-independent antennas, Chap. 18 in *Antenna Engineering Handbook,* edited by H. Jasik, McGraw-Hill Book Company, New York, 1961.

5. **Isbell, D. E.** A log-periodic reflector feed, *Proc. IRE,* Vol. 47, No. 6, pp. 1152–1153, June, 1959.

6. **DuHamel, R. H. and D. G. Berry.** Logarithmically periodic antenna arrays, *IRE Wescon Record,* part 1, Vol. 2, pp. 161–174, 1958.

7. **Isbell, D. E.** Log-periodic dipole arrays, *IRE Trans. Antennas and Propagation,* Vol. AP-8, No. 3, pp. 260–267, May, 1960.

8. **Carrel, R. L.** The design of log-periodic dipole antennas, *IRE National Convention Record,* part 1, Vol. 9, pp. 61–75,1961.

9. **Berry, D. G. and F. R. Ore.** Log-periodic monopole arrays, *IRE National Convention Record,* part 1, Vol. 9, pp. 76–85, 1961.

10. Mittra, R. Log-periodic antennas, Chap. 22 in *Antenna Theory,* Part II, edited by R. E. Collin and F. J. Zucker, McGraw-Hill Book Company, New York, 1969.

11. **Carrel, R. L.** Analysis and design of the log-periodic dipole antenna, *University of Illinois Technical Report No. 52,* Urbana, 1961.

12. **Guillemin, E. A.** *Communication Networks,* Chaps. II and IV, John Wiley and Sons, Inc., New York, 1953.

13. **Jordan, E. C.** *Electromagnetic Waves and Radiating Systems,* Prentice-Hall, Inc., Englewood Cliffs, N. J., 1950, p. 464.

14. **Schelkunoff, S. A. and H. T. Friis.** *Antennas: Theory and Practice,* John Wiley and Sons, Inc., New York, 1952, p. 425.

15. **Ma, M. T., P. W. Arnold, and L. C. Walters.** Ground effect on the behavior of a vertically polarized log-periodic dipole used in an HF communication system, presented at 1969 IEEE International Conference on Communications, June 9–11, Boulder, Colo.

16. **Ma, M. T. and L. C. Walters.** Power gains for antennas over lossy plane ground, *ESSA Technical Report No. ERL 104-ITS 74,* April, 1969, Boulder, Colo.

17. **Carr, J. W.** Some variations in log-periodic antenna structures, *IRE Trans. Antennas and Propagation,* Vol. AP-9, No. 2, pp. 229–230, March, 1961.

18. **Dyson, J. D.** The characteristics and design of the conical log-spiral antenna, *IEEE Trans. Antennas and Propagation,* Vol. AP-13, No. 4, pp. 488–498, July, 1965.

19. **Kieburtz, R. B.** Analysis and synthesis of aperture fields of log-periodic antennas, *Proceedings of the International Symposium on Electromagnetic Wave Theory,* Delft, The Netherlands, 1965.

20. **Mei, K. K. and D. Johnstone** A broadside log-periodic antenna, *Proc. IEEE,* Vol. 54, No. 6, pp. 889–890, June, 1966.

21. **Mei, K. K., M. W. Moberg, V. H. Rumsey, and Y. S. Yeh.** Directive frequency independent arrays, *IEEE Trans. Antennas and Propagation,* Vol. AP-13, No. 5, pp. 807–809, September, 1965.

ADDITIONAL REFERENCES

Barbano, N. Phase center distributions of spiral antennas, *IRE Wescon,* part 1, Vol 4, pp. 123–130, 1960.

Bawer, R. and J. J. Wolfe. The spiral antenna, *IRE International Convention Record,* part 1, Vol. 8, pp. 84–95, 1960.

Bell, R. L., C. T. Elfving, and R. E. Franks. Near-field measurements on a log-periodic antenna, *IRE Trans. Antennas and Propagation,* Vol. AP-8, No. 6, pp. 559–567, November, 1960.

Brillouin, L. *Wave Propagation in Periodic Structures,* Dover Publications, New York, 1953.

Burdine, B. H. and R. M. McElvery. Spiral antennas, *MIT Research Laboratory of Electronics Report No. 1,* March, 1955.

Burdine, B. H. and H. J. Zimmerman. The spiral antenna, *MIT Research Laboratory of Electronics Report No. 2,* April, 1955.

Cheo, B. R. S., V. H. Rumsey and W. J. Welch. A solution to the frequency-independent antenna problem, *IRE Trans. Antennas and Propagation,* Vol. AP-9, No. 6, pp. 527–534, November, 1961.

Copeland, J. R. Radiation from the balanced conical equi-angular spiral antenna, *Ohio State University Report No. 903–12,* September, 1960.

Curtis, W. L. Spiral antennas, *IRE Trans. Antennas and Propagation,* Vol. AP-8, No. 3, pp. 298–306, May, 1960.

Deschamps, G. A. Impedance properties of complementary multiterminal planar structures, *IRE Trans. Antennas and Propagation,* Vol. AP-7, pp. s371–s378, December, 1959.

DuHamel, R. H. Log-periodic antennas and circuits, in *Electromagnetic Theory and Antennas,* part 2, edited by E. C. Jordan, Pergamon Press, New York, 1962.

DuHamel, R. H. and F. R. Ore. Logarithmically periodic antenna designs, *IRE National Convention Record,* part 1, Vol. 6, pp. 139–151, 1958.

DuHamel, R. H. and F. R. Ore. Log-periodic feeds for lens and reflectors, *IRE National Convention Record,* part 1, Vol. 7, pp. 128–137, 1959.

Dyson, J. D. The equiangular spiral antenna, *IRE Trans. Antennas and Propagation,* Vol. AP-7, No. 2, pp. 181–187, April, 1959.

Dyson, J. D. The unidirectional equiangular spiral antenna, *IRE Trans. Antennas and Propagation,* Vol. AP-7, No. 4, pp. 329–334, October, 1959.

Dyson, J. D. Multi-mode logarithmic spiral antennas—possible applications, *Proc. NEC,* Vol. 17, pp. 206–213, 1961.

Dyson, J. D. and P. E. Mayes. New circularly-polarized frequency-independent antennas with conical beam or omni-directional patterns,

IRE Trans. Antennas and Propagation, Vol. AP-9, No. 4, pp. 334–342, July, 1961.

Dyson, J. D. A survey of very wideband and frequency-independent antennas—1945 to present, *J. Res. NBS,* Vol. 66D, pp. 1–6, 1962.

Dyson, J. D. Frequency-independent antennas: Survey of development, *Electronics,* Vol. 35, No. 16, pp. 39–44, April, 1962.

Elfving, C. T. Design criterion for log-periodic antennas, *IRE Wescon,* paper No. 1-2, Vol. 5, 1961.

Elfving, C. T. Array theory principles applied to log-periodic arrays, *GAP Symposium Digest,* pp. 52–57, 1964.

Greiser, J. W. and P. E. Mayes. Vertically polarized log-periodic zigzag antennas, *Proc. NEC,* Vol. 17, pp. 193–204, 1961.

Greiser, J. W. and P. E. Mayes. The bent backfire zigzag—a vertically polarized frequency independent antenna, *IEEE Trans. Antennas and Propagation,* Vol. AP-12, No. 3, pp. 281–290, May, 1964.

Hudock, E. Near-field investigation of uniformly periodic monopole arrays, *Technical Report No. 1,* Antenna Lab., University of Illinois, Urbana, June, 1963.

Isbell, D. E. Nonplanar logarithmically periodic antenna structures, *Technical Report No. 30,* Antenna Lab., University of Illinois, Urbana, February, 1958.

Isbell, D. E. Log-periodic dipole arrays, *IRE Trans. Antennas and Propagation,* Vol. AP-8, No. 3, pp. 260–267, May, 1960.

Jordan, E. C., G. A. Deschamps, J. D. Dyson, and P. E. Mayes. Developments in broadband antennas, *IEEE Spectrum,* Vol. 1, No. 4, pp. 58–71, April, 1964.

Karjala, D. S. and R. Mittra. Radiation from some periodic structures excited by a waveguide, *Technical Report No. 65-15,* Antenna Lab., University of Illinois, Urbana, October, 1965.

Mast, P. E. A theoretical study of the equiangular spiral antenna, *Technical Report No. 35,* Antenna Lab., University of Illinois, Urbana, September, 1958.

Mayes, P. E., G. A. Deschamps, and W. T. Patton. Backward-wave radiation from periodic structure and application to the design of frequency-independent antennas, *Proc. IRE,* Vol. 49, No. 5, pp. 962–963, May, 1961.

Mittra, R. Theoretical study of a class of logarithmically periodic circuits, *Technical Report No. 59,* Antenna Lab., University of Illinois, Urbana, July, 1962.

Mittra, R. and K. E. Jones. A study of continuously scaled and log-periodically loaded transmission lines, *Technical Report No. 73,* Antenna Lab., University of Illinois, Urbana, September, 1963.

Mittra, R. and K. E. Jones. Theoretical k-β diagram for monopole and dipole arrays and their application to log-periodic antennas, *IEEE Trans. Antennas and Propagation*, Vol. AP-12, No. 5, pp. 533–540, September, 1964.

Mittra, R. and K. E. Jones. How to use k-β diagrams in log-periodic antenna design, *Microwaves*, p. 18, June, 1965.

Tang, C. H. A class of modified log-spiral antennas, *IEEE Trans. Antennas and Propagation*, Vol. AP-11, No. 4, pp. 422–427, July, 1963.

Wheeler, M. S. On the radiation from several regions in spiral antennas, *IRE Trans. Antennas and Propagation*, Vol. AP-9, No. 1, pp. 100–102, January, 1961.

CHAPTER 6

TRAVELING-WAVE ANTENNAS ABOVE LOSSY GROUND

In the previous two chapters, we analyzed arrays of standing-wave and log-periodic antennas. With an assumed form for current distribution on the antenna, important characteristics such as far-fields, input impedance, and power gain from those arrays above different lossy grounds were presented. In this chapter, we treat still another kind, namely, the traveling-wave type of antenna. Basically, the antenna involved is terminated with an approximately matching resistance such that the currents in the antenna wires are substantially traveling waves. Of course, considerable power may be lost in the terminating resistance. This loss is apparently the price paid for desirable features such as simplicity of construction, relatively wide bandwidth of operation, and higher directive gain offered by the radiator. Specific antennas to be discussed are the elevated sloping vee antenna, the sloping rhombic, and the side-terminated vertical half-rhombic. These antennas are very useful for HF and VHF point-to-point communications.

Since the mutual impedance between antennas in these classes has not been systematically investigated and is difficult to formulate, we are not considering it here. For this reason, only the analysis of characteristics for a single antenna above lossy ground is treated. The overall radiation pattern of an array consisting of identical traveling-wave antennas will, then, when the coupling effect between antennas is ignored, be the product of the radiation pattern of a single antenna and the array factor studied in Chapter 1.

6.1 Elevated Sloping Vee Antenna

This antenna is made of two diverging wires, as shown in Fig. 6.1, forming an apex angle of 2γ. The feed and termination points are, respectively, H and H' meters above the ground. The angle between the antenna wires and the ground plane is α'. The projection of the two wires on the ground surface makes an angle β' with the x axis. The length of each wire is l meters.

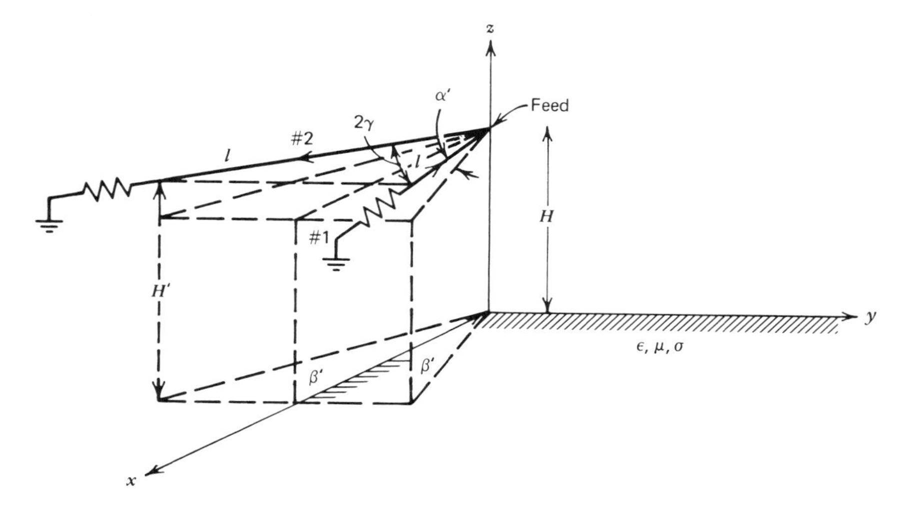

Fig. 6.1 Elevated sloping vee antenna above flat lossy ground.

Although the structure of the antenna is simple, the mathematical derivation of an exact expression for the input impedance with a finite l, even in free space, is quite involved. An approximate formula for the average impedance, when $l \to \infty$, has been obtained elsewhere[1,2]:

$$Z_{in} = 120\left(\ln \frac{\lambda}{2\pi a} - 0.60 + \ln \sin \gamma\right) - j170 \text{ ohms}, \tag{6.1}$$

where a is the radius of the wire and λ is the wavelength at the operating frequency.

If we take, for example, a typical set of parameters, say $\gamma = 20°$ and $a = 0.002$ m, then

$$\underline{\lambda = 10 \text{ m } (f = 30 \text{ MHz})}, \qquad \underline{20 \text{ m } (15 \text{ MHz})}, \qquad \underline{30 \text{ m } (10 \text{ MHz})};$$

$$R_{in} = 600.77, \qquad 683.94, \qquad 732.61 \text{ ohms}. \tag{6.2}$$

When $\gamma = 15°$ and $a = 0.002$ m, we have

$$\underline{\lambda = 10 \text{ m}}, \qquad \underline{20 \text{ m}}, \qquad \underline{30 \text{ m}};$$

$$R_{in} = 567.32, \qquad 650.50, \qquad 699.16 \text{ ohms}. \tag{6.3}$$

Since the values obtained above are valid only for $l \to \infty$ and also vary with f, a, and γ, a compromise value of 600 ohms for the input resistance of all vee antennas with finite lengths has been traditionally considered adequate by practical engineers.[3] For this reason, a resistance of 300 ohms is usually used for termination of each leg in Fig. 6.1. Under this condition, even though the antenna is not perfectly matched, the currents in the antenna wires are still very much traveling waves in nature. Thus, we can write

$$I_1(s) = \text{current distribution on antenna wire No. 1} = -I_m e^{-jks} \quad (6.4)$$

and

$$I_2(s) = \text{current distribution on antenna wire No. 2} = I_m e^{-jks},$$

where I_m is the current amplitude and s is the distance from the feed point.

Referring to Fig. 4.2, we have

for wire No. 1, $\theta' = 90° - \alpha'$, $\varphi' = \beta'$,

$$\cos\psi_1 = \cos\theta \sin\alpha' + \sin\theta \cos\alpha' \cos(\varphi - \beta'), \quad (6.5)$$

and for wire No. 2, $\theta' = 90° - \alpha'$, $\varphi' = -\beta'$,

$$\cos\psi_2 = \cos\theta \sin\alpha' + \sin\theta \cos\alpha' \cos(\varphi + \beta'), \quad (6.6)$$

where

$$\alpha' = \sin^{-1}\left(\frac{H' - H}{l}\right), \qquad -(90° - \gamma) \leqslant \alpha' \leqslant 90° - \gamma, \quad (6.7)$$

$$\beta' = \sin^{-1}\left(\frac{\sin\gamma}{\cos\alpha'}\right). \quad (6.8)$$

For the special case of a horizontal vee antenna, $H' = H$, $\alpha' = 0$, and $\beta' = \gamma$.

The height of the current element ds above the ground is

$$H_s = z = H + (\sin\alpha')s. \quad (6.9)$$

Substituting (6.4) and (6.9) into (4.59) and (4.60), we obtain, respectively,

$E_{\theta 1} = \theta$ component of the far-field contributed by wire No. 1

$$= 30 I_m \frac{e^{-jkr}}{r} \{ -\cos\alpha' \cos\theta \cos(\varphi - \beta') [F_1 - F_3 R_v \exp(-j2kH\cos\theta)]$$

$$+ \sin\alpha' \sin\theta [F_1 + F_3 R_v \exp(-j2kH\cos\theta)] \}, \quad (6.10)$$

$E_{\varphi 1} = \varphi$ component of the far-field contributed by wire No. 1

$$= 30 I_m \frac{e^{-jkr}}{r} \cos\alpha' \sin(\varphi - \beta')[F_1 + F_3 R_h \exp(-j2kH\cos\theta)], \quad (6.11)$$

$E_{\theta 2} = \theta$ component of the far-field contributed by wire No. 2

$$= 30 I_m \frac{e^{-jkr}}{r} \{\cos\alpha' \cos\theta \cos(\varphi + \beta')[F_2 - F_4 R_v \exp(-j2kH\cos\theta)]$$

$$- \sin\alpha' \sin\theta [F_2 + F_4 R_v \exp(-j2kH\cos\theta)]\}, \quad (6.12)$$

and

$E_{\varphi 2} = \varphi$ component of the far-field contributed by wire No. 2

$$= -30 I_m \frac{e^{-jkr}}{r} \cos\alpha' \sin(\varphi + \beta')[F_2 + F_4 R_h \exp(-j2kH\cos\theta)], \quad (6.13)$$

where r is the distance from the feed point to the far-field point,

$$F_i = \frac{1 - \exp[-jkl(1 - \cos\psi_i)]}{1 - \cos\psi_i}, \qquad i = 1, 2, 3, \text{ and } 4, \quad (6.14)$$

$$\cos\psi_3 = -\cos\theta \sin\alpha' + \sin\theta \cos\alpha' \cos(\varphi - \beta'), \quad (6.15)$$

$$\cos\psi_4 = -\cos\theta \sin\alpha' + \sin\theta \cos\alpha' \cos(\varphi + \beta'), \quad (6.16)$$

and R_h and R_v are, respectively, given in (4.53) and (4.58). The total field components become

$$E_\theta = E_{\theta 1} + E_{\theta 2} = 30 I_m \frac{e^{-jkr}}{r} F_\theta$$

and

$$E_\varphi = E_{\varphi 1} + E_{\varphi 2} = 30 I_m \frac{e^{-jkr}}{r} F_\varphi, \quad (6.17)$$

where

$$F_\theta = \cos\alpha' \cos\theta \{-\cos(\varphi - \beta')[F_1 - F_3 R_v \exp(-j2kH\cos\theta)]$$

$$+ \cos(\varphi + \beta')[F_2 - F_4 R_v \exp(-j2kH\cos\theta)]\}$$

$$+ \sin\alpha' \sin\theta [F_1 - F_2 + (F_3 - F_4) R_v \exp(-j2kH\cos\theta)] \quad (6.18)$$

and

$$F_{\varphi} = \cos\alpha' \{ \sin(\varphi - \beta')[F_1 + F_3 R_h \exp(-j2kH\cos\theta)] - \sin(\varphi + \beta')[F_2 + F_4 R_h \exp(-j2kH\cos\theta)] \}. \quad (6.19)$$

The expression for the power gain will then be

$$G(\theta,\varphi) = \frac{30[|F_\theta|^2 + |F_\varphi|^2]}{R_{\text{in}}}, \quad (6.20)$$

where R_{in} is the real part of (6.1) or is assumed to be 600 ohms according to the general practice.[3] Note that the free-space approximation is implied if the following simple relation is used:

$$R_{\text{in}} = 120\left(\ln\frac{\lambda}{2\pi a} + \ln\sin\gamma - 0.60\right). \quad (6.21)$$

In all of the numerical examples presented later, we will use (6.21) because an expression for the mutual impedance between the vee antenna and its ground image has yet to be derived.

Before we present numerical results for the antenna considered in this section, we note that the expressions $E_{\theta 2}$ and $E_{\varphi 2}$ in (6.12) and (6.13) with $\beta' = 0$ should represent the fields radiated by a traveling-wave sloping long-wire antenna.[4] Further, if the free-space ($R_v = R_h = 0$) pattern of the wire in the forward direction ($\varphi = 0$) is desired (as is generally considered important), the field expression simplifies to $\cos(\alpha' + \theta)F_2$, yielding a series of conical lobes around the antenna wire. The lobe nearest to the antenna wire is the largest in field strength (the main lobe) and all others progressively diminish in strength (sidelobes). For this simple antenna, the level of the largest sidelobe is only about 5.0–6.0 dB below that of the main lobe, depending on the wire length l. This is why the usefulness of the long-wire antenna is very restricted. An example of free-space field pattern for a horizontal traveling-wave long-wire antenna ($\alpha' = \beta' = 0$) in the forward direction ($\varphi = 0$) is shown in Fig. 6.2.

The goal of designing an elevated sloping vee antenna is, then, to select the wire length (l), the antenna heights (H and H'), and the apex angle (2γ) so that the maxima of the main lobe of each wire and their reflected parts from the ground will reinforce each other in a desired direction. The sidelobes of the individual wires can only be combined arbitrarily. In this manner, the final ratio of the levels of the main lobe to sidelobes may be improved.

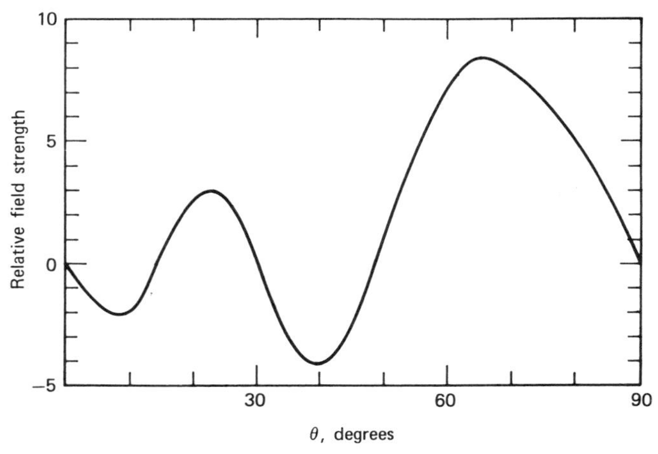

Fig. 6.2 Free-space field pattern for a horizontal traveling-wave long-wire antenna ($\alpha'=\beta'=0$) in the forward direction ($\varphi=0$) with $l=4\lambda$.

Numerical results for a typical sloping vee antenna in the forward direction ($\varphi=0$) are given in Figs. 6.3(a) and 6.3(b), where $H=15$ m, $H'=33.75$ m, $a=0.0016$ m, $\gamma=15°$, $f=10$ MHz, $\epsilon_r=4$, and $\sigma=0.001$ mho/m (poor ground). We choose poor ground in this example because it is closer to the condition assumed in (6.21). The corresponding results for a ground with higher conductivity such as sea water should be approximately 1–2 dB better. Note that, with respect to the frequency of 10 MHz used there, the power gain changes only slightly from $l=150$ m (5λ) to 225 m (7.5λ). This reveals that for a fixed leg length, say $l=150$ m, the performance of a sloping vee antenna in free space should also remain practically unchanged within the frequency range between $f=10$ and 15 MHz. This is the reason we claimed at the beginning of this chapter that this type of antenna can operate within "a relatively wide bandwidth." Of course, when the ground effect is also included, the final characteristics in terms of sidelobe level may change considerably, depending on the type of ground involved. This change is mainly because of the height parameters H and H'. For the values of l in Figs. 6.3(a) and 6.3(b), the position of the maximum gain remains approximately around 76°–78° and that of the first sidelobe is around 41°–51°. These results can be compared with those obtained elsewhere.[3,5]

Details of changes in levels of the maximum gain and the first sidelobe

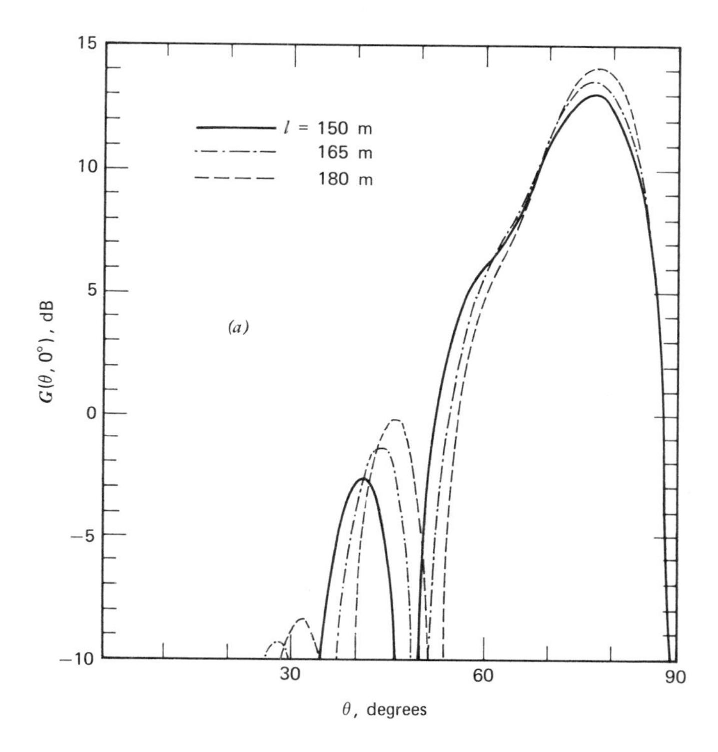

Fig. 6.3 (*a*) Power gain (in the vertical plane) of an elevated sloping vee antenna above poor ground with $H = 15$ m, $H' = 33.75$ m, $a = 0.0016$ m, $\gamma = 15°$, $f = 10$ MHz, and $l = 150$ m, 165 m, and 180 m.

with γ, H, or H' are, respectively, shown in Figs. 6.4, 6.5, and 6.6. From Fig. 6.4 we see clearly that, with a given set of H, H', a, f, and ground constants, the maximum of $G(\theta, 0)_{\max}$ can be realized with $\gamma = 20°$, $18.5°$, and $16.5°$, respectively, for $l = 150$ m (5λ), 180 m (6λ), and 210 m (7λ). Actually, these optimum values for γ can also be derived theoretically. To simplify the mathematics involved, let us consider the field expressions in free space. From (6.18) and (6.19) with $R_v = R_h = 0$, we obtain

$$F_\theta^0 = \cos\alpha' \cos\theta \left[-F_1 \cos(\varphi - \beta') + F_2 \cos(\varphi + \beta') \right] + \sin\alpha' \sin\theta (F_1 - F_2) \tag{6.22}$$

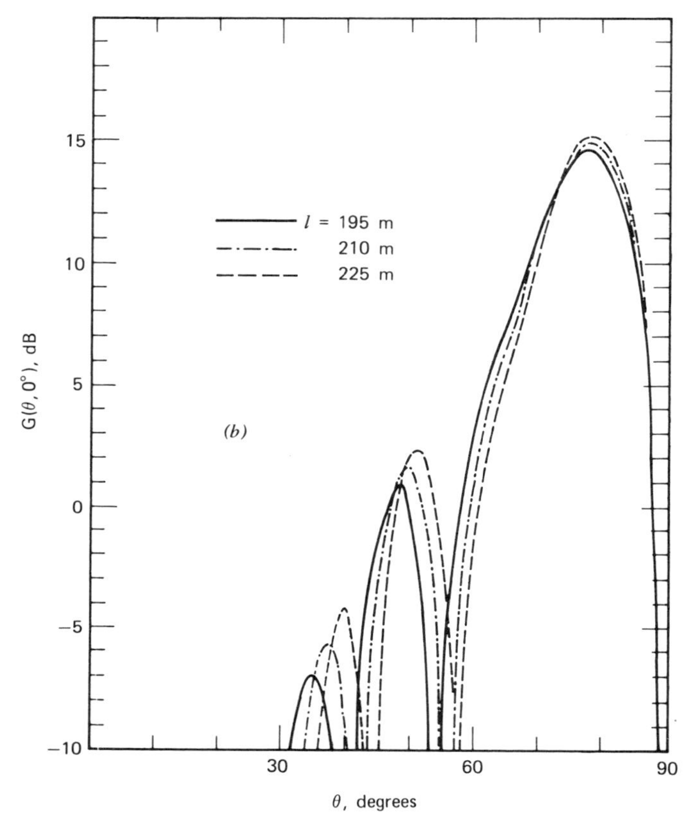

Fig. 6.3 (*b*) Continuation of Fig. 6.3(*a*) with $l = 195$ m, 210 m, and 225 m.

and

$$F_{\varphi}^{0} = \cos\alpha'[F_1 \sin(\varphi - \beta') - F_2 \sin(\varphi + \beta')], \tag{6.23}$$

where the superscript 0 denotes the fields in free space. In the forward direction $\varphi = 0°$ and $\theta = 90° - \alpha'$, we have

$$\cos\psi_1 = \cos\psi_2 = \sin^2\alpha' + \cos^2\alpha' \cos\beta', \tag{6.24}$$

$$F_1 = F_2, \qquad F_{\theta}^{0} = 0, \qquad \text{and} \qquad F_{\varphi}^{0} = -2F_1 \cos\alpha' \sin\beta'.$$

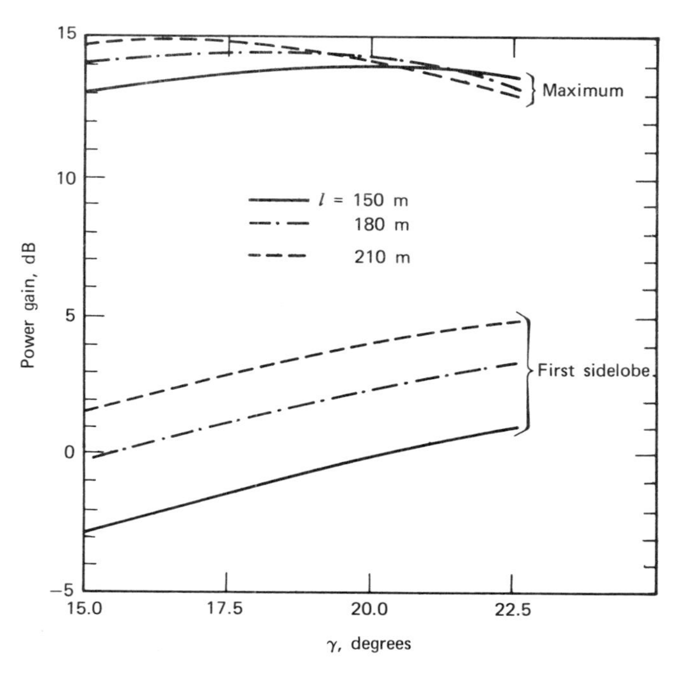

Fig. 6.4 Variation of power gain with γ for the same antenna as in Fig. 6.3.

Thus,

$$|F^0|^2 = |F_\varphi^0|^2 = 4\cos^2\alpha' \sin^2\beta' |F_1|^2$$

$$= 16\cos^2\alpha' \sin^2\beta' \frac{\sin^2[\frac{1}{2}kl(1-\cos\psi_1)]}{(1-\cos\psi_1)^2}. \quad (6.25)$$

For a fixed set of H and H', α' is constant. Setting $d|F^0|^2/d\beta' = 0$ yields

$$\tfrac{1}{2}kl\sin^2\beta' \frac{\cos^2\alpha' - \cos\psi_1}{1-\cos\beta'} = \cos^2\alpha' \tan[\tfrac{1}{2}kl(1-\cos\psi_1)]. \quad (6.26)$$

When $H' - H \ll l$, $\cos\alpha' \cong 1$, $\sin\alpha' \cong 0$, $\psi_1 \cong \beta' \cong \gamma$, Eq. (6.26) simplifies to

$$\tfrac{1}{2}kl\sin^2\gamma = \tan[\tfrac{1}{2}kl(1-\cos\gamma)]. \quad (6.27)$$

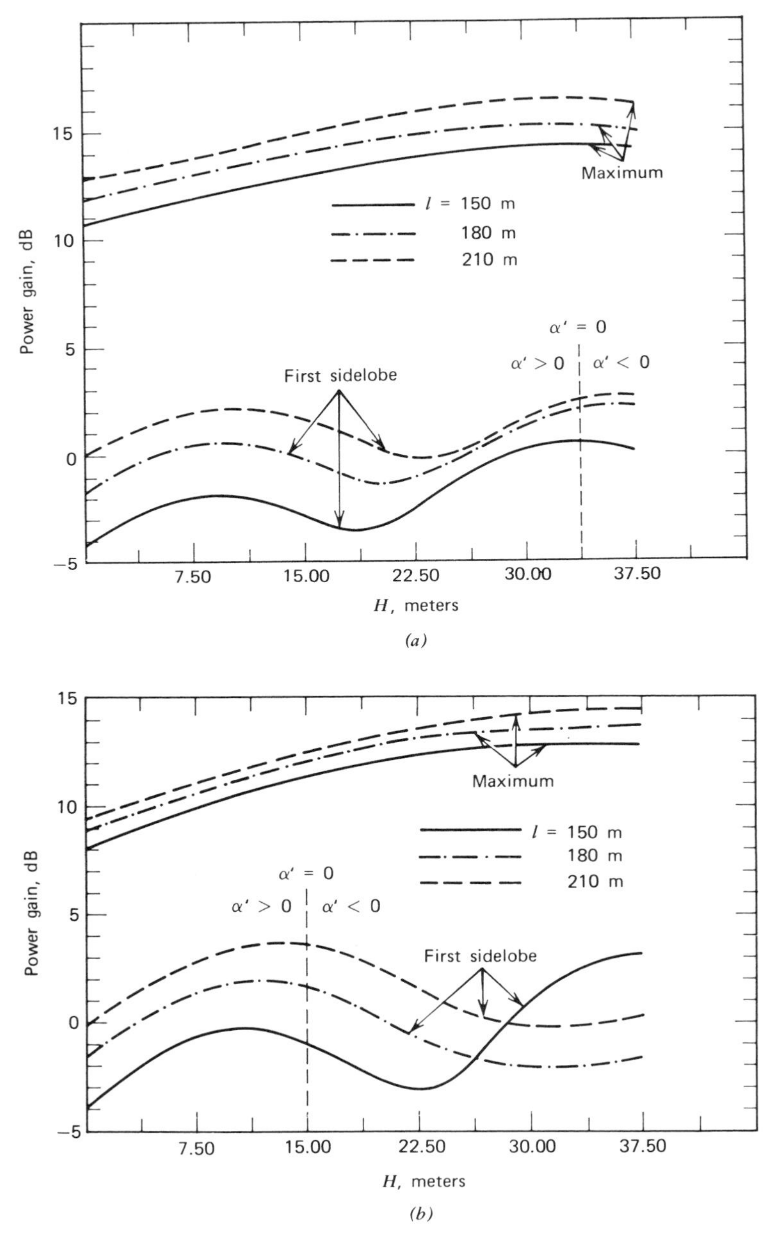

Fig. 6.5 Variation of power gain with H for an elevated sloping vee antenna above poor ground with $a=0.0016$ m, $\gamma=15°$, $\varphi=0$, and $f=10$ MHz: (a) $H'=33.75$ m and (b) $H'=15.00$ m.

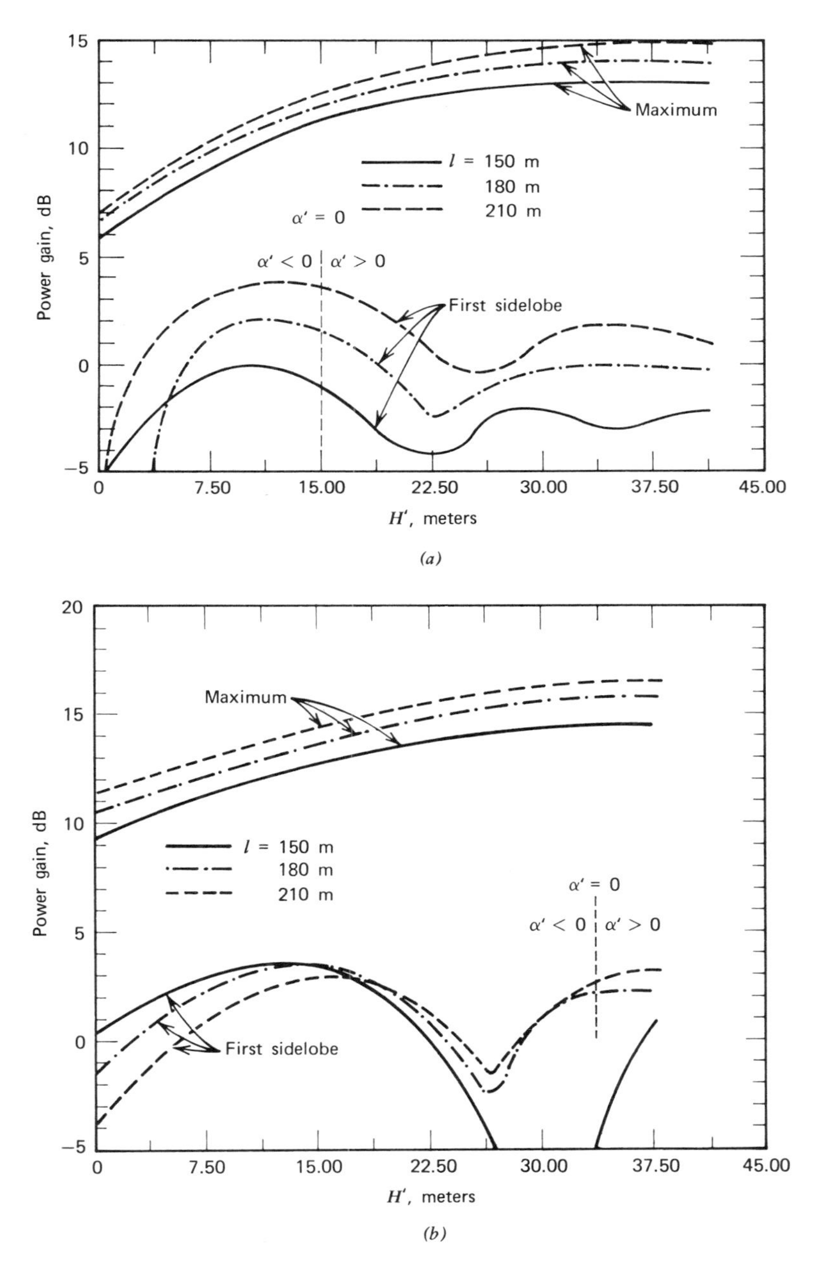

Fig. 6.6 Variation of power gain with H' for an elevated sloping vee antenna above poor ground with $a = 0.0016$ m, $\gamma = 15°$, $\varphi = 0$, $f = 10$ MHz: (*a*) $H = 15.00$ m and (*b*) $H = 33.75$ m.

Solving for γ_{opt}, we obtain

l/λ	γ_{opt}
2	34.6°
3	28.4°
4	24.4°
5	21.9°
6	20.2°
7	18.6°

(6.28)

which are slightly larger than those shown in Fig. 6.4. These deviations are, naturally, due to the ground effect and the positive slope of the vee antenna in Fig. 6.4.

In Fig. 6.5(*a*) where $H'=33.75$ m, the maximum of $G(\theta,0)_{max}$ occurs approximately at $H=33.75$ m (or $\alpha'=0$) for all the three leg lengths considered there. On the other hand, Fig. 6.5(*b*) reveals that, with $H'=15.00$ m, the maximum of $G(\theta,0)_{max}$ is obtainable at $H>15$ m ($\alpha'<0$). Figures 6.6(*a*) and 6.6(*b*) with respective feed heights (H) of 15.00 and 33.75 m show that, for both cases the maximum of $G(\theta,0)_{max}$ occurs when $\alpha'>0$. The information displayed in these figures, as far as the maximum gain is concerned, indicates that, with the range of parameters (H',H,α') considered, the effect of feed and termination heights is more important than the slope angle. The first sidelobe reaches, however, its highest level near $\alpha'=0$ in all of these figures, except in Fig. 6.6(*b*) where the level of the first sidelobe is the highest when $\alpha'<0$.

To complete the presentation, we also give examples for power gains in an azimuthal surface, $G(76°,\varphi)$, in Figs. 6.7(*a*) and 6.7(*b*) with $\alpha'>0$ and $\alpha'<0$, respectively. Corresponding results for $\alpha'=0$ are shown in Fig. 6.8. General shapes of these figures are all similar. Note that the curves in Figs. 6.7 and 6.8 are constructed from data computed for every five degrees in φ.

From the economic point of view, the elevated sloping vee antenna is generally constructed with $H'<H$ (or $\alpha'<0$) so that only one high pole is required for supporting the feed point and two shorter poles are installed for the termination points. Results for $G(\theta,0)$ with $H=33.75$ m, $H'=15.0$ m, $a=0.0016$ m, $\gamma=15°$, $f=10$ MHz, $\epsilon_r=4$, and $\sigma=0.001$ mho/m (poor

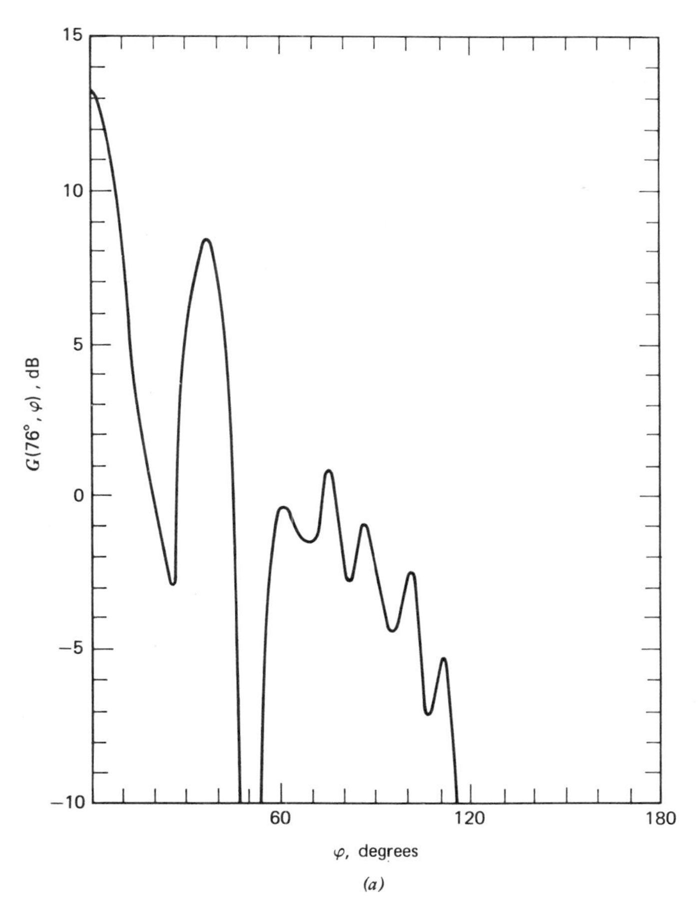

Fig. 6.7 (*a*) Power gain (in an azimuthal surface) of an elevated sloping vee antenna above poor ground with $l = 150$ m, $a = 0.0016$ m, $H = 15$ m, $H' = 33.75$ m ($\alpha' > 0$), $\gamma = 15°$, and $f = 10$ MHz.

ground) are shown in Fig. 6.9, which may be compared with those shown in Fig. 6.3.

Finally, we should also mention that some limited scaled-model measurements of an elevated sloping vee antenna have been performed.[6] Satisfactory agreement between these measured results and the computed results obtained elsewhere[3] has been noted in general.

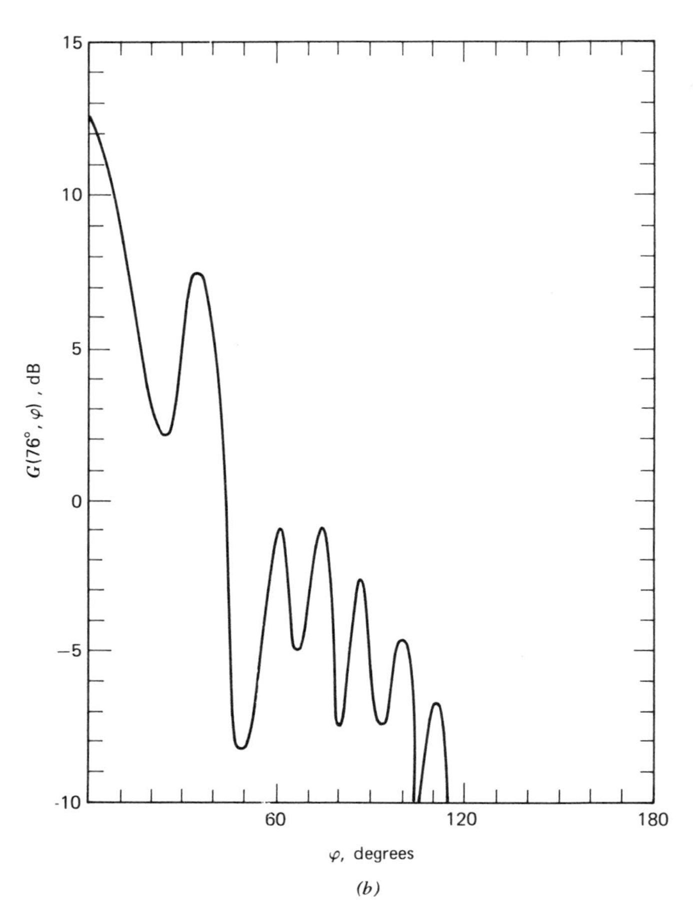

Fig. 6.7 (*b*) Continuation of Fig. 6.7(*a*) but with $H = 33.75$ m and $H' = 15$ m ($\alpha' < 0$).

6.2 Sloping Rhombic Antenna

Since the rhombic was first introduced by Bruce,[7,8] it has been extensively used for short-wave communications. The antenna consists of four straight wires of the same length l arranged in the form of a rhombus (see Fig. 6.10). It can be considered as an extension of the vee antenna studied in the previous section.

If H and H' are designated as the respective heights of the feed point

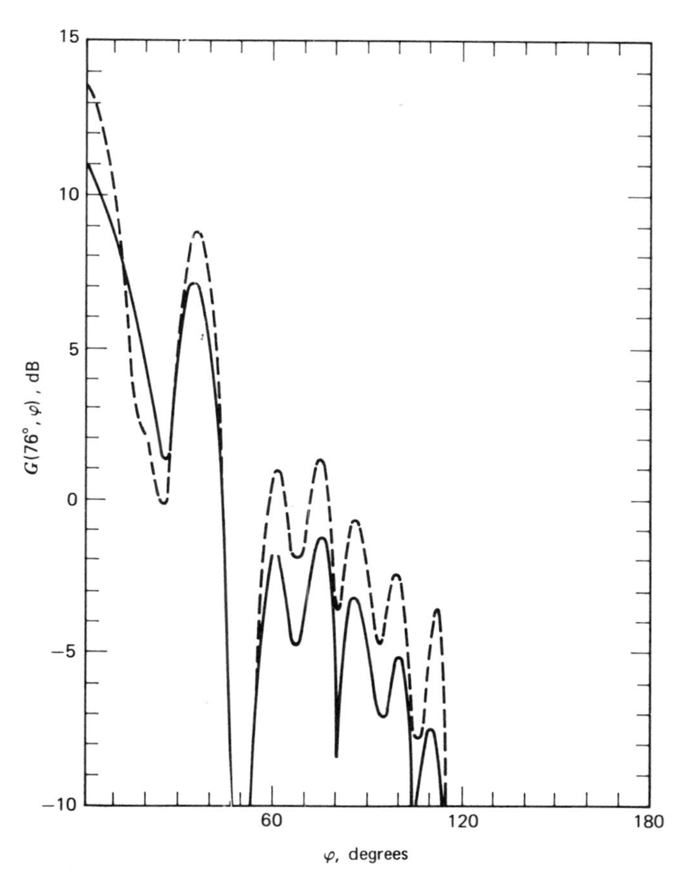

Fig. 6.8 Continuation of Fig. 6.7 with $H = H' = 15$ m ($\alpha' = 0$, solid curve) and $H = H' = 33.75$ m ($\alpha' = 0$, dashed curve).

and termination above ground, the height of the other two vertexes can be expressed as

$$H'' = \tfrac{1}{2}(H + H'). \tag{6.29}$$

The angle between No. 1 wire and the ground plane now becomes

$$\alpha' = \sin^{-1}\left(\frac{H'' - H}{l}\right) = \sin^{-1}\left(\frac{H' - H}{2l}\right). \tag{6.30}$$

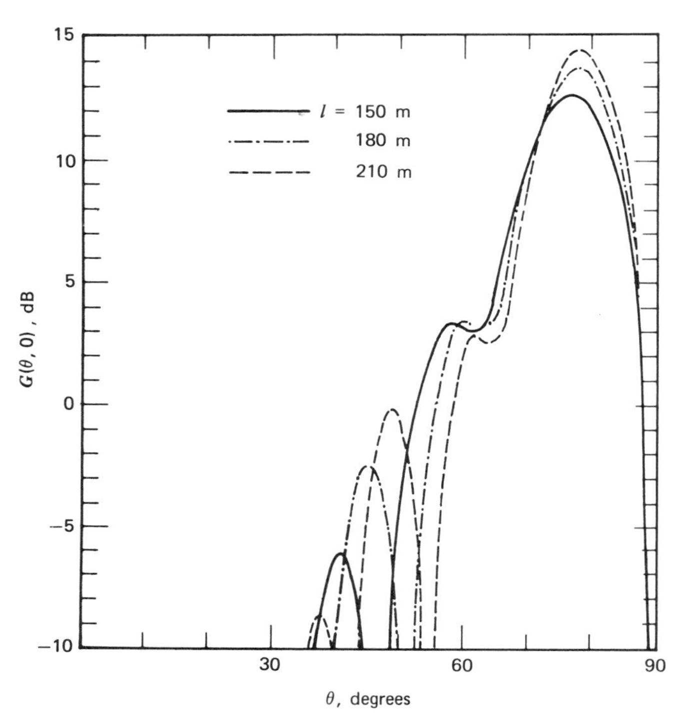

Fig. 6.9 Power gain (in the vertical plane) of an elevated sloping vee antenna above poor ground with $H = 33.75$ m, $H' = 15.00$ m, $a = 0.0016$ m, $\gamma = 15°$, $f = 10$ MHz, and $l = 150$ m, 180 m, and 210 m.

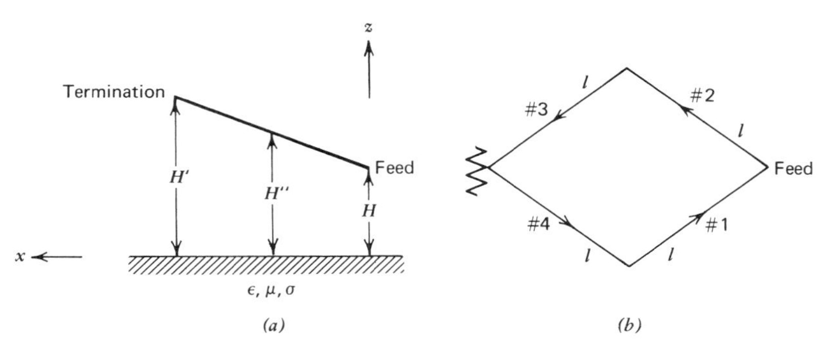

Fig. 6.10 Sloping rhombic antenna above flat lossy ground: (*a*) side view, and (*b*) top view.

The parameters β' and γ defined in the previous section still remain valid here.

Following the same notation used in the previous section, we have the same field expressions for wires No. 1 and No. 2 as those given in (6.10) through (6.13). For wires No. 3 and No. 4, the associated relations are

$$H_s = z = H'' + (\sin\alpha')s, \tag{6.31}$$

$$I_3 = I_m e^{-jkl} e^{-jks} = -I_4, \tag{6.32}$$

$E_{\theta 3} = \theta$ component of the far-field contributed by wire No. 3

$$\begin{aligned} &= 30 I_m e^{-jkl} \frac{e^{-jkr'}}{r'} \{\cos\alpha' \cos\theta \cos(\varphi - \beta')[F_1 - F_3 R_v \exp(-j2kH'' \cos\theta)] \\ &\qquad - \sin\alpha' \sin\theta [F_1 + F_3 R_v \exp(-j2kH'' \cos\theta)]\} \\ &= 30 I_m \frac{e^{-jkr}}{r} \exp[-jkl(1 - \cos\psi_2)] \\ &\quad \times \{\cos\alpha' \cos\theta \cos(\varphi - \beta')[F_1 - F_3 R_v \exp(-j2kH'' \cos\theta)] \\ &\quad - \sin\alpha' \sin\theta [F_1 + F_3 R_v \exp(-j2kH'' \cos\theta)]\}, \end{aligned} \tag{6.33}$$

$E_{\varphi 3} = \varphi$ component of the far-field contributed by wire No. 3

$$\begin{aligned} &= -30 I_m \frac{e^{-jkr}}{r} \exp[-jkl(1 - \cos\psi_2)] \\ &\quad \times \cos\alpha' \sin(\varphi - \beta')[F_1 + F_3 R_h \exp(-j2kH'' \cos\theta)], \end{aligned} \tag{6.34}$$

$E_{\theta 4} = \theta$ component of the far-field contributed by wire No. 4

$$\begin{aligned} &= 30 I_m \frac{e^{-jkr}}{r} \exp[-jkl(1 - \cos\psi_1)] \\ &\quad \times \{-\cos\alpha' \cos\theta \cos(\varphi + \beta')[F_2 - F_4 R_v \exp(-j2kH'' \cos\theta)] \\ &\quad + \sin\alpha' \sin\theta [F_2 + F_4 R_v \exp(-j2kH'' \cos\theta)]\}, \end{aligned} \tag{6.35}$$

and

$E_{\varphi 4} = \varphi$ component of the far-field contributed by wire No. 4

$$\begin{aligned} &= 30 I_m \frac{e^{-jkr}}{r} \exp[-jkl(1 - \cos\psi_1)] \\ &\quad \times \cos\alpha' \sin(\varphi + \beta')[F_2 + F_4 R_h \exp(-j2kH'' \cos\theta)], \end{aligned} \tag{6.36}$$

where $\cos\psi_1$ and $\cos\psi_2$ are, respectively, given in (6.5) and (6.6) and F_i can be found in (6.14). Note that the distance from the vertex at the connection of wires No. 2 and No. 3 to the far-field point, r', has been replaced by the usual approximation $r' = r - l\cos\psi_2$ in the phase term in (6.33) and (6.34). A similar relation, $r'' = r - l\cos\psi_1$, has been used in (6.35) and (6.36).

The total field components will then be

$$E_\theta = E_{\theta 1} + E_{\theta 2} + E_{\theta 3} + E_{\theta 4}$$

$$= 30 I_m \frac{e^{-jkr}}{r} F'_\theta$$

and

$$E_\varphi = E_{\varphi 1} + E_{\varphi 2} + E_{\varphi 3} + E_{\varphi 4}$$

$$= 30 I_m \frac{e^{-jkr}}{r} F'_\varphi, \tag{6.37}$$

where

$$\begin{aligned} F'_\theta = F_\theta &+ \exp[jkl(1-\cos\psi_2)]\{F_1[\cos\alpha'\cos\theta\cos(\varphi-\beta') - \sin\alpha'\sin\theta] \\ &- F_3 R_v \exp(-j2kH''\cos\theta)[\cos\alpha'\cos\theta\cos(\varphi-\beta') + \sin\alpha'\sin\theta]\} \\ &+ \exp[-jkl(1-\cos\psi_1)]\{F_2[-\cos\alpha'\cos\theta\cos(\varphi+\beta') + \sin\alpha'\sin\theta] \\ &+ F_4 R_v \exp(-j2kH''\cos\theta)[\cos\alpha'\cos\theta\cos(\varphi+\beta') + \sin\alpha''\sin\theta]\} \end{aligned} \tag{6.38}$$

and

$$\begin{aligned} F'_\varphi = F_\varphi &+ \exp[-jkl(1-\cos\psi_1)]\cos\alpha'\sin(\varphi+\beta') \\ &\times [F_2 + F_4 R_h \exp(-j2kH''\cos\theta)] \\ &- \exp[-jkl(1-\cos\psi_2)]\cos\alpha'\sin(\varphi-\beta') \\ &\times [F_1 + F_3 R_h \exp(-j2kH''\cos\theta)], \end{aligned} \tag{6.39}$$

with F_θ and F_φ given, respectively, in (6.18) and (6.19).

The power gain of the sloping rhombic thus becomes

$$G(\theta,\varphi) = \frac{30[|F'_\theta|^2 + |F'_\varphi|^2]}{R_{\text{in}}}. \tag{6.40}$$

According to the general practice,[1] the expression given in (6.21) is still a good approximation for the input resistance R_{in} in (6.40). Numerical results presented later are actually obtained on this basis. The termination resistance for $f=10$ MHz and $\gamma=20°$ is then approximately 700 ohms to ensure a substantial traveling wave for the currents in antenna wires.

Assuming $R_v=R_h=0$ (free space) and $\alpha'\cong 0$ ($H'-H \ll 2l$), we also can derive an expression similar to (6.27) for the optimum apex angle yielding maximum power gain in the forward direction ($\varphi=0°$ and $\theta=90°$). Under these conditions, we have

$$F_\theta'^0=0 \qquad \text{and} \qquad F_\varphi'^0=-2\{1-\exp[-jkl(1-\cos\gamma)]\}F_1\sin\gamma.$$

Thus,

$$|F'^0|^2=|F_\varphi'^0|^2=64\frac{\sin^2\gamma\sin^4[\frac{1}{2}kl(1-\cos\gamma)]}{(1-\cos\gamma)^2}. \tag{6.41}$$

Equating the derivative of $|F'^0|^2$ with respect to γ to zero, we find

$$kl\sin^2\gamma=\tan[\tfrac{1}{2}kl(1-\cos\gamma)], \tag{6.42}$$

yielding

l/λ	γ_{opt}
2	38.5°
3	31.4°
4	27.1°
5	24.2°
6	22.1°
7	20.5°

(6.43)

which may be compared with those obtained in (6.28) for the corresponding vee antenna.

A set of numerical results for a typical sloping rhombic antenna above sea water is presented in Figs. 6.11 through 6.15. We choose sea water as the ground here so that we can compare our results with those obtained elsewhere for a horizontal rhombic above perfect ground.[4,9,10] Figure 6.11 gives power gain in the vertical plane ($\varphi=0°$) with $H=10$ m, $H'=20$ m ($\alpha'>0$), $\gamma=20°$, $a=0.0016$ m, and $f=10$ MHz for $l=90$ m (3λ) and 120 m

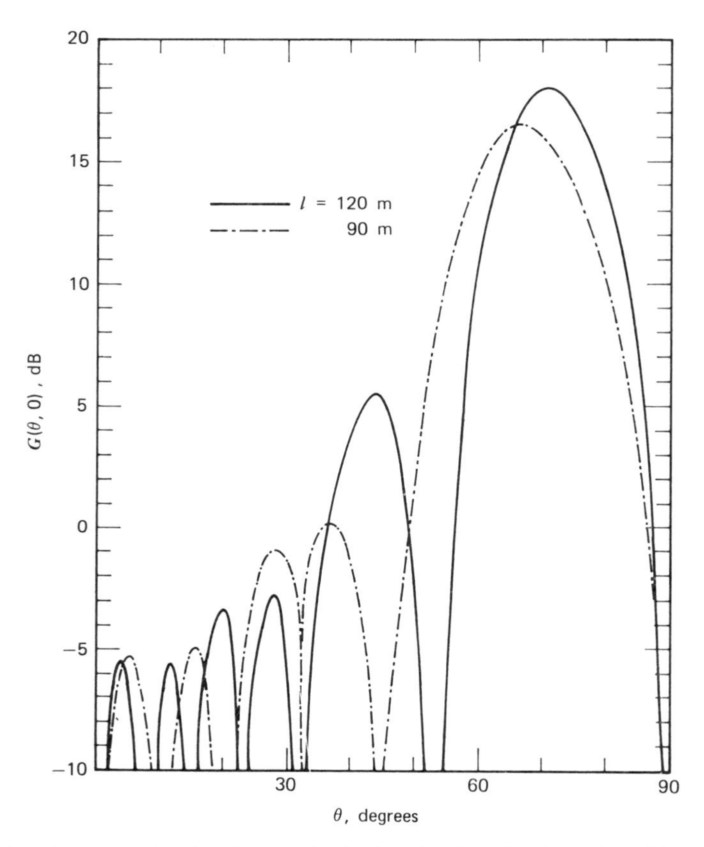

Fig. 6.11 Power gain (in the vertical plane) of a sloping rhombic antenna above sea water with $H=10$ m, $H'=20$ m, $a=0.0016$ m, $\gamma=20°$, $l=90$ m and 120 m, and $f=10$ MHz.

(4λ). It is also clear that both the maximum power gain and the first sidelobe, in level and position, depend on l and ground constants. As generally recognized by practical engineers,[11] the level of the first sidelobe is always high. Variations of the maximum power gain with respect to γ for $H=10$ m, $H'=20$ m, $l=90$ and 120 m, and $f=10$ MHz are shown in Fig. 6.12. Evidently, the maximum of $G(\theta, 0°)_{\max}$ occurs, respectively, at $\gamma=26.25°$ and $\gamma=22.50°$ for $l=90$ m (3λ) and $l=120$ m (4λ). These optimum values are smaller than the corresponding values obtained in (6.43) for a horizontal rhombic in free space.

The influence of H on levels of the maximum power gain and the first

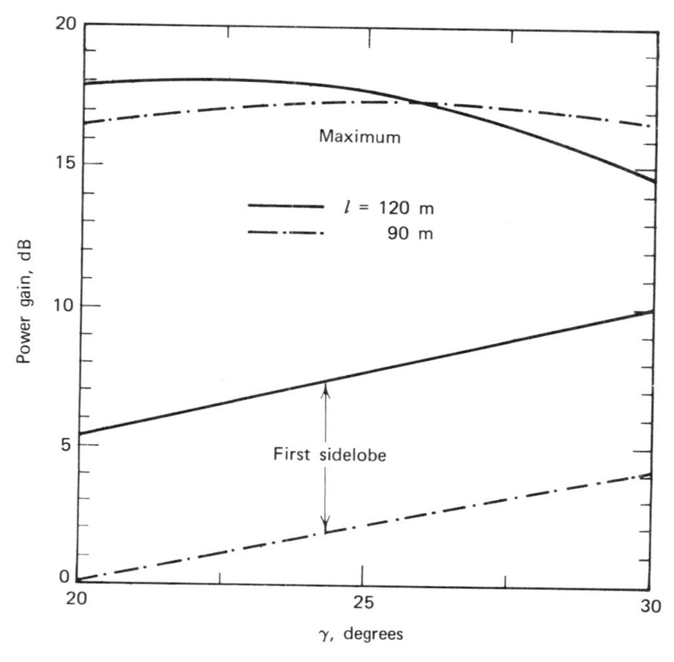

Fig. 6.12 Variation of power gain with γ for the same rhombic antennas as in Fig. 6.11.

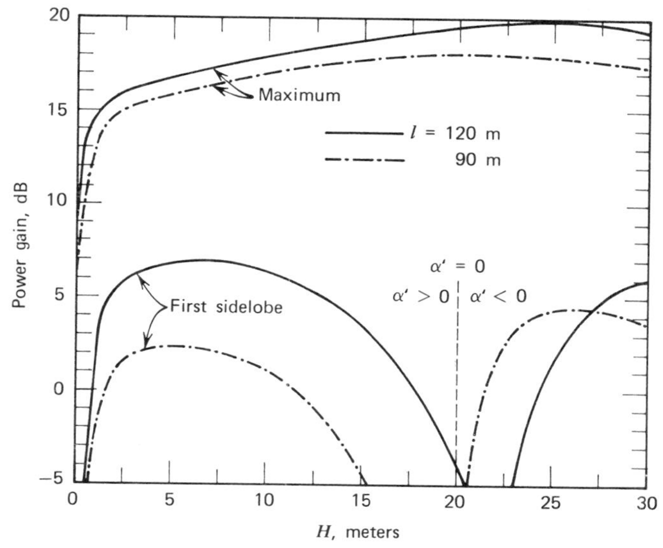

Fig. 6.13 Variation of power gain with H for a sloping rhombic antenna above sea water with $H'=20$ m, $a=0.0016$ m, $\gamma=22.5°$, and $f=10$ Mchz.

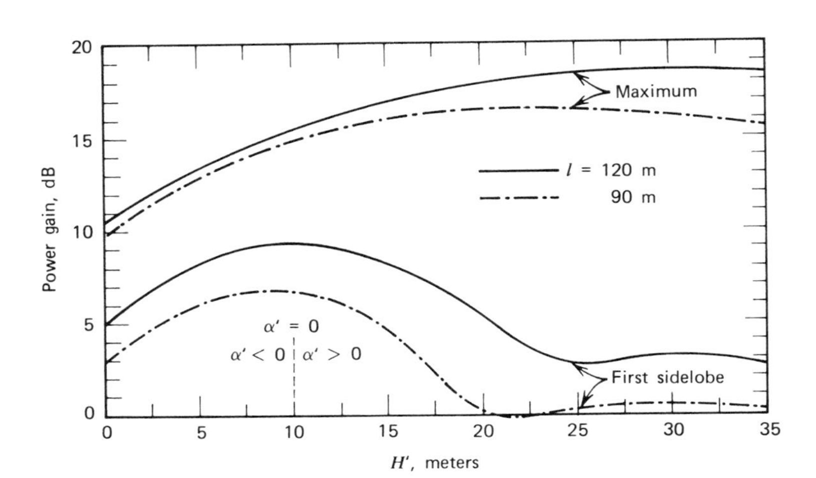

Fig. 6.14 Variation of power gain with H' for the same antenna as in Fig. 6.11.

sidelobe with $H'=20$ m, $\gamma=22.5°$, $\varphi=0°$, $f=10$ MHz, and sea water as the ground is presented in Fig. 6.13. For this particular geometry and arrangement, it is apparent that $H \cong H'=20$ m ($\alpha' \cong 0$) should be a good choice for both the maximum power gain and the first sidelobe.

On the other hand, when the parameter H is fixed at 10 m, $H'=H=10$ m ($\alpha'=0$) will yield the worst first-sidelobe level for the rhombic with $\gamma=20°$ and $f=10$ MHz above sea water, as clearly shown in Fig. 6.14. From the same figure, we also see that $H'=22.5$ m and $H'=30.0$ m should be chosen as the respective termination heights for $l=90$ m and $l=120$ m to produce best results for the maximum power gain and the first sidelobe.

Finally, power gain in an azimuthal surface such as $G(70°,\varphi)$ is calculated for every five degrees in φ with $H=10$ m, $H'=20$ m ($\alpha'>0$), $a=0.0016$ m, $l=120$ m, $\gamma=20°$, $f=10$ MHz, and sea water. The result of these calculations is given in Fig. 6.15, which is very similar to that shown in Fig. 6.7(a) for a sloping vee antenna. Since the parameter α' produces little effect on $G(70°,\varphi)$, corresponding results with $\alpha'=0$ and $\alpha'<0$ are omitted here.

Fair agreement between computed and measured results for a single horizontal rhombic above ground has been noted in some scaled-model measurements,[12] although the agreement in power gains has not been as close as that in patterns. This is, of course, expected in view of the

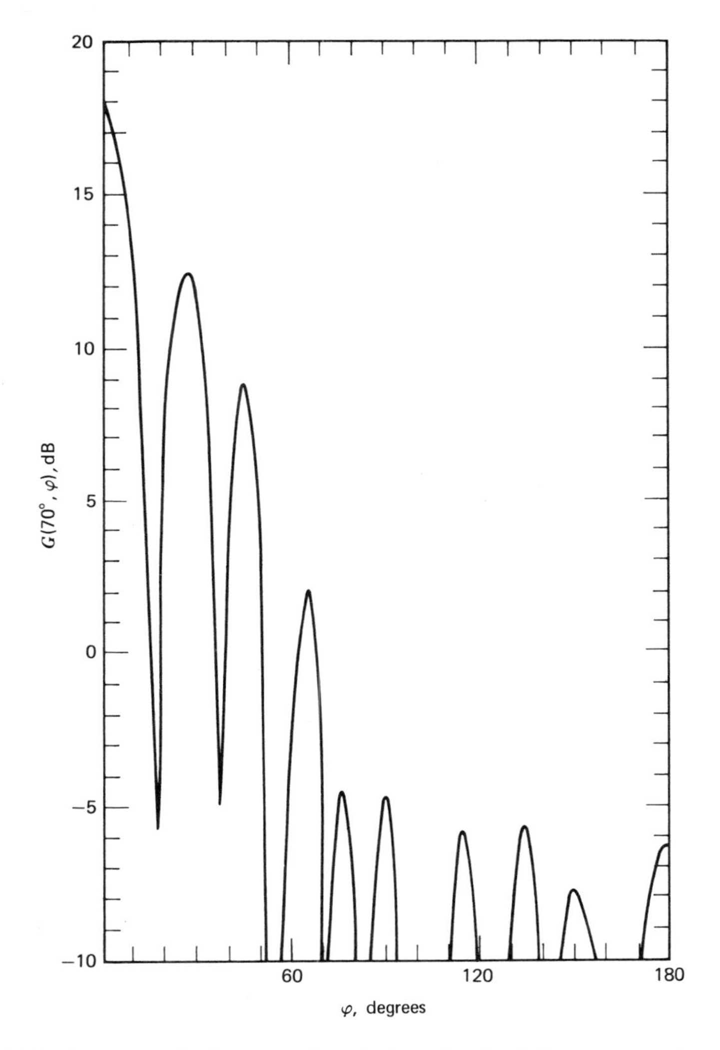

Fig. 6.15 Power gain (in an azimuthal surface) of the same antenna as in Fig. 6.11 with $l = 120$ m.

approximation for R_{in} used in (6.40) and the assumption of pure traveling waves for currents in our theoretical model. The computed power gains are, on the average, 1 or 2 dB higher than the measured ones.

Before concluding this section, we should note that an approximate expression for the radiation resistance of a single rhombic antenna in free space is also available[13,14]:

$$R_{rad} = 240(\ln 2kl \sin^2\gamma + 0.577) \text{ ohms}, \qquad l \geqslant 2\lambda, \tag{6.44}$$

which may be used to replace R_{in} in (6.40). The result is then, according to the definition in (1.7), directive gain. The ratio R_{rad}/R_{in} should represent radiation efficiency of the rhombic. A set of R_{rad} is presented as follows: for $\gamma = 20°$,

l/λ	R_{rad} (ohms)
2	397.27
3	494.58
4	563.64
5	617.20
6	660.94
7	697.96

(6.45)

and for $\gamma = 15°$,

l/λ	R_{rad} (ohms)
2	263.47
3	360.78
4	430.53
5	483.36
6	527.13
7	564.12

(6.46)

which may be compared, respectively, with the values for R_{in} given in (6.2) and (6.3) to gain some feeling about the radiation efficiency. However, we must bear in mind that a "strict" comparison of R_{rad} and R_{in} is not too meaningful, since (6.21) and (6.44) were derived by different approaches and are both approximate in nature.

6.3 Side-Terminated Vertical Half-Rhombic Antenna

This antenna is constructed from two long wires in series, forming an inverted vee antenna,[4] as shown in Fig. 6.16. The antenna, together with its ground image, constitutes a rhombic antenna in the vertical plane. If the termination is properly matched, the currents in both wires are again substantially of the traveling-wave type.

With a treatment similar to the previous two sections, we have in this case for wire No. 1,

$$\theta' = 90° - \alpha', \qquad \varphi' = 0°, \qquad H_{s1} = z = (\sin\alpha')s, \tag{6.47}$$

$$\cos\psi_1 = \cos\theta\sin\alpha' + \sin\theta\cos\alpha'\cos\varphi, \tag{6.48}$$

and for wire No. 2,

$$\theta' = 90° + \alpha', \qquad \varphi' = 0°, \qquad H_{s2} = (l-s)\sin\alpha', \tag{6.49}$$

$$\cos\psi_2 = -\cos\theta\sin\alpha' + \sin\theta\cos\alpha'\cos\varphi. \tag{6.50}$$

It is also clear that

$$\begin{aligned} I_1(s) &= I_m e^{-jks}, \\ I_2(s) &= I_m e^{-jkl} e^{-jks}, \end{aligned} \tag{6.51}$$

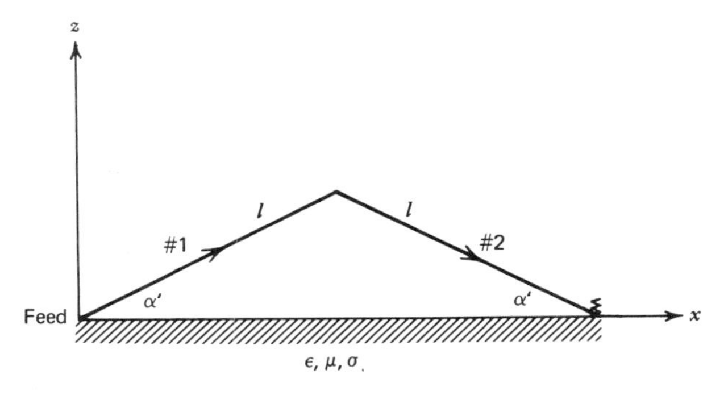

Fig. 6.16 Side-terminated vertical half-rhombic antenna above flat lossy ground.

and

$$\gamma = \alpha'. \tag{6.52}$$

The field components contributed by the individual wires are then

$$E_{\theta 1} = 30 I_m \frac{e^{-jkr}}{r} [\cos\alpha' \cos\theta \cos\varphi (F_1 - R_v F_2)$$

$$- \sin\alpha' \sin\theta (F_1 + R_v F_2)], \tag{6.53}$$

$$E_{\theta 2} = 30 I_m \frac{e^{-jkr}}{r} \exp[-jkl(1 - \cos\psi_1)]$$

$$\times \{\cos\alpha' \cos\theta \cos\varphi [F_2 - R_v F_1 \exp(-j2kl \sin\alpha' \cos\theta)]$$

$$+ \sin\alpha' \sin\theta [F_2 + R_v F_1 \exp(-j2kl \sin\alpha' \cos\theta)]\}, \tag{6.54}$$

$$E_{\varphi 1} = -30 I_m \frac{e^{-jkr}}{r} \cos\alpha' \sin\varphi (F_1 + R_h F_2), \tag{6.55}$$

and

$$E_{\varphi 2} = -30 I_m \frac{e^{-jkr}}{r} \exp[-jkl(1 - \cos\psi_1)]$$

$$\times \cos\alpha' \sin\varphi [F_2 + R_h F_1 \exp(-2kl \sin\alpha' \cos\theta)], \tag{6.56}$$

where

$$F_i = \frac{1 - \exp[-jkl(1 - \cos\psi_i)]}{1 - \cos\psi_i}, \qquad i = 1 \text{ and } 2, \tag{6.57}$$

with $\cos\psi_1$ and $\cos\psi_2$ given, respectively, in (6.48) and (6.50). Combining appropriate expressions, we obtain

$$E_\theta = E_{\theta 1} + E_{\theta 2} = 30 I_m \frac{e^{-jkr}}{r} F''_\theta \tag{6.58}$$

and

$$E_\varphi = E_{\varphi 1} + E_{\varphi 2} = -30 I_m \frac{e^{-jkr}}{r} F''_\varphi, \tag{6.59}$$

where

$$F''_\theta = \cos\alpha' \cos\theta \cos\varphi \big(F_1 + F_2 \exp[-jkl(1-\cos\psi_1)]$$

$$- R_v \{ F_2 + F_1 \exp[-jkl(1-\cos\psi_2)] \} \big)$$

$$- \sin\alpha' \sin\theta \big(F_1 - F_2 \exp[-jkl(1-\cos\psi_1)]$$

$$+ R_v \{ F_2 - F_1 \exp[-jkl(1-\cos\psi_2)] \} \big), \tag{6.60}$$

$$F''_\varphi = \cos\alpha' \sin\varphi \big(F_1 + F_2 \exp[-jkl(1-\cos\psi_1)]$$

$$+ R_h \{ F_2 + F_1 \exp[-jkl(1-\cos\psi_2)] \} \big). \tag{6.61}$$

The power gain of the side-terminated vertical half-rhombic antenna is then

$$G(\theta,\varphi) = \frac{30\big[|F''_\theta|^2 + |F''_\varphi|^2\big]}{R''_{\text{in}}}, \tag{6.62}$$

where

$$R''_{\text{in}} = \tfrac{1}{2} R_{\text{in}}, \tag{6.63}$$

with R_{in} given in (6.21). Note that (6.63) is only approximately true if the ground involved is finitely conducting. It becomes exact when the ground is perfect ($R_v = 1$ and $R_h = -1$). Under this latter condition, the vertical half-rhombic antenna should behave much as a corresponding rhombic in free space. Thus, optimum values for γ given in (6.43) are still valid for a vertical half-rhombic above perfect ground.

A set of numerical results for a typical vertical half-rhombic is now presented. In Fig. 6.17, $G(\theta,0)$ is given with $l = 120$ m, $a = 0.0016$ m, $\alpha' = 27.5°$, $f = 10$ MHz, and sea water as the ground. While the general characteristics, such as the realizable level of the maximum power gain and relatively high levels of the sidelobes, are essentially similar to those associated with the other two antennas already discussed in this chapter, the position of the maximum power gain now occurs at $\theta = 88°$, much closer to the ground ($\theta = 90°$) than that with the sloping vee or rhombic antenna. This rather striking feature is a direct result of the antenna being

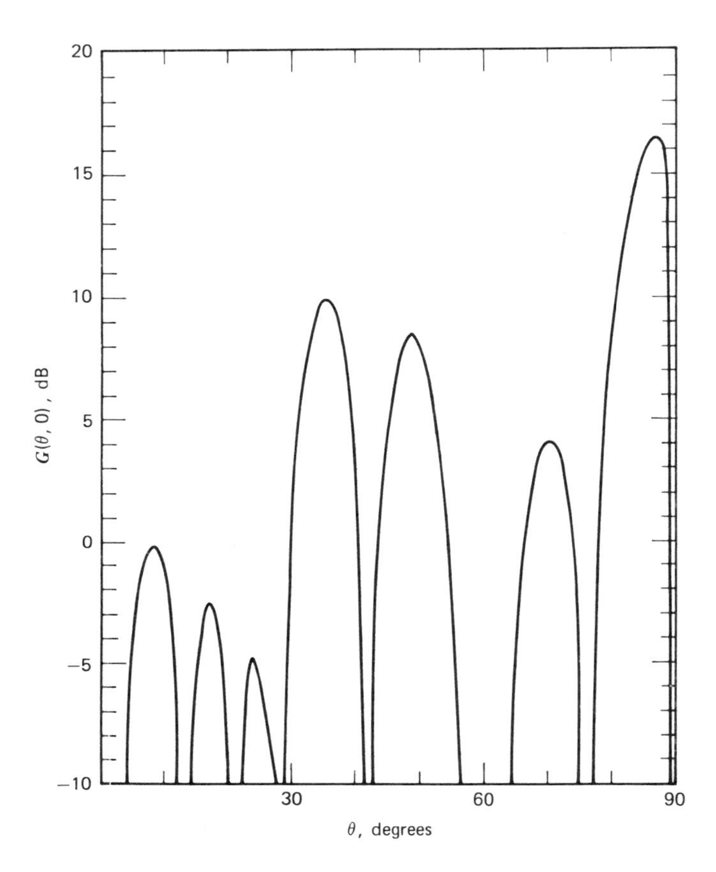

Fig. 6.17 Power gain (in the vertical plane) of a side-terminated half-rhombic above sea water with $l = 120$ m, $a = 0.0016$ m, $\alpha' = 27.5°$, and $f = 10$ MHz.

vertical. Figure 6.18 shows the variation of power gain with α' for the same antenna in Fig. 6.17. It is clear that the highest level of the maximum power gain occurs approximately at $\alpha' = 26.6°$, which is very close to the theoretical value predicted in (6.43). Finally, corresponding power gains in azimuthal surfaces identified by $\theta = 88°$ and $70°$ are presented in Fig. 6.19, which do not differ much from those shown in Figs. 6.7 and 6.8 for the elevated sloping vee antenna and those shown in Fig. 6.15 for the sloping rhombic antenna.

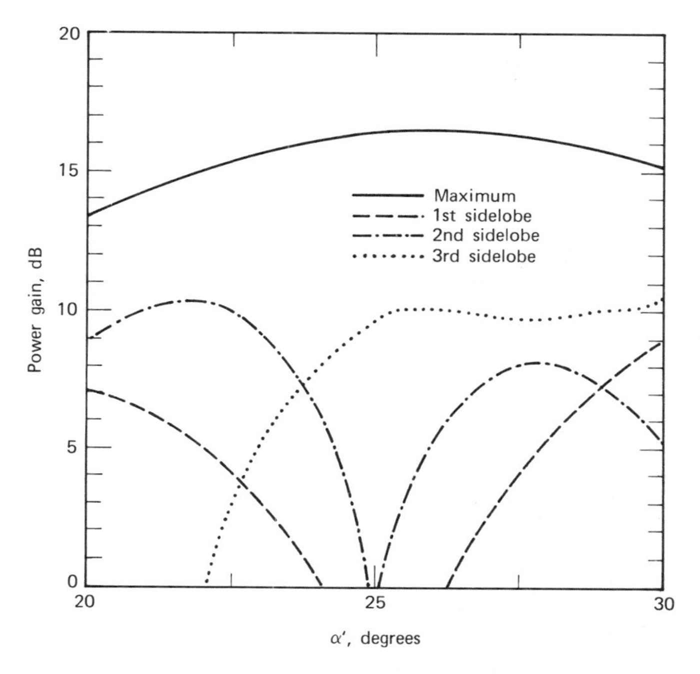

Fig. 6.18 Variation of power gain with α' for the same antenna described in Fig. 6.17.

6.4 Concluding Remarks

In this chapter, we have analyzed characteristics of elevated sloping vee, rhombic, and vertical half-rhombic antennas, above lossy ground. Generally speaking, higher gains can be achieved with relatively simpler structure by these antennas than by the ones discussed in Chapters 4 and 5. Antenna wires and terminations of the three antennas analyzed here are assumed, respectively, lossless and perfectly matched, so that the currents flowing in the antenna wires take the form of a traveling wave. Of course, in reality, none of these assumptions is correct. A way of estimating more accurately the input impedance of a traveling-wave antenna is perhaps by actual measurement. When the attenuation along antenna wires is known, a suitable form for the current such as $I(s) = \pm I_m e^{-\alpha s} e^{-jks}$, with α representing the attenuation constant, may be used to replace that assumed in (6.4).

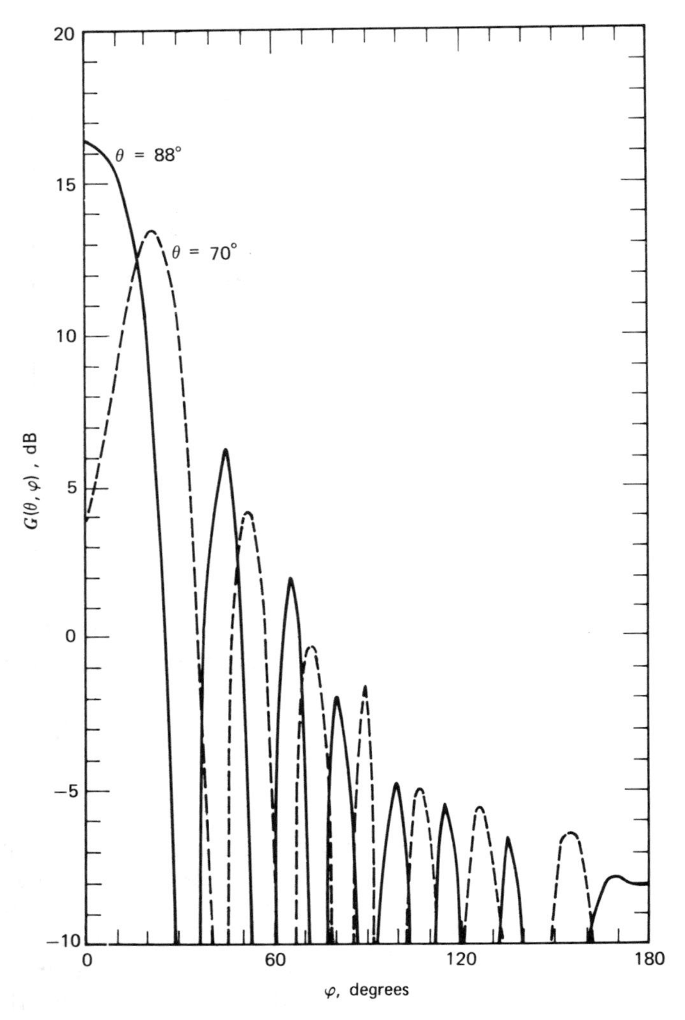

Fig. 6.19 Power gain (in an azimuthal surface) of the same antenna described in Fig. 6.17; solid and dashed curves are for $\theta = 88°$ and $\theta = 70°$, respectively.

In this case, the resultant analysis should not be much more complicated than that outlined in this chapter. Corresponding numerical results may then be one or a fraction of a decibel lower than those presented, depending on the actual attenuation and wire lengths.

Due to lack of existing formulation for calculating the mutual impedance between traveling-wave antennas, only the analysis for a single antenna above ground is given. Characteristics of arrays consisting of identical traveling-wave atennas can be analyzed through those of a single antenna (as presented) and those of the array factor studied in Chapter 1. For example, referring back to Fig. 6.7(a), where power gain at $\theta=76°$ of an elevated sloping vee antenna was presented, we find that the level of the sidelobe at $\varphi=35°$ is unusually high for most of the practical applications. In order to reduce it, we certainly can synthesize a uniform broadside array along the y axis such that the array factor will have a null at or near $\varphi=35°$. According to (2.10) and (2.12), the power pattern associated with a uniformly excited three-element array should take the form of $P_0(y)=(y+1)^2$, where $y=2\cos u$, $u=kd\sin\theta\sin\varphi$ $k=2\pi/\lambda$, and d is the spacing between consecutive elements. Obviously, the only null due to the array factor occurs at $y=-1$ or $u=120°$. If we wish to place this null at $\varphi=35°$ and $\theta=76°$, we require $d=0.5989\lambda$, or 17.9674 m at $f=10$ MHz, which is the actual frequency used in Fig. 6.7(a). This means that an array of three identical elevated sloping vee antennas along the y axis with an approximate element spacing of 18 m will reduce the level of the concerned sidelobe very substantially. With a similar consideration, we can also synthesize arrays to reduce the level of a sidelobe in Fig. 6.3, where power gains in the vertical plane ($\varphi=0$) of a sloping vee antenna were presented. Of course, the same principle applies equally well to the sloping rhombic and the vertical half-rhombic, although we choose the sloping vee antenna for the purpose of illustration. In addition, interlaced arrays or other combinations of rhombic antennas have also been investigated for enhancing the maximum power gain.[4,15] In any case, synthesis techniques presented in Chapter 2 should be useful for approximately meeting specific criteria.

REFERENCES

1. **Schelkunoff, S. A. and H. T. Friis.** *Antennas—Theory and Practice*, John Wiley and Sons, Inc., New York, 1952.

2. **Schelkunoff, S. A.** *Electromagnetic Waves*, D. Van Nostrand Co., New York, 1943.

3. **Perlman, S.** Extended spectrum response with sloping vee antenna, *IEEE International Convention Record*, Vol. 13, part 5, pp. 121–127, 1965.

4. **Laport, E. A.** Long-wire antennas, Chap. 4 in *Antenna Engineering Handbook*, edited by H. Jasik, McGraw-Hill Book Co., New York, 1961.

5. **Harrison, C. W., Jr.** Radiation from vee antennas, *Proc. IRE*, Vol. 31, No. 7, pp. 362–364, July, 1943.

6. **Kilpatrick, E. L. and H. V. Cottony.** Scaled model measurements of performance of an elevated sloping vee antenna, *ESSA Technical Report No. ERL 72-ITS 62*, Boulder, Colo., May, 1968.

7. **Bruce, E.** Developments in short-wave directive antennas, *Proc. IRE*, Vol. 19, No. 8, pp. 1406–1433, August, 1931.

8. **Bruce, E., A. C. Beck, and L. R. Lowry.** Horizontal rhombic antennas, *Proc. IRE*, Vol. 23, No. 1, pp. 24–46, January, 1935.

9. **Christiansen, W. N.** Directional patterns of rhombic antenna, A.W.A. *Tech. Rev.*, Vol. 7, No. 1, pp. 33–51, September, 1946.

10. **LaPort, E. A. and A. C. Veldhuis.** Improved antennas of the rhombic class, *RCA Rev.*, Vol. 21, No. 1, pp. 117–123, January, 1960.

11. **Harper, A. E.** *Rhombic Antenna Design*, D. Van Nostrand Co., New York, 1941.

12. **Proctor, L. L.** Scaled-model, experimental verification of computed performance of a single rhombic antenna and an array of two rhombic antennas, *ESSA Technical Report No. IER 52-ITSA 50*, Boulder,Colo., July, 1967.

13. **Lewin, L.** Discussion of "Radiation from rhombic antenna, by Foster," *Proc. IRE*, Vol. 29, No. 9, p. 523, September, 1941.

14. **Foster, D.** Radiation from rhombic antenna, *Proc. IRE*, Vol. 25, No. 10, pp. 1327–1353, October, 1937.

15. **Ma, M. T. and L. C. Walters.** Power gains for antennas over lossy plane ground, *ESSA Technical Report No. ERL 104-ITS 74,* Boulder, Colo., April, 1969.

ADDITIONAL REFERENCES

Altshuler, E. E. The traveling-wave linear antenna, *IRE Trans. Antennas and Propagation*, Vol. AP-9, No. 4, pp. 324–329, July, 1961.

Colebrook, F. M. The electric and magnetic fields of a linear radiator carrying a progressive wave, *J. IEE* (London), Vol. 86, No. 518, pp. 169–178, February, 1940.

De Carvalho Fernandes, A. A. On the design of some rhombic antenna arrays, *IRE Trans. Antennas and Propagation*, Vol. AP-7, No. 1, pp. 39–46, January, 1959.

Hall, J. S. Non-resonant sloping -V aerial, *Wireless Engineer*, Vol. 30, No. 9, pp. 223–226, September, 1953.

Iampol'skii, V. G. The V-shaped inclined antenna, *Telecommunications,* Vol. 13, No. 4, pp. 403–411, April, 1959.

Kilpatrick, E. L. and H. V. Cottony. Optimum phasing of nested rhombic antennas connected in parallel, *ESSA Technical Report No. ERL 67-ITS 59*, Boulder, Colo., March, 1968.

Wolff, E. A. *Antenna Analysis*, Chap. 8, John Wiley and Sons, Inc., New York, 1966.

AUTHOR INDEX

SUBJECT INDEX